Immunological
Adjuvants and
Vaccines

NATO ASI Series

Advanced Science Institutes Series

A series presenting the results of activities sponsored by the NATO Science Committee, which aims at the dissemination of advanced scientific and technological knowledge, with a view to strengthening links between scientific communities.

The series is published by an international board of publishers in conjunction with the NATO Scientific Affairs Division

A	**Life Sciences**	Plenum Publishing Corporation
B	**Physics**	New York and London
C	**Mathematical and Physical Sciences**	Kluwer Academic Publishers Dordrecht, Boston, and London
D	**Behavioral and Social Sciences**	
E	**Applied Sciences**	
F	**Computer and Systems Sciences**	Springer-Verlag
G	**Ecological Sciences**	Berlin, Heidelberg, New York, London,
H	**Cell Biology**	Paris, and Tokyo

Recent Volumes in this Series

Volume 174—Cell and Molecular Biology of Artemia Development
edited by Alden H. Warner, Thomas H. MacRae,
and Joseph C. Bagshaw

Volume 175—Vascular Endothelium: Receptors and Transduction Mechanisms
edited by John D. Catravas, C. Norman Gillis, and Una S. Ryan

Volume 176—Processing of Sensory Information in the
Superficial Dorsal Horn of the Spinal Cord
edited by F. Cervero, G. J. Bennett, and P. M. Headley

Volume 177—Prostanoids and Drugs
edited by B. Samuelsson, F. Berti, G. C. Folco,
and G. P. Velo

Volume 178—The Enzyme Catalysis Process: Energetics, Mechanism,
and Dynamics
edited by Alan Cooper, Julien L. Houben, and Lisa C. Chien

Volume 179—Immunological Adjuvants and Vaccines
edited by Gregory Gregoriadis, Anthony C. Allison,
and George Poste

Volume 180—European Neogene Mammal Chronology
edited by Everett H. Lindsay, Volker Fahlbusch,
and Pierre Mein

Series A: Life Sciences

Immunological Adjuvants and Vaccines

Edited by

Gregory Gregoriadis

Royal Free Hospital School of Medicine
London, United Kingdom

Anthony C. Allison

Syntex Research
Palo Alto, California

and

George Poste

Smith Kline French Laboratories
Philadelphia, Pennsylvania

Plenum Press
New York and London
Published in cooperation with NATO Scientific Affairs Division

Proceedings of a NATO Advanced Study Institute on
Immunological Adjuvants and Vaccines,
held June 24–July 5, 1988,
in Cape Sounion Beach, Greece

Library of Congress Cataloging in Publication Data

NATO Advanced Study Institute on Immunological Adjuvants and Vaccines
(1988: Ájra Sounion, Greece)
  Immunological adjuvants and vaccines / edited by Gregory Gregoriadis, An-
thony C. Allison, and George Poste.
    p.   cm.—(NATO ASI series. Series A, Life sciences; v. 179)
  "Proceedings of a NATO Advanced Study Institute on Immunological Ad-
juvants and Vaccines, held June 24–July 5, 1988, in Capa Sounion Beach,
Greece"—T.p. verso.
  "Published in cooperation with NATO Scientific Affairs Division."
  Includes bibliographical references.

ISBN 978-1-4757-0285-9   ISBN 978-1-4757-0283-5 (eBook)
DOI 10.1007/978-1-4757-0283-5

  1. Immunological adjuvants—Congresses. 2. Vaccines—Congresses. I.
Gregoriadis, Gregory. II. Allison, Anthony C. (Anthony Clifford), 1925-   III.
Poste, George. IV. North Atlantic Treaty Organization. Scientific Affairs Division.
V. Title. VI. Series.
  [DNLM: 1. Adjuvants, Immunologic—congresses. 2. Vaccines—congresses.
QW 800 N279i 1988]
QR187.3.N37   1988
615'.372—dc20
DNLMÇDLC   89-25561
for Library of Congress   CIP

© 1989 Plenum Press, New York
Softcover reprint of the hardcover 1st edition 1989

A Division of Plenum Publishing Corporation
233 Spring Street, New York, N.Y. 10013

PREFACE

Vaccination, chiefly responsible for the eradication of smallpox and the control of poliomyelitis and German measles in man and of foot-and-mouth, Marek's and Newcastle disease in domestic animals, remains the best answer to infectious diseases. Early vaccines were live wild type organisms but these have been largely replaced by attenuated or killed organisms or by purified components (subunits) thereof. More recently, developments in recombinant DNA techniques, the advent of monoclonal antibodies and progress in our understanding of the immunological structure of proteins, have laid the foundations for a new generation of vaccines. For instance, subunit vaccines have been produced through gene cloning and a number of peptides mimicking small regions of proteins on the outer coat of viruses and capable of eliciting virus neutralizing antibodies, have been synthesized. Such vaccines are defined at the molecular level, can elicit immune responses controlling specific infectious organisms and are, thus, potentially free of the problems inherent in conventional ones. However, because subunit and peptide vaccines are only weakly or non-immunogenic, they require the presence of immunological adjuvants. These are a diverse array of agents that promote specific humoural and/or cell-mediated immunity responses to antigens.

This book contains the proceedings of the 1st NATO Advanced Studies Institute "Immunological Adjuvants and Vaccines" held in Cape Sounion Beach, Greece during 24 June-5 July, 1988. It deals with traditional and modern immunological adjuvants as applied to a variety of conventional and new generation vaccines, mechanisms of adjuvanticity and related immune responses as well as optimization of such responses by the use of appropriate adjuvant formulations. We express appreciation to Professors Ruth Arnon and J.H.L. Playfair for their advice in the planning of the ASI, to Dr. G. Deliconstantinos who, as Chairman of the local committee, contributed most effectively to its success, and to Mrs. A. Massaro for her help with practical aspects of the ASI. We are particularly grateful to Mrs. Susan Gregoriadis for her invaluable input in the editing of the book. The ASI was held under the sponsorship of NATO Scientific Affairs Division and co-sponsored and generously financed by Smith Kline and French Laboratories (Philadelphia). Financial assistance was also provided by Syntex (Palo Alto), Biophor (College Station), Sclavo (Sienna), Institut Merieux (Lyons), Hoffmann-La-Roche (Basle), Boehringer (Mannheim), Merck Sharp and Dohme (West Point), Organon (Oss), Connaught Laboratories (Willowdale), Schering (W. Berlin), Ciba Geigy (Horsham), Merz and Dade (Dudingen), Johnson and Johnson (La Jolla) and Northumbria Biologicals (Cramlington).

June 1989

Gregory Gregoriadis
Anthony C. Allison
George Poste

CONTENTS

ANTIGENS AND ADJUVANTS FOR A NEW GENERATION OF VACCINES

Anthony C. Allison

Syntex Research
Palo Alto
California, USA

INTRODUCTION

More than a decade has passed since vaccination made possible the global eradication of smallpox. Vaccination has also been remarkably successful in reducing morbidity and mortality due to yellow fever virus in Africa as well as Central and South America. These vaccines, as well as the vaccines developed after the last World War against measles and rubella, contain live attenuated viruses. While there is no question about the efficacy of such vaccines and their safety in the great majority of recipients, they can produce encephalitis or other complications in humans with immunodeficiency. Many persons in developing countries are immuno-deficient because of infections and poor nutrition (Dowd and Heatley, 1984). The high prevalence of human immunodeficiency virus (HIV) in parts of Africa is now recognized. Live bacterial vaccines, e.g. Mycobacterium bovis BCG, can also produce generalized infections in immunocompromised persons. Hence the desirability of developing vaccines based on inactivated viruses or bacteria, or purified antigens, is generally recognized. The question is whether such development is feasible: there is still a widespread belief that only living vaccines can elicit cell-mediated immunity and protection against some infections.

Viral nucleic acids or their DNA homologues can become incorporated into host cell DNA, and are known to trans-activate protooncogenes, which could have undesirable long-term effects. Hence antigens lacking nucleic acids are preferred for the new generation of vaccines. Some viral subunit antigens are readily prepared from the viruses themselves, e.g. influenza virus haemagglutinin (HA), or from naturally occurring materials, e.g. hepatitis B virus surface antigen (HBsAg) from the serum of carriers. However, two procedures have revolutionized antigen production during the past decade.

Recombinant Antigens

One is production by recombinant DNA technology. The prototype is the surface antigen of hepatitis B virus (HBsAg) cloned and expressed in yeast in a form physically and antigenically resembling the 22 nm particles in serum (Valenzuela et al, 1983; Murray et al, 1984). Recombinant HBsAg has been shown to have immunogenicity in humans comparable to that of HBsAg derived from serum (Scolnick et al, 1984), and it is now authorized for human

1

use in the USA and several other countries. In principle it is possible to produce a wide range of antigens by recombinant DNA technology, although optimal expression systems vary with different antigens. For example, HBsAg is not readily produced in Escherichia coli, which is however a good expression system for the nucleocapsid antigen (HBcAg) of the same virus (Murray et al, 1984). HBcAg from E. coli is in the form of particles physically and antigenically resembling those produced during natural infection; conversion of this p21 polypeptide to a non-particulate form in detergent changes its antigenicity to that of HBeAg (MacKay et al, 1981), illustrating the importance of conformation for antigenicity. When the three-dimensional structure of antigens is known from X-ray crystallography and monoclonal antibodies defining protective epitopes, many are found to be conformational rather than linear. This is true of influenza virus haemagglutinin (Wiley et al, 1981) and poliovirus (Hogle et al, 1985), for example. Using suitable expression systems, e.g. E. coli, yeast, baculovirus or mammalian cells, it is possible to produce protein antigens with conformations similar to those naturally occurring in many infectious agents. For example, herpes simplex virus (HSV-2) glycoproteins B and D expressed in mammalian cells are typically glycosylated and, inoculated with suitable adjuvants, efficiently protect guinea pigs from genital HSV-2 infections (Berman et al, 1985).

Many microbial antigens are being expressed by recombinant technology, including pertussis and tetanus toxoids, bacterial fimbriae, respiratory syncytial virus fusion protein and HIV surface and nucleocapsid antigens. Recombinant technology can also be used to express peptides as fusion proteins with bacterial flagella, fimbriae, outer membrane proteins and other potential carriers. All of these recombinant antigens are candidates for inclusion in a new generation of vaccines.

Synthetic Peptide Antigens

The second strategy, which is becoming a major industry, is identification of peptides in antigens recognized by T-lymphocytes and by antibodies and using the corresponding synthetic (or occasionally recombinant) peptides in vaccines. The finding that quite small peptides bind to class II MHC glycoproteins of antigen-presenting and stimulate T-lymphocytes (discussed by Howard Gray; see below) has encouraged this approach. The concept of synthetic peptide vaccines is of course appealing, and some interesting candidates are emerging, including peptides of foot and mouth disease virus and the common cold virus (discussed by Fred Brown, below). However, two potential difficulties should not be overlooked. First, it is seldom possible with synthetic peptides to reproduce conformational epitopes (although disulphide-constrained loops and some other specific peptide conformations can be synthesized). Second, immune responses to peptides are frequently genetically restricted. For example, mice of the $H-2^r$ haplotype do not respond to S region peptides of HBV although they can respond to pre-S peptides of the virus (Milich et al, 1985). Hence it seems likely that in outbred populations such as humans there will be a higher proportion of low responders to peptides than to multi-epitope recombinant subunits. No synthetic peptide vaccine is yet in use; however, there is no shortage of experimental work on the subject, and efficacious peptide vaccines may emerge during the next few years.

IMPROVEMENT OF CURRENTLY USED VACCINES

To vaccinate against hepatitis B virus (HBV), three doses of HBsAg adjuvanted with alum are given. It would be convenient to reduce the dose of expensive HBsAg and to use two injections instead of three. It would also be helpful to improve responses to HBV vaccine, especially in newborn children in developing countries and in special groups, such as intravenous drug

users, in developed countries. Alum is the only adjuvant currently authorized for human use. While it is a good adjuvant for bacterial toxins, it is
ineffective with some other antigens. Antibody responses to influenza virus
haemagglutinin are not increased by alum (Nicholson et al, 1979). Influenza
vaccine is used mainly in humans aged 65 years or over, who are particularly
susceptible to the disease. In such individuals antibody responses to influenza vaccine are inconsistent (Arden et al, 1986) and an improved vaccine
is urgently needed. The currently used inactivated pertussis vaccine is
crude and occasionally produces neurological complications. It should be
possible to produce an efficacious vaccine without risk of such side effects. Hence there is room for improvement of vaccines already in use.

VACCINES FOR THE FUTURE

The need for new vaccines is even greater: topping the list of those
urgently required are vaccines against the human immunodeficiency virus
(HIV-1) and malaria. However, other vaccines that may be more readily
developed within a few years include vaccines for hepatitis A virus, respiratory syncytial virus and common bacterial infections of childhood. The
development of vaccines against herpesviruses also seems feasible.

How can existing vaccines be improved and new ones developed? To real-
ize the full potential of recombinant and synthetic peptide antigens they
will have to be used with an adjuvant that elicits a protective immune
response. For vaccine design it is useful to know what type of immune
response is required for protection. Is the formation of antibodies sufficient or is cell-mediated immunity required? Is any antibody response
sufficient or are certain subclasses of antibodies, for example those able
to activate complement or mediate antibody-dependent cellular cytotoxicity,
required for optimal protection? Is secretory IgA required for protection
against mucosal infections?

As dicussed below, some adjuvants more consistently elicit cell-
mediated immunity than others, and they can also favour the formation of
antibodies of protective isotypes. Adjuvants are therefore useful probes to
analyse immune responses. For a long time adjuvants such as alum and
Freund's complete adjuvant were used empirically without any knowledge of
how they exert their effects. With the great advances that have taken place
in immunology during the last decade, at least some of the effects of adjuvants are being defined at the cellular and molecular levels. There is
widespread misunderstanding about the distinction between adjuvants and
inducers of non-specific resistance to infections, as there is about the
ambiguous uses of terms such as "carrier". It is therefore useful to begin
with a few relevant definitions.

DEFINITIONS

Elsewhere I have discussed the distinction between adjuvants, which
increase specific immune responses, and agents which non-specifically increase resistance to infections and tumors (Allison, 1978). This they do by
activating cells of the monocyte-macrophage lineage or natural killer cells.
Lipopolysaccharide endotoxins (LPS) and some muramyl dipeptides (MDP), or
related compounds such as muramyl tripeptide phosphatidylethanol-amine (MTP-
PE), can have both activities. For reasons that will be discussed compounds
that are pure adjuvants, lacking capacity to increase non-specific resistance, are preferable for vaccine administration. Major differences between
the two effects are observed in timing and sensitivity to radiation. Non-
specific resistance to infectious agents following administration of Mycobacterium bovis BCG, Propionibacterium acnis, or active polymers or small

3

molecules, including LPS and MDP, takes two to three weeks to become fully established and is then radioresistant. Adjuvants must be given about the time of antigen administration, and their effects are eliminated by immuno-suppressive doses of X-radiation.

The term adjuvant can be applied to an agent that augments specific immune responses to antigens. The term vehicle can be used for the mineral oil emulsion of Freund's adjuvant, liposomes and the squalane-Pluronic[R] polymer emulsion which we have developed (Allison and Byars, 1986). Thus a vehicle is distinguished from a carrier, which is immunogenic and when bound to a non-immunogenic or weakly immunogenic hapten (e.g. a peptide) increases antibody formation against the latter. An adjuvant in a suitable vehicle can be termed an adjuvant formulation. Pharmacists know that the successful use of a drug depends critically on effective formulation, and the same is true of an adjuvant. The adjuvant formulation should be efficacious, safe, stable and easy to use.

AFFINITIES AND ISOTYPES OF ANTIBODIES

Traditionally the efficacy of adjuvants has been judged by the levels of antibodies elicited, using a convenient test such as ELISA or haema-gglutination (HA). While these assays have provided useful information, they should now be supplemented by other measures of the quantity and qual-ity of antibodies elicited. Preferably antibody levels should be quantified by tests relevant to function, such as neutralization of bacterial toxins or viruses. Because of potential problems with solid-phase assays, at least some measurements of antibody levels using fluid-phase assays should be made. In addition to the quantities of antibodies elicited by a vaccine, two properties of the antibodies are likely to be important for protection: their affinities for antigen and their isotype. To neutralize a virus or bacterial toxin, antibodies should bind them with sufficiently high affin-ity. If the complexes are not removed by phagocytic cells, antibodies must bind to a virus or toxin with an affinity of at least the same order as that of the natural receptor. We have developed an economical and convenient method for measuring affinities of antibodies in solution which provides the required information (see paper by Kenney, below).

Another important property of antibodies is their isotype. Antibodies of the IgG class pass from the vascular to the extravascular compartment more easily than those of the IgM class, and only the former are trans-ferred across the placenta or by milk to foetuses and newborn animals. Antibodies of some isotypes efficiently activate complement, bind to high-affinity receptors on monocytes, and act synergistically with antibody-dependent effector cells to produce cytotoxicity. Examples are IgG2a anti-bodies in mice and IgG1 antibodies in humans, both of which bind to high affinity to $Fc\gamma I$ receptors (Unkeless et al, 1988). Studies with isotype-switch variants of murine monoclonal antibodies (which have the same Fab regions, so binding to antigen is comparable) show that IgG2a antibodies confer better protection against tumors than those of other isotypes (Steplewski et al, 1985; Kaminski et al, 1986). Antibodies of the IgG2a isotype are also involved in protection against at least some infectious agents (Wechsler et al, 1986). Studies with "reshaped" human antibodies, geneticaly constructed so as to have antigen-binding hypervariable regions like those of rodent monoclonals, confirm the superiority of the human IgG1 isotype in complement-mediated and ADCC-mediated lysis (Reichmann et al, 1988). The desirability of developing an adjuvant that preferentially elicits high-affinity antibodies of the IgG2a isotype in mice and IgG1 in humans is therefore apparent. Interactions of antibodies and effector cells are certainly important in defence against infections in vivo. An obvious example is the opsonization of bacteria for phagocytosis by neutrophils, monocytes or macrophages. Both antibodies and functionally mature macrophages are required for protection against viruses (Allison, 1988), so antibodies which facilitate this interaction are likely to be particularly effective.

One of the properties of Freund's adjuvants is that they change the isotypes of antibodies elicited in the guinea pig. The incomplete adjuvant, like alum and LPS, elicits antibodies mainly of the γ1 isotype whereas the complete adjuvant elicits antibodies of the γ2 isotype (White, 1976). While alum, LPS and Quil A elicit antibodies mainly of the IgG1 isotype in the mouse, the adjuvant formulation that we have developed (SAF-1) augments the production of IgG2a antibodies (see below). For reasons discussed below this may be related to the elicitation of cell-mediated immunity.

ANTIBODIES AGAINST EPITOPES OF NATIVE AND DENATURED PROTEINS

During the course of preparation of vaccines in some adjuvants (e.g. emulsification in Freund's complete adjuvant) a substantial amount of denaturation takes place. While this may increase immune responses, e.g. denatured gamma globulin is a much better immunogen than the native molecule (Dresser and Mitchison, 1968), it may also change the structure of relatively unstable antigens such as pertussis toxoid. It is therefore useful to have a measure of the proportion of antibodies binding to epitopes in the native and denatured molecule, which can provide an indication of whether the adjuvant formulation is suitable for use with unstable antigens. We have assessed such responses using human serum albumin as an antigen (Kenney et al, 1989), and found the adjuvant formulation that we have developed to be much less denaturing than Freund's complete adjuvant.

ELICITATION OF CELL-MEDIATED IMMUNITY (CMI)

The importance of eliciting T-cell immune responses to vaccines and T-cell memory is obvious. Helper T-lymphocytes are required for the formation of antibodies against most antigens. In addition cytotoxic T-lymphocytes can lyse infected cells or produce mediators such as interferon-γ following interaction with antigen in a genetically restricted situation (Morris et al, 1982). Selective deficiencies of antibody formation and of CMI in experimental animals and humans provide useful information about the principal mechanisms of immunity against different groups of organisms. Mice repeatedly injected neonatally with antibodies against the μ-chain of immunoglobulin do not produce antibodies and are highly susceptible to bacterial infections. Children with selective deficiencies of IgG have recurrent bacterial infections as well as an increased risk of paralytic poliovirus and echovirus meningoencephalitis (Hayward, 1982). Children with selective deficiencies of T-cell functions have severe herpesvirus and poxvirus infections as well as generalized BCG infections. Passive protection experiments confirm the importance of T-lymphocytes in protection against herpesviruses (Rager-Zisman and Allison, 1976); the protective T-cells are of the Lyt-1+ phenotype (Nash and Gell, 1983). Cell-mediated immunity is also important for protection against tumour viruses (Allison, 1980).

Cytotoxic T-lymphocytes able to lyse autologous cells expressing several antigens of human immunodeficiency virus (HIV) are demonstrable in infected persons (Walker et al, 1988), although it is not yet known whether they have a protective role. In this and other virus infections cytotoxic T-cells frequently recognize internal (nucleoprotein and/or polymerase) antigens of the virus, and this may be a strategy by which broad protection rather than variant-specific immunity might be achieved (Allison, 1988).

5

It is therefore likely that for optimal protection against some infectious agents, e.g. herpesviruses and possibly HIV, the elicitation of cell-mediated immunity (CMI) is desirable. Tests for CMI should include not only delayed hypersensitivity but also proliferative responses to the antigen and the release of IL-2 (Byars and Allison, 1987). Cytotoxicity for autologous or syngeneic infected target cells should also be studied. If mice or rats are used, syngeneic target cells are readily available. With outbred species such as humans and subhuman primates, B-cells transformed by Epstein-Barr virus, transfected with a vaccinia virus vector expressing the antigen under study (e.g. HIV antigens, Walker et al, 1988), can provide autologous target cells for studies of genetically-restricted cell-mediated cytotoxicity.

UNDESIRABLE EFFECTS OF ADJUVANTS

Desirable attributes of adjuvants have been outlined in previous sections. In summary, they should elicit cell-mediated immune responses and adequate levels of antibodies of sufficient affinity and of isotypes to confer protection against the infectious agent in question. They should also elicit both T-cell and B-cell memory. The question is whether these can be obtained without undesirable side effects. Regulatory authorities will have to be satisfied on these points before authorizing new adjuvants for human use. Alum was introduced in 1926 before strict control by regulatory authorities was practiced; whether it would be allowed by regulatory authorities today is far from certain. It is therefore necessary to have a clear understanding of the known side effects of adjuvants so as to avoid them as far as possible.

Reactions at Injection Sites

The first complication of the use of an adjuvant is acute or subacute tissue damage at the injection site, or a later granulomatous reaction. Many surface-active adjuvants produce tissue damage at injection sites. A convenient test is measurement of creatine phosphokinase in the circulation after intramuscular injection (Byars and Allison, 1987). Limits of circulating creatine phosphokinase acceptable to regulatory authorities for injections of drugs and vitamins have been defined. Granulomas can be assessed by histological examination at various times following vaccine injection, or by measurement of leukocyte enzymes following intramuscular injection (Davies and Allison, 1976). Even alum produces an appreciable granulomatous response at the injection site (Turk and Parker, 1977). Freund's complete adjuvant elicits a severe granulomatous reaction at the injection site which precludes use in humans, domestic and farm animals, and may prevent future use in at least some laboratory animals on humanitarian grounds. Trehalose dimycolate, which is a component of some adjuvant formulations, including monophosphoryl lipid A, elicits granulomatous reactions at injection sites (Bloch, 1955). Even mineral oil adjuvants can induce granulomatous responses, and in some strains of mice plasmacytomas (Potter and Boyce, 1962), which is why their use in humans is not authorized. For a long time it was thought that slow release of antigen from depots and macrophages in granulomas at injection sites were required for optimal antigen presentation and adjuvant activity, but experimental excision of injection sites after a few days does not impair immune responses.

Pyrogenicity

A second undesirable effect of adjuvants is pyrogenicity. A regulatory requirement of biological products introduced into humans is that they should not be pyrogenic. Some adjuvants, such as LPS and MDP, are pyrogenic. It was believed that adjuvants exert their effects by inducing the formation of interleukin-1 by accessory cells (Oppenheim et al, 1980) and that IL-1 is the same as endogenous pyrogen (Dinarello, 1986). If this were true, it would not be possible to separate adjuvant activity from pyrogenicity. However, synthetic analogues of MDP, including murabutide (Chedid et al, 1982) and N-acetylmuramyl-L-threonyl-D-isoglutamine (Allison and Byars, 1986), are potent adjuvants with greatly reduced pyrogenicity compared with naturally occurring MDP. The monophosphoryl derivative of lipid A retains adjuvant activity with reduced pyrogenicity (Ribi et al, 1987). Some of the properties previously attributed to IL-1, including pyrogenicity and induction of acute-phase protein synthesis in the liver, are now thought to be due to IL-6 (Helle et al, 1988; Gauldie et al, 1987). Hence dissociation between adjuvant activity and pyrogenicity is possible and desirable. Candidate components for adjuvant formulations should be tested for pyrogenicity in sensitive animal models, e.g. intravenous administration in rabbit.

Anterior uveitis

Small doses of LPS or MDP injected intravenously in the rabbit increase vascular permeability in the eye, as shown by passage of fluoresceinated macromolecules into the anterior chamber (Waters et al, 1986). Histological examination shows leukocyte emigration into the anterior urea that can lead to irreversible changes. We have found that repeated administration of lipophilic MDP derivatives can produce blindness in cynomolgus monkeys; incorporating these derivatives into liposomes does not prevent uveitis but actually enhances that complication. Anterior uveitis is a component of Reiter's syndrome, a complication of Yersinia and some other gram-negative infections in genetically predisposed persons, who are often of the HLA-B27 haplotype (Geczy et al, 1983). We have correlated uveitis produced by LPS and MDP with activation of cells of the monocyte/macrophage lineage to produce prostaglandins and sulfidopeptide leukotrienes. Hence macrophage activation, as shown by these properties or by increased resistance to infections, is not a desirable attribute of an adjuvant. It may induce or aggravate uveitis in only a small minority of predisposed recipients, but even that risk is unacceptable. It is estimated that one in 300,000 recipients of Bordetella pertussis vaccine has neurological complications, but that is enough to give vaccination in general a bad name. Even occasional, complications of adjuvant usage could defer for decades regulatory approval of much needed adjuvant formulations. For that reason we suggest that it is essential to select for human use adjuvant components that show good separation between potency as adjuvants and as monocyte/macrophage activators and inducers of anterior uveitis.

Arthritis

Arthritis is also a component of Reiter's syndrome and might be produced in genetically predisposed humans by analogy with the arthritis produced by FCA in rats. MDP can also sensitize guinea pigs for arthritis. It is therefore essential that any potential adjuvant formulation be tested for capacity to induce arthritis in predisposed experimental animals such as rats and guinea pigs. The threonine analog of MDP does not induce arthritis in rats (Byars and Allison, 1987).

TARGETING OF ANTIGENS TO ANTIGEN-PRESENTING CELLS (APC)

Since a granuloma at the injection site is not required for adjuvant activity, the question arises how adjuvant formulatioons optimize contact of antigens with APC. It is now believed that while cells of the monocyte/macrophage lineage can function as APC when activated, for example during infections, three cell types function as the principal APC under physiological conditions, for example in an animal responding to vaccine antigens.

Langerhans cells

Cells of the Langerhans cell lineage originate in the bone marrow, migrate through the blood to the skin where they remain for about a week and then migrate through afferent lymphatics to the T-dependent areas of lymph nodes, where they are termed interdigitating cells (Balfour et al, 1981). Dendritic cells isolated from the spleen (Steiniman and Nussenzweig, 1980) have similar properties and may be of the same lineage. Because of possible confusion with follicular dendritic cells, which have a different location and properties, the term Langerhans cells is used in this presentation. Cells of this lineage efficiently present antigens associated with their surfaces, for example contact-sensitizing chemicals and myelin basic protein, to elicit T-lymphocyte-dependent immune responses (Knight et al, 1983, 1985).

Follicular dendritic cells (FDC)

As their name implies, these cells are found in lymphoid follicles in lymph nodes, spleen and other sites, where their branching cytoplasmic extensions are closely related to B-lymphocytes. FDC express CD4 and high-affinity C3b receptors (CRI). Immune complexes activating complement injected into mice became localized on FDC, and this process appears to be required for the generation of B-lymphocyte memory, in other words proliferation of clones of B-lymphocytes responding to antigen with consequent priming for a secondary response (Klaus et al, 1980). Immune complexes binding FDC become associated with beaded cell membrane extensions which are readily taken up by follicular B-lymphocytes expressing class II major histocompatbility antigens (Szakal et al, 1988). The antigen can be demonstrated for at least one week by immunocytochemistry in endocytic vacuoles within B-cells; in such a compartment they may be partially digested for presentation to T-lymphocytes.

B-lymphocytes

During the past few years evidence has accumulated that B-lymphocytes efficiently present antigens to T-lymphocytes (Ron and Sprent, 1987; Abbas, 1988). In fact, depletion of B-cells by repeated injections of antibody against the µ-chain of immunoglobulin markedly decreases responses to antigens of T-lymphocytes in peripheral lymphoid tissues. A major role of surface membrane immunoglobulin receptors for antigens on B-cells may be to bind the antigen for subsequent presentation to T-cells.

Targeting of antigens to FDC may be a crucial factor in the efficient presentation to B-lymphocytes and, through them, to T-lymphocytes. In secondary immune responses this occurs through the formation of complement-activating immune complexes. Adjuvant formulations can facilitate such localization by themselves activating complement. This is true of lipopolysaccharides and Syntex Adjuvant Formulation-1; liposomes of compositions that activate complement are better adjuvants than those that do not (Allison and Byars, 1986). Such complement activation should be moderated so as to have enough C3b on the antigen-bearing micelles, emulsions or liposomes to allow targeting of associated antigens, but not sufficient complement activation at injection sites to elicit inflammatory lesions.

SELECTION BY ADJUVANTS FOR THE PRODUCTION OF ANTIBODIES OF HIGH AFFINITY AND PROTECTIVE ISOTYPES

For reasons discussed above it is frequently desirable to elicit isotype antibodies of high affinities and protective isotypes, e.g. IgG2a in the mouse. It has long been known that the use of particular adjuvants can influence the isotypes of antibodies. An example is the use of low doses of antigen with alum, Bordetella pertussis or saponin to produce IgE antibodies in the mouse. Antigens administered to guinea pigs in Freund's incomplete adjuvant elicit mainly antibodies of the γ_1 isotype whereas with the complete adjuvant γ_2 antibodies are formed (White, 1976). We have compared antibodies elicited by human serum albumin and recombinant human interleukin-1α administered to mice in different adjuvants by the intraperitoneal and subcutaneous routes (Kenney et al, 1989). Considerable differences were observed. Freund's complete adjuvant elicited good levels of antibodies, but these were not of very high affinity; many were directed to epitopes not exposed on the native molecule. Syntex Adjuvant Formulation-1 (SAF-1) elicited the highest proportion of antibodies of the IgG2a isotype. The antibodies against IL-1 were potent in neutralizing the biological activity of the molecule, and cells from the mice were used for production of monoclonal antibodies. Aluminium hydroxide and Quil A elicited antibodies largely of the IgG1 isotype.

Thus adjuvants can select for the isotype of the antibodies formed. Moreover, production of hybridomas does not require the barbarous traditional procedure of immunizing intraperitoneally with Freund's complete adjuvant. Subcutaneous or intramuscular immunization with SAF-1 or Quil A, depending on the desired isotype, is equally effective. The use of FCA for laboratory animal immunization is already restricted in several large research centres and, since more human adjuvants are equally effective, they should be used everywhere.

ROLE OF CYTOKINES IN ISOTYPE SELECTION

Until recently the mechanisms by which the formation of antibodies of particular isotypes are favoured were unknown. Now evidence is accumulating that cytokines play a role in isotype selection in the mouse, together with indications that the same is true in cultured human cells. When IL-4 is added to cultures of purified mouse B-lymphocytes, it stimulates the secretion of IgG1 and IgE and suppresses secretion of IgG2a (Snapper et al, 1988). In contrast, recombinant IFN-γ increases IgG2a secretion, decreases IgG1 and IgE secretion and has no effect on IgM secretion (Finkelman et al, 1988a). Antibodies against IL-4 in the mouse strongly suppress the formation of IgE antibodies and less strongly suppress IgG1 antibodies (Finkelman et al, 1988b). IgE production by normal human B-lymphocytes induced by alloreactive T-cell clones is likewise increased by IL-4 and suppressed by IFN-γ (Pène et al, 1988). In the mouse IL-5 selectively increases the formation of IgA by splenic and Peyer's patch B-lymphocytes (Lehman and Coffman, 1988).

These findings explain why adjuvants that are designed to increase cell-mediated immunity, such as FCA and SAF-1, concurrently select for antibodies of the IgG2a isotype. Potent T-cell-mediated responses to antigenic stimulation release IFN-γ (which can occur from both the helper and cytotoxic subset of T-cells), and IFN-γ augments the formation of IgG2a antibodies. Adjuvants which less consistently stimulate T-cell responses, e.g. aluminium hydroxide and Quil A, favour the production of IgG1 and IgE antibodies, presumably by stimulating the release of more IL-4 than IFN-γ.

For most immunizations production of IgE antibodies is undesirable. Veterinarians have long known that vaccines containing saponin can elicit antibodies giving acute hypersensitivity in cattle. When rhesus monkeys were immunized with inactivated SIV in ISCOM form they showed acute hypersensitivity reactions following challenge with virus. Monkeys immunized with the same antigen in SAF-1 did not show any allergic reactions when receiving the same challenge (Letvin et al, 1987). Production ofIgE may be required for immunity to schistosmula and some other parasites, but usually it should be avoided.

Thus the mode of action of adjuvants is being elucidated in terms of contemporary immunology. The next few years should see further application of immunological principles to vaccine development. To ensure exploitation of the exciting prospect of developing a new generation of adjuvants will also be required. Since the ill effects of adjuvants are now well defined, rational decisions on that important issue should be possible.

REFERENCES

Abbas, A.K., 1988, A reassessment of the mechanisms of antigen-specific T-cell-dependent B-cell activation, Immunol.Today, 9:89.

Allison, A.C., 1978, Macrophage activation and nonspecific immunity, Int.Rev.Exp.Pathol., 18:303.

Allison, A.C., 1980, Immune responses to polyoma virus and polyoma virus-induced tumors, in: "Viral Oncology", G. Klein, ed., Raven Press, New York.

Allison, A.C., 1988, The role of monocytes, macrophages, Langerhans cells and follicular dendritic cells in persistent virus infections, in: "Persistent Virus Infections", C. Lopez and H. Margolis, eds., Alan R. Liss, New York.

Allison, A.C. and Byars, N.E., 1986, An adjuvant formulation that selectively elicits the formation of antibodies of protective isotypes and cell-mediated immunity, J.Immunol.Methods, 95:157.

Arden, N.H., Patriarca, P.A. and Kendal, A.P., 1986, Experiences in the use and efficacy of influenza vaccine in nursing homes, in: "Options for Control of Influenza", A.P. Kendal and P.A. Patriarca, eds., Alan R. Liss, New York.

Balfour, B.M., Drexhage, H.A., Kamperdijk, E.W.A. and Hoefsmid, E.C., 1981, Antigen-presenting cells, including Langerhans cells, veiled cells and interdigitating cells, Ciba Foundation Symposium, 84:281.

Berman, P.W., Gregory, T., Crase, P. and Lasky, L.A., 1985, Protection from genital herpes simplex type 2 infection by vaccination with cloned glycoprotein D, Science, 227:1490.

Bloch, H., 1955, Virulence of mycobacteria, Adv.Tuberc.Res., 6:49.

Byars, N.E. and Allison, A.C., 1987, Adjuvant formulation for use in vaccines to elicit both cell-mediated and humoral immunity, Vaccine, 5:223.

Davies, P. and Allison, A.C., 1976, Secretion of macrophage enzymes in relation to the pathogenesis of chronic inflammation, in: "Immunobiology of the Macrophage", D.S. Nelson, ed., Academic Press, New York.

Dinarello, C.A., 1986, Multiple biological properties of recombinant human interleukin 1 (beta), Immunobiology, 172:301.

Dowd, P.S. and Heatley, R.V., 1984, The influence of undernutrition on immunity, Clin.Sci., 66:241.

Dresser, D. and Mitchison, A., 1968, The mechanism of immunologic paralysis, Adv. Immunol., 8:129.

Finkelman, F.D., Katona, I.M., Mosmann, T.R. and Coffman, R.L., 1988a, Interferon- regulates the isotypes of immunoglobulin secreted during in vivo humoral responses, J.Immunol., 140:1022.

Finkelman, F.D., Katona, I.M., Urban, J., Holmes, J., Ohara, J., Tung, A.S., Sample, J.G. and Paul, W.E., 1988b, IL-4 is required to generate and sustain in vivo IgE responses, J.Immunol., 141:2335.

Gauldie, J., Richards, C. Harnish, D., Lansdorf, P. and Baumann, H., 1987, Interferon β_2/B-cell stimulatory factor type 2 shares identity with monocyte-derived hepatocyte-stimulating factor and regulates the major acute-phase response in liver cells, Proc.Natl.Acad.Sci.USA, 84:7251.

Geczy, A.F., Alexander, K., Bashir, H.V., Edmonds, J.P., Upfold, L. and Sullivan, J., 1983, HLA-B27, Klebsiella and ankylosing spondylitis: biological and chemical studies, Immunol.Rev., 70:23.

Hayward, A.J., 1982, Immunodeficiency, in: "Clinical Aspects of Immunology", P.J. Lachmann and D.K. Peters, eds., Blackwell Scientific Publications, 2:1658, Oxford.

Helle, M., Boeije, L. and Aarden, L.A., 1988, Functional discrimination between interleukin-6 and interleukin-1, Eur.J.Immunol., 18:1535.

Hogle, J.M., Chow, M. and Tilman, D.J., 1985, Three-dimensional structure of poliovirus at 2.9A resolution, Science, 29:1358.

Kaminski, M.S., Kitamura, K., Maloney, D.G., Campbell, M.J. and Levy, R., 1986, Importance of antibody isotype in monoclonal anti-idiotype therapy of murine B cell lymphoma. A study of hybridoma class-switch variants, J.Immunol., 136:1123.

Kenney, J.S., Hughes, B.W., Masada, M. and Allison, A.C., 1989, Evaluation of adjuvants for the production of murine monoclonal antibodies, J.Immunol.Methods (in press).

Klaus, G.G., Humphrey, J.H., Kunkl, A. and Dongworth, D.W., 1980, The follicular dendritic cell: its role in antigen presentation in the generation of immunological memory, Immunol.Rev., 53:3.

Knight, S.C., Krejci, J., Malkovsky, M., Colizzi, V., Gautam, A. and Asherson, G.L., 1985, The role of dendritic cells in the initiation of immune responses to contact sensitizers. I. In vivo exposure to antigen, Cell.Immunol., 94:427.

Knight, S.C., Mertin, J., Stackpole, A. and Clarke, J., 1983, Induction of immune responses in vivo with small numbers of veiled (dendritic) cells, Proc.Natl.Acad.Sci.USA, 80:6032.

Lehman, D.A. and Coffman, R.L., 1988, The effects of IL-4 and IL-5 on the IgA response by murine Peyer's patch B cell subpopulations, J.Immunol., 141:2050.

Letvin, N.L., Daniel, M.D., King, N.W., Kiyotaki, M., Kannagi, N., Chalifoux, L.V., Sehgal, P.K., Desrosiers, R.C., Arthur, L.O. and Allison, A.C., 1987, AIDS-like disease in macaques induced by simian immunodeficiency virus: a vaccine trial, in: "Vaccines", R.M. Chanock, R.A. Lener, F. Braun and H.S. Ginasky, eds., Cold Spring Harbor, New York.

MacKay, P., Lees, J. and Murray, K., 1981, The conversion of hepatitis B core antigen synthesized in E. coli into e antigen, J.Med.Virol., 8:237.

Milich, D.R. and Chisari, F.V., 1982, Genetic regulation of the immune response to hepatitis B surface antigen (HBsAg). I. H-2 restriction of the murine humoral immune response to the a and d determinants of HBsAg, J.Immunol., 129:320.

Morris, A.G., Lin, Y-L. and Askonas, B.A., 1982, Immune interferon release when a cloned cytotoxin T-cell line meets its correct influenza-infected target, Nature, 295:150.

Murray, K., Bruce, S.A., Hinnen, A., Wingfield, P., van Erd, P.M., de Reus, A. and Schellekens, H., 1984, Hepatitis B virus antigens made in microbial cells immunize against viral infection, EMBO J., 3:645.

Nash, A.A. and Gell, P.H.G., 1983, Membrane phenotype of murine effector and suppressor cells involved in delayed hypersensitivity and protective immunity to herpes simplex virus, Cell.Immunol., 75:348.

Nicholson, K.G., Tyrrell, D.A.J., Harrison, P., Potter, C.W., Jennings, R.,

Clark, A., Schild, G.C., Wood, J.M., Yells, R., Seagrott, V., Huggens, A. and Anderson, S.G., 1979, Clinical studies of monovalent inactivated whole virus and subunit A/USSR/77 (H_1N_1) vaccine: serological and clinical reactions, J.Biol.Stand., 7:123.

Oppenheim, J.J., Togawa, A., Chedid, L. and Mizel, S., 1980, Components of mycobacteria and muramyl dipeptide with adjuvant activity induce lymphocyte-activating factor, Cell.Immunol., 50:71.

Pene, J., Rousset, F., Briere, F., Chretien, I., Paliard, X., Banchereau, J., Spits, H. and DeVries, J.E., 1988, IgE production by normal human B cells induced by alloreactive T cell clones is mediated by IL-4 and suppressed by IFN-γ, J.Immunol., 141:1218.

Potter, M. and Boyce, C.R., 1962, Induction of plasma cell neoplasms in Balb/c strain mice with mineral oil and mineral oil adjuvants, Nature, 193:1086.

Rager-Zisman, B. and Allison, A.C., 1976, Mechanism of immunologic resistance to herpes simplex virus (HSV-1) infection, J.Immunol., 116:35.

Reichmann, L., Clark, M., Waldmann. H. and Winter, G., 1988, Reshaping human antibodies for therapy, Nature, 332:323.

Ron, Y. and Sprent, J., 1987, T-cell priming in vivo: a major role for B-cells in presenting antigen to T-cells in lymph nodes, J.Immunol., 138:2848.

Scolnick, E.M., McLean, A.A., West, D.J., McAleer, W.J., Miller, W.J. and Bunyak, E.B., 1984, Clinical evaluation in healthy adults of a hepatitis B vaccine made by recombinant DNA, J.A.M.A., 251:2812.

Snapper, C.M., Finkelman, F.D. and Paul, W.E., 1988, Differential regulation of IgG1 and IgE synthesis by interleukin-4, J.Exp.Med., 167: 183.

Steinman, R.M. and Nussenzweig, M.C., 1980, Dendritic cells: features and functions, Immunol.Rev., 53:127.

Steplewski, Z., Spira, G., Blasczyc, M., Lubeck, M.D., Radlmuch, A., Illges, H., Herlyn, D., Rajewsky, K. and Scharff, M., 1985, Isolation and characterization of anti-monosialoganglioside monoclonal antibody 19-9S class switch variants, Proc.Natl.Acad.Sci., USA, 82:3653.

Szakal, A.K., Kosco, M.H. and Tew, J.G., 1988, A novel in vivo follicular dendritic cells - its role in antigen presentation and in the generation of immunological memory, J.Immunol., 140:341.

Turk, J.L. and Parker, D., 1977, Granuloma formation in normal guinea pigs injected intradermally with aluminum and zirconium compounds, J.Invest.Dermatol., 68:336.

Unkeless, J.C., Scigliano, E. and Freedman, V.H., 1988, Structure and function of human and murine receptors for IgG, Ann.Rev.Immunol., 6:251.

Valenzuela, P., Medina, A., Rutter, W.J., Annerer, A. and Hall, B.D., 1983, Synthesis and assembly of hepatitis B virus surface antigen particles in yeast, Nature, 291:503.

Waters, R.V., Terrell, T.G. and Jones, G.H., 1986, Uveitis induction in the rabbit by muramyl dipeptides, Infect.Immun., 51:816.

Walker, B.D., Fiemer, C., Paradis, T.J., Fuller, T.C., Hirsch, M.S., Schooley, R.T. and Moss, B., 1988, HIV-1 reverse transcriptase is a target for cytotoxic T-lymphocytes in infected individuals, Science, 240:64.

Wechsler, D.S. and Konghshavn, P.A.L., 1986, Heat-labile antibodies effects cure of Trypanosoma musculi infection in C57BL/6 mice, J.Immunol., 137:2968.

White, R.G., 1976, The adjuvant effect of microbiol products on the immune response, Ann.Rev.Microbiol., 30:579.

Wiley, D.C., Wilson, I.A. and Skehel, J.J., 1981, Structural identification of the antibody binding sites of influenza haemagglutinin and their involvement in antigenic variation, Nature, 289:373.

STRUCTURAL AND FUNCTIONAL STUDIES ON MHC-PEPTIDE ANTIGEN INTERACTIONS

Howard M. Grey*, Søren Buus[+] and Alessandro Sette*

*Cytel, San Diego, CA, USA
[+]University of Copenhagen, Copenhagen, Denmark

INTRODUCTION

T cells recognize protein antigens on the surface of antigen presenting cells (APC), but only after the antigen has been "processed" (physically altered by denaturation or fragmentation) by an APC (Shimonkevitz et al, 1983) and subsequently "displayed" in association with MHC molecules on the APC surface (Allen et al, 1987; Buus et al, 1987b). Previous studies on the mechanism of antigen recognition have established that T cells recognize a complex formed between MHC and peptide fragments of protein antigens (Werdelein 1982; Watts et al, 1984; Buus et al, 1986). Moreover, a strong correlation exists between the capacity to form complexes with peptides and the capacity to serve as the MHC restriction element used in the immune response to the same peptides (Babbitt et al, 1985; Buus et al, 1987a).

This paper considers two major questions arising from the concept of T cell recognition of MHC-antigen complexes: 1) What is the structural basis of the interaction between peptides and MHC that allows the few MHC molecules expressed by an individual to bind a large universe of antigenic peptides? and 2) What is the relevance of this determinant selection by MHC in defining immune responses to protein antigens?

RESULTS AND DISCUSSION

Structural Requirements for the Interaction between Ia and Antigen

To understand how a presumably large universe of immunogenic peptides can bind to and be restricted by the very small number of Ia specificities present within an individual, we have undertaken to characterize the requirements for the binding of peptides to a given MHC molecule. These studies were predominantly carried out utilizing the class II MHC restriction element IA^d, and more recently IE^d, and a set of peptides that were capable of binding to those MHC molecules. Much of the work has concentrated on the immunogenic peptide from chicken ovalbumin (Ova 323-339). The first approach taken was to compare the structure of the binding region of Ova 323-339 with the binding regions of other unrelated peptides that were also capable of binding to IA^d with a high affinity. To define the IA^d interacting region within the peptide Ova 323-339, we synthesized a series of N- and C-terminal truncated analogs and tested these analogs for binding to purified IA^d molecules. By this procedure we were able to define a critical core region within Ova 323-339 that was involved in binding to IA^d. It consisted of the hexapeptide 327-332 and had the sequence VHAAHA.

In the last few years we have also identified many other peptides from unrelated proteins that also bind strongly to IA^d. Since each of these peptides can competitively inhibit the binding of Ova 323-339 to IA^d, it is most likely that they all bind to the same site on IA^d as the Ova peptide and therefore should share the critical structural features required for such binding. We developed a computerized procedure which analyzes protein sequences, searching for hexapeptides structurally similar to Ova 327-332. Using the core binding region of Ova as the "master" sequence, we were able to identify in each of the other peptides the region that was most similar to this core IA^d binding region from the Ova peptide. These alignments are shown in Table 1. The common structural motif that emerges from this set of alignments is: position 1 appears to be the most similar in the six peptides being occupied in most peptides by the hydrophobic residues V, I, or L; the next position is usually occupied either by a basic residue (H, R, or K) or the polar residues serine or threonine; this is often followed by two hydrophobic residues in positions 3 and 4; position 5 is more variable but tends to be occupied by polar or charged amino acids; finally, position 6 is occupied either by A or by another residue with a short side chain (S). Although this motif could be discerned in the great majority of good IA^d binding peptides, we have also identified IA^d binding peptides that apparently do not share this motif. In a recent analysis of over 80 peptides, approximately two-thirds of the strong binders to IA^d contained the above motif, while it was contained in only 10% of the peptides that failed to bind IA^d.

We have been able to experimentally confirm the involvement of these predicted core binding regions in two of the peptides listed in Table 1, influenza virus hemaglutinin (Ha) 130-142 and sperm whale myoglobin (Myo) 106-118 (Sette et al, 1988). A similar N- and C-terminal truncation analysis we had previously performed for Ova 323-339 revealed a core region within the Ha peptide that consists of the sequence NGVTAACS as being critical for binding to IA^d. This sequence contains the predicted binding region based on the structural similarity analysis, VTAACS. Similarly, for the myoglobin peptide, N- and C-terminal truncation analysis indicated that the IA^d binding region was contained in the sequence AIIHVLHS. Again, the predicted region of similarity with the Ova peptide, IHVLHS, is contained within this IA^d binding region of the myoglobin peptide. Thus, the structural similarities between these unrelated peptides that we have discerned on the basis of sequence analysis are indeed critical for the expression of IA^d binding capacity by these peptides. A further indication that this is the case is that we have been able to predict other peptides that can bind to IA^d on the basis of their possessing this motif and have been able to experimentally verify this potential IA^d binding capacity by synthesizing the peptide and confirming its IA^d binding capacity (Sette et al, 1988).

To further characterize the relative contribution to IA^d binding of individual residues within the Ova 323-339 peptide, we synthesized a series of 55 single substitution analogs of the peptide Ova 323-336, which is highly stimulatory for certain T cells and binds equally well as Ova 323-339 to IA^d. For each of the 11 positions, 325-335, we synthesized five analogs, each carrying a single amino acid substitution (two conservative, one semi-conservative, and two non-conservative). When these analogs were tested for their capacity to bind to IA^d, it was found that most of the substitutions had little or no effect on Ia binding. Only nine out of 55 (16%) had a significant effect (more that three-fold reduction), and only seven substitutions led to more than a five-fold reduction in IA^d binding capacity (Fig. 1). The most dramatic changes were seen at residues V_{327} and A_{332}. These

Table 1. Structural Similarities Between Unrelated Peptides with Strong IA^d Binding Capacity

	Residue Number					
Peptide	1	2	3	4	5	6
Ova 323-339	V	H	A	A	H	A
Ha 130-142	V	T	A	A	C	S
Ha 187-206	V	G	T	Y	V	S
Myo 63-78	V	T	V	L	T	A
Myo 108-118	I	H	V	L	H	S
A^d α 6-20	I	T	V	Y	Q	S
Nase 101-120	V	R	Q	G	L	A
Nase 1-20	A	T	L	I	K	A
Ova 317-327	L	K	I	S	Q	A
p cyt 88-104	L	K	Q	A	T	A
CS 388-393	A	K	M	E	K	A
λ rep 12-26	A	R	R	L	K	A
HEL 74-86	C	S	A	L	L	S
Ova 213-322	L	S	G	I	S	S

results are in agreement with those obtained with the series of truncated peptides and indicate that V_{327} and A_{332} are the most critical residues involved in determining IA^d binding. Significant but less dramatic effects were detected at H_{328} and E_{333}. Although we have interpreted the decrease in IA^d binding capacity associated with substitutions at these positions as being caused by alterations of IA^d contact residues, we cannot exclude the possibility that the decreased binding could be secondary to a change in peptide conformation caused by the substitutions.

We have also recently performed a similar analysis of the interaction between IE^d molecules and its peptide ligands. In this case the motif

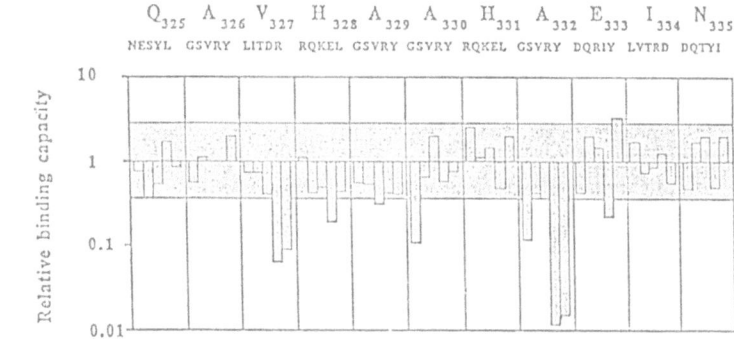

Fig. 1. The relative capacity of analog OVA 323-336 peptides to bind to IA^d. IA^d-interacting capacity of single substitution analogs of Ova 323-336. Five different substitutions (two conservative, one semi-conservative, two non-conservative) were introduced into each of the eleven positions 325-335. These peptides were then tested for IA^d binding. Each bar represents the arithmetic mean of two to four independent determinations. Shaded area represents 99% confidence limits. a, parent sequence: b, single substitutions with:

Table 2. IE^d Motifs in Unrelated Peptides with Good IE^d Binding Capacity

Peptide	Residue number 1	2	3	4	5	6
Myo 108-118	H	V	L	H	S	R
Nase 111-130	H	E	Q	H	L	R
HSV 8-23		R	F	R	G	K
HEL 105-120		R	N	R	C	K
HIV p 17, 17-32		K	Y	K	L	K
Dynorphin 1-13		R	I	R	P	K
HIV p 17, 9-24		K	I	R	L	R
λ rep 12-26			R	R	L	K
CS 325-341			K	K	I	K

Underlined residues designate the IE^d motif (basic residue, basic residue, non-charged residue, basic residue) described in the text.

recognized is different and based on the presence of a basic residue either in position 1, 2 or 3, and two more basic residues at positions 4 and 6. Position 5 is non-charged and tends to be hydrophobic (S, L, G, C, P, I, have been found in their position) (Table 2). Analogously to what is shown in the IA^d system we have been able to experimentally confirm the involvement of these predicted core binding regions in four of these IE^d binding peptides (Sette et al, 1988; Adorini et al, 1988, Sette et al, manuscript in preparation). We have also analyzed for their IE^d binding capacity a large panel of single amino acid substitution analogs of HEL 107-116. Significant effects were seen only at the three basic residues implicated as critical by the sequence and truncation analysis. They also confirm the very permissive nature of peptide binding to Ia, in that about 90% of the substitutions tested had little or no effect on the IE^d binding capacity of HEL 107-116 (Sette et al, manuscript in preparation).

In summary, these findings are in keeping with the determinant selection hypothesis and suggest the possible mechanism by which Ia molecules could bind many seemingly unrelated peptides.

The Relative Contribution of "Determinant Selection" and "Holes in the T Cell Repertoire" in Defining T Cell Responses to a Protein Antigen

Two main theories have been proposed to explain the MHC control of T cell immune responses. According to one line of reasoning ("determinant selection"), MHC molecules act as specific antigen receptors, thereby allowing some, but not all, antigens to interact with a particular MHC and form potentially T cell stimulatory complexes (Rosenthal, 1978). The other line of reasoning ("holes in the T cell repertoire") claims that MHC control of T cell responses is at the level of the T cell itself and not at the level of the interaction between antigen and MHC (Klein, 1982). In the broadest sense such "holes in the T cell repertoire" may be due to either an absence of the relevant T cell receptor (caused by limitations of the T cell receptor gene repertoire or by deletion of T cells) or, alternatively, by the unresponsiveness of antigen-specific T cells (caused by anergy or by suppression) (Paul, 1984).

Table 3. Binding of a Panel of Fourteen Nase Peptides to Four Ia Molecules

Nase peptide	Concentration (μM) of Nase peptide needed to obtain 50% inhibition of binding of:			
	^{125}I-Ova 323-339 to IA^d	^{125}I-repr 12-26 to IE^d	^{125}-HEL 46-61 to IA^k	^{125}I-Nase 101-120 to IE^k
Nase 1-20	25	–	–	300
Nase 11-30	150	–	–	450
Nase 21-40	–	–	–	–
Nase 31-50	–	–	–	500
Nase 41-60	–	–	–	–
Nase 51-70	–	–	–	–
Nase 61-80	225	–	–	500
Nase 71-90	–	–	–	–
Nase 81-100	–	–	–	25
Nase 91-110	125	–	65	–
Nase 101-120	20	500	400	35
Nase 112-130	–	150	550	350
Nase 121-140	–	–	–	125
Nase 131-149	–	–	–	–

50% inhibition >600 μM.

We therefore performed a series of functional and biochemical experiments in an attempt to evaluate the extent to which determinant selection influences the immune response to a foreign protein. For this study we selected a panel of 14 overlapping peptides representing the entire sequence of the protein staphylococcal nuclease (Nase) to estimate the frequency of Ia binding sites within a natural protein antigen, and to evaluate the relative contribution of "determinant selection" vs "holes in the T cell repertoire" in the generation of T cell responses to these peptides. Each of the 20 amino acid-long peptides overlapped the sequence of the adjacent peptides by 10 amino acids (Finnegan et al, 1986).

The ability of this set of peptides to bind to the Ia molecules expressed by H-2^d and H-2^k mice was measured by their capacity to inhibit the binding of radiolabeled peptides that had been previously characterized for their capacity to bind to particular Ia specificities. Table 3 shows the capacity of these four different affinity purified Ia molecules to bind the panel of 14 Nase peptides. Five peptides, 21-40, 41-60, 51-70, 71-90 and 131-149, failed to bind any of the four Ia molecules tested, whereas one peptide, 101-120, bound to all four Ia specificities. The remaining peptides bound to one to three of the four Ia molecules. Five peptides were found to bind to IA^d, two to IE^d, three to IA^k and eight to IE^k. Due to the overlapping nature of the peptides, the binding of two adjacent peptides to a particular MHC molecule may be due to a single Ia binding site present on both peptides. If this were the case in all instances of adjacent peptide binding to the same MHC, then there would be three peptide binding regions for IA^d, one for IE^d, two for IA^k, and six for IE^k.

We next studied the capacity of each of the 14 Nase peptides to immunize Balb/c (IA^d, IE^d) and CBA/J (IA^k, IE^k) mice (Table 4). For those peptides that were immunogenic, the MHC restriction element used in the immune response was determined by anti-Ia inhibition or by immunization of congenic

Table 4. Immunogenicity of the Nase Peptides

Nase peptide	Mean 3H-TdR incorporation x 10-3 by T cells from		
	Balb/c	CBA/J	BIO.A(4R)
Nase 1-20	57	25	1
Nase 11-30	-1	15	NT
Nase 21-40	3	4	NT
Nase 31-50	-1	9	NT
Nase 41-60	0	0	NT
Nase 51-70	0	1	NT
Nase 61-80	90	-1	NT
Nase 71-90	0	-2	NT
Nase 81-100	6	107	0
Nase 91-110	2	70	33
Nase 101-120	61	71	27
Nase 112-130	5	110	32
Nase 121-140	1	50	1
Nase 131-149	1	-3	NT

Mice were immunized with 25 µg of the different Nase peptides in complete Freunds adjuvant. One week later lymph node cells were obtained and T cells nylon wool purified. In triplicate microtiter plates, 4 x 10^5 T cells/well were incubated with 2 x 10^5 4000R irradiated spleen cell/wells and 1 µg of peptide. 3H-TdR incorporation was measured five days later by scintillation spectrometry. The mean responses of two or more experiments are reported with the background (T cells + spleen cells without antigen) subtracted.

strains of mice. In Balb/c mice, three peptides (1-20, 61-80 and 101-120) were found to be immunogenic. All three peptides were restricted to IA^d, and for one peptide, Nase 101-120, IE^d was also used as a restriction element. For CBA/J mice, seven peptides, 1-20, 11-30, 81-100, 91-110, 101-120, 112-130 and 121-140 were immunogenic. By their failure to be immunogenic in BIO.A(4R) mice, it could be concluded that three of these peptides were restricted solely by IE^K (Nase 1-20, 81-100 and 121-140). Both IE^K and IA^K were used as restriction elements, roughly equally, for Nase 91-110 and 112-130; and for 101-120, IE^K was used as the major restriction element, together with a minor IA^K component. Due to the relatively low response, the restriction element used by Nase 11-30 could not be determined.

Finally, the data on the capacity of the four Ia molecules to bind the 14 Nase peptides were combined with the data on the ability of the different Ia specificities to serve as restriction elements for Nase specific T cell responses. The level at which binding of a peptide to Ia was significant with respect to immunogenicity was not known, but empirically it was observed that all combinations of Nase peptides and Ia that resulted in very weak or undetectable binding (50% inhibitory concentration >600 uM) did not stimulate a T cell response. Using this as a lower limit of "significant binding", we assessed the relationship between binding and immunogenicity. Sixteen of the 54 combinations tested (30%) showed significant binding to Ia; of those, a T cell response was elicited in 10 (62%) instances. In no instance did a peptide that failed to bind to Ia at this level induce a T cell response.

Table 5. Correlation Between Binding to Ia and Immunogenicity

Binding to Ia (50% inhibition dose)	Immunogenicity		
	Yes	No	Total
5-100 uM	5	0	5
101-500 uM	5	7	12
>500 uM	0	37	37
Total	10	44	54

The finding of a significant correlation between peptide binding to Ia and immunogenicity strongly supports the determinant selection hypothesis, i.e. that Ia serves as receptor that selects antigenic determinants and that binding of antigen to Ia is a necessary, albeit not sufficient prerequisite for T cell recognition. If staphylococcal nuclease peptides are representative of other protein antigens, the data would suggest that only about 30% of peptide/Ia interactions are of an affinity compatible with immunogenicity. That is, at the level of any single MHC specificity, less than one-third of the potential antigenic determinant are selected for presentation. Furthermore, of those peptides that have been selected on the basis of their capacity to interact with MHC, only about 60% induce a T cell response, leaving the other 40% as possible "holes in the T cell repertoire". Thus, we conclude that both determinant selection and holes in the T cell repertoire act in concert to define the immune responsiveness of an individual.

The significance of peptide/Ia interaction in terms of what strength of binding is needed to allow presentation to T cells has been hitherto unknown. In a previous study of a panel of peptides that had been selected for their immunogenicity, we found that most of these peptides bound to their restriction element with a 50% inhibitory concentration of 5-100 μM (Buus et al, 1987a). In Table 5, the binding of the Nase peptides to Ia has been divided into three groups: 50% inhibition at 5-100 μM, at 101-600 μM, and >600 μM, and the T cell stimulatory capacity of these groups determined. Five of five peptides that bound to Ia at a 50% inhibitory concentration of 5-100 μM were capable of eliciting T cell responses. Whereas only five of the 12 peptides (42%) that bound to Ia at 50% inhibitory concentrations between 101-600 μM were capable of eliciting T cell responses, none of the peptided with a 50% inhibitory concentration >600 μM elicited an immune response. Thus, the affinity of a peptide for Ia had a profound influence on its T cell stimulatory capacity. The great majority of the best interactions are productive, dropping off to less than one-half of the interactions with intermediate affinity and disappearing completely with the lowest affinity. These data suggest that defects in the T cell repertoire are most pronounced for peptides with intermediate affinity for Ia and are rarely seen for the best Ia binding peptides. We have previously observed one instance of a strong binding peptide to a given Ia that failed to be immunogenic in the context of that restriction element (Guillet et al, 1987).

On a population basis, Ia molecules are among the most polymorphic proteins known; however, each individual possesses only a few Ia alleles. Studies of binding of peptides to Ia have suggested that each Ia molecule possesses a single binding site. How does the immune system achieve a sufficient T cell repertoire despite the requirements for specific binding of antigen to Ia? We have previously found that each Ia can bind many seemingly different peptides and that Ia is very permissive in its capacity to bind antigen, probably because it recognizes abroadly defined "motifs" within antigens (Sette et al, 1988). Indeed, the data presented herein demonstrate that in the set of 14 peptides representing the Nase protein, a minimum of three sites were detected for Ia^d, one for IE^d, one for IA^k, and six for IE^k. Thus, on average, an Ia specificity bound three peptide regions within the Nase protein. By extrapolation it can be estimated that each Ia specificity will bind approximately 18 sites on a protein antigen of 100,000 MW. Thus, using an unbiased panel of peptides, it does not appear likely that even the smallest micro-organism with only a few proteins could escape the immune system due to the absence of Ia binding sites.

Acknowledgements

This work was supported in part by NIH grants AIO9758 and AI18634. The expert secretarial assistance of Joyce Joseph is gratefully acknowledged.

REFERENCES

Adorini, L., Sette, A., Buus, S., Grey, H.M., Darsley, M., Lehmann, P.V., Doria, G., Nagy, Z.A. and Appella, E., 1988, Interaction of an immunodominant epitope with Ia molecules in T-cell activation, Proc.Natl.Acad.Sci.USA., 85:5181.
Allen, P.M., Babbitt, B. and Unanue, E., 1987, T cell recognition of lysozyme - the biochemical basis of presentation, Immunol.Rev., 98:172.
Babbittt, B., Allen, P., Matsueda, G., Haber, E. and Unanue, E., 1985, The binding of immunogenic peptides to Ia histocompatibility molecules, Nature, 317:359.
Buus, S., Sette, A., Colon, S.M., Jenis, D.M. and Grey, H.M., 1986, Isolation and characterization of antigen-Ia complexes involved in T cell recognition, Cell, 47:1071.
Buus, S., Sette, A., Colon, S.M., Miles, C. and Grey, H.M., 1987a, The relation between major histocompatibility complex (MHC) restriction and the capacity of Ia to bind immunogenic peptides, Science, 235:1352.
Buus, S., Sette, A. and Grey, H.M., 1987b, The interaction between protein-derived immunogenic peptides and Ia, Immunol.Rev., 98:115.
Finnegan, A., Smith, M.A., Smith, J.A., Berzofsky, J., Sachs, D.H. and Hodes, R.J., 1986, The T cell repertoire for recognition of a phylogenetically distant protein antigen, J.Exp.Med., 164:897.
Guillet, J., Lai, M., Briner, T.J., Buus, S., Sette, A., Grey, H.M., Smith, J.A. and Gefter, M., 1987, Immunological self, nonself discrimination, Science, 235:1353.
Klein, J., 1982, "Immunology", John Wiley & Sons, New York.
Paul, W., 1984, "Fundamental Immunology", Raven Press, New York.
Rosenthal, A.S., 1978, Determinant selection and macrophage function in genetic control of the immune response, Immunol.Rev., 40:136.
Sette, A., Buus, S., Colon, S., Miles, C. and Grey, H.M., 1988, Structural analysis of peptides capable of binding to more than one Ia antigen, J.Immunol, in press.
Shimonkevitz, R., Kappler, J., Marrack, P. and Grey, H., 1983, Antigen recognition by H-2 restricted T cells, J.Exp.Med., 158:303.
Watts, T.H., Brian, A.A., Kappler, J.W., Marrack, P. and McConnell, H.M., 1984, Antigen presentation by supported planar membranes containing affinity purified I-A^d, Proc.Natl.Acad.Sci.USA., 81:7564.
Werdelin, O.J., 1982, Chemically related antigens compete for presentation by accessory cells to T cells, J.Immunol., 129:1883.

ROLE OF INTERFERON GAMMA IN THE PRIMING OF HUMAN ACCESSORY CELLS FOR

AUTOCRINE SECRETION OF MONOKINES (IL-1 AND TNF) AND HLA CLASS II GENE

EXPRESSION

Virelizier, J.L.[1], Arenzana-Seisdedos, F.[1] and
Mogensen, S.C.[2]

[1]Laboratoire d'Immunologie Virale, Institut Pasteur
75724 Paris Cedex 15, France and [2]Institute of Medical
Microbiology, University of Aarhus, Aarhus, Denmark

INTRODUCTION

Vaccine preparations of the future are likely to use as antigens recombinant proteins or synthetic peptides rather than whole micro-organisms.
This implies the use of adjuvants to increase the poor immunogenicity of
such molecules for both antibody and T lymphocyte responses. The mechanism
of action of adjuvants is complex and not entirely understood, but is likely
to be associated with increased antigen-presenting ability of specialized
cell types such as monocyte-macrophage or dendritic cells. It is apparent
that an essential step in the induction of responsiveness to foreign proteins is the activation of helper T lymphocytes after specific recognition
of relevant epitopes on the membrane of antigen-presenting cells (APC). The
aim of the present communication is not to review the complex events associated with antigen presentation, and the reader is referred to excellent
reviews by Unanue et al, 1984 and Germain, 1986. Rather, we will discuss
regulation of some properties of APC by exogenous lymphokines such as interferon (IFN) gamma, and the role of autocrine secretion of monokines in IFN
gamma-induced macrophage activation and HLA class II gene expression, two
essential requirements for efficient antigen presentation (Kurt-Jones et al,
1985).

Resting and activated T lymphocytes exhibit distinct stimulatory
(antigen-presenting cells) requirements for growth and lymphokine release
(Inaba and Steinman, 1984). APC provide two types of signals to T lymphocytes. One is to present on the membrane a processed form of foreign antigens in association with class II major histocompatibility complex (MHC)
antigens for MHC-restricted T cell receptors to recognize and bind specifically such self-plus-non-self complex. This signal may be sufficient to
induce specific responses in T cells already activated in vitro, such as T
cell clones or hybridomas. The second signal is non-specific, and consists
in the induction of a state of activation in resting T lymphocytes. Such
activation will enable resting T cells to leave the GO state, enter the Gl
state, respond to growth factors and proliferate clonally to perform their
immune effector functions. There is evidence that the latter signal is
associated with the production of interleukin 1 (IL-1) by APC. We will now
discuss evidence that IFN-gamma, a lymphokine produced after specific stim-

21

ulation of T lymphocytes, may regulate these two human APC functions by increasing the IL-1 secretory potential and, in synergy with tumor necrosis factor (TNF), the expression of HLA class II genes in such cells.

ENHANCED IL-1 SECRETORY FUNCTION OF HUMAN PERIPHERAL BLOOD MONOCYTES INCUBATED WITH IFN GAMMA

When human circulating monocytes are separated by adherence onto plastic surfaces and cultured in vitro, very little if any IL-1 activity can be detected in supernatants, provided that the culture medium does not contain bovine serum and is not contaminated by endotoxins (LPS). Addition of LPS or Poly IC to the culture induces in a few hours secretion of IL-1 in supernatants, measurable in standard murine thymocyte proliferation assays. We have investigated the ability of the three types of IFN (alpha, beta and gamma) to modulate the IL-1 secretory potential of human monocytes in vitro (Arenzana-Seisdedos et al, 1985). Monocytes cultured in medium containing human serum did not secrete detectable IL-1. Overnight incubation of such cultures with recombinant IFN alpha, beta or gamma resulted in a much enhanced secretion of IL-1, which was observed after washing and stimulation with either LPS or Poly IC. Incubation with IFN without further stimulation of cells did not induce detectable IL-1 secretion, indicating that IFN does not behave as a direct IL-1 inducer. Instead, incubation with IFN induced an enhanced potential for IL-1 secretion that could be revealed by a triggering signal provided by endotoxins or Poly IC. In such freshly explanted monocyte cultures, IFN gamma consistently showed preferential priming effects over IFN alpha or beta, since the levels of LPS-induced IL-1 secretion was greater, and obtained with lower concentrations of IFN-gamma.

When monocyte cultures were aged for 4 to 12 days in vitro, the enhancing effects of IFN-gamma on IL-1 secretory potential was found to be selective, not shared by IFN alpha or beta. We have observed a progressive and profound loss of the ability of LPS or Poly IC to induce IL-1 activity in supernatants of aged cultures. The mechanism of this loss is not understood. It did not appear to be due to a regulatory effect of secreted prostaglandins, since dialysing the supernatants or incubating the cells with indomethacin did not modify the results. IFN-gamma, but not IFN-alpha or beta, was found to maintain (when added at the onset of the cultures) or reverse the loss (when added on the fourth day of culture) of the IL-1 secretory function (Arenzana-Seisdedos et al, 1985). Others have shown that IFN-gamma-primed human monocytes accumulate more IL-1 alpha and IL-1 beta steady state mRNA in response to LPS (Burchett et al, 1988). These authors also showed that, in the absence of LPA, IFN-gamma does not increase IL-1 alpha or IL-1 beta mRNA. The exact mechanism whereby IFN-gamma enhances IL-1 gene expression is not yet understood. Collart et al (1986) have suggested that IFN-gamma inhibits the production of a putative labile transcription repressor, thus prolonging the period of increased transcription induced by LPS stimulation. Whatever the causative mechanism(s), it is clear that IFN-gamma modulates IL-1 production in monocytes by inducing a potential rather than ongoing secretory function. In both fresh and aged monocyte cultures, IFN gamma primes monocytes for enhanced IL-1 gene expression that will eventually be triggered by a second signal.

MODULATION OF IL-1 SECRETION IN LANGERHANS CELLS BY IFN-GAMMA

Monocyte-macrophages may not be the most efficient antigen-presenting cells of the body. Dendritic cells such as Langerhans cells of the skin appear to be much more efficient for antigen presentation than T lymphocytes (Bjerck et al, 1984). Since separation of Langerhans cells from the skin is difficult, we have taken advantage of their OKT6 (CD1) membrane positivity to sort them from histiocytosis X granuloma biopsy (where such cells largely predominate) in order to investigate their IL-1 secretory potential (Arenzana-Seisdedos et al, 1986). We found that Langerhans cells spontaneously secrete IL-1, and that secretion is increased after LPS stimulation in culture. When Langerhans cells were incubated with recombinant IFN-gamma, there was no increase in background level of IL-1 secretion. Similarly to what was previously observed with monocyte cultures, IFN-gamma-treated cultures showed increased levels of IL-1 activity in their supernatants after simulation with LPS.

RESPECTIVE ROLE OF IFN-GAMMA AND TNF IN HLA CLASS II GENE INDUCTION IN HUMAN MONOCYTES

All three species (alpha, beta and gamma) of IFN induce HLA class I gene expression in human cells, but only IFN-gamma is able to induce HLA class II gene expression (Rosa and Fellous, 1984). Thus pure preparations of IFN-gamma were shown to induce HLA-DR expression on a large repertoire of cell types, including human monocytes (Sztein et al, 1984) and monocytic cell lines (Virelizier et al, 1984). The question arises, however, of whether IFN-gamma is the sole cytokine able to induce class II expression, since induction of such genes by IFN-gamma is variable in different cell types. Furthermore, there is evidence that crude lymphokine supernatants contain a mediator, different from IFN-gamma, which participates in the full induction of class II antigens in monocytes (Walker et al, 1984).

With these observations in mind, we have investigated (Arenzana-Seisdedos et al, 1988) whether TNF could have a role in IFN-gamma-induced HLA class II induction in human monocytic cell lines, taking advantage of our previous observation that one such cell line (THP-1) is constitutively positive for membrane DR expression, whereas another (U937) is not (Virelizier et al, 1984). Recombinant IFN-gamma was shown to be very efficient at enhancing DR expression in THP-1 cells, whereas de novo induction of DR antigens in U937 cells was very modest, even when high concentrations of IFN-gamma were used. When recombinant TNF was used alone, a clear induction of DR mRNA and antigens was observed in THP-1, but not in U937 cells. When IFN-gamma and TNF were used in combination, both U937 and THP-1 cells showed clear induction of class II expression. This synergism was observed at both cytoplasmic mRNA and membrane DR and DQ antigen level.

The lack of detectable DR gene response to TNF stimulation in U937 cells was not due to deficient TNF receptor expression in this cell line. Indeed incubation of U937 cells in TNF-containing medium resulted in a clear induction of HLA class I and TNF gene expression, ruling out defective responsiveness of U937 cells to TNF. Alternatively, the different sensitivity of the two cell lines to IFN-gamma-induced DR expression could be due to distinct effects of IFN-gamma on TNF secretion in vitro. This was indeed shown to be the case, since IFN-gamma induced TNF activity in supernatants of THP-1, but not U937 cells. Direct demonstration of a role for endogenous TNF secretion in this system was obtained by neutralization of TNF molecules by a specific antibody. In the presence of antibody to TNF, IFN-gamma was found to be less efficient at inducing DR antigen in THP-1 cells, whereas the modest induction seen in U937 cells was not decreased by TNF neutralization (Arenzana-Seisdedos et al, 1988).

In the light of our finding that TNF synergizes with IFN-gamma for HLA class II gene induction in the two monocytic cell lines used, but has no detectable effects on the gene when used alone in U937 cells, it appears that TNF does not behave as a direct HLA class II gene inducer, but rather amplifies an ongoing expression of such genes, whether constitutive (as in THP-1 cells) or IFN-gamma-induced (as in both U937 and THP-1 cells). This concept may help reconcile apparently conflicting results showing that TNF does not induce HLA class II genes in fibroblasts (Collins et al, 1986), but does so in a murine myelomonocytic cell line (Chang and Lee, 1986). The different level of basal class II antigen expression in fibroblasts (class II negative) and macrophages is likely to account for such differences in the effect of TNF used alone.

The concept that IFN-gamma is able to induce its own amplifier cytokine has interesting implications. This phenomenon may provide a selective advantage in terms of HLA class II expression, to cell types such as monocyte-macrophages able to produce TNF under the influence of very low concentrations of IFN-gamma, and also respond to TNF in an autocrine manner by amplifying class II gene expression. This may be an efficient "double-lock" system selecting, among many cell types with IFN-gamma receptors, specialized antigen-presenting cells with a high sensitivity to the HLA class II-inducing effects of IFN-gamma.

CONCLUSION: ROLE OF AUTOCRINE SECRETION OF MONOKINES IN IFN-GAMMA-DEPENDENT REGULATION OF ACCESSORY CELL FUNCTIONS

From the observations described above, there is clear evidence that IFN-gamma enhances IL-1 secretion in human monocytes and Langerhans cells, and also induces TNF secretion in monocytes, which in turn amplifies the HLA class II-inducing effects of IFN-gamma. The signals provided to APC by IFN-gamma and monokines appear to be different, acting at different stages of the activation process, and may thus not be redundant. It appears that IFN-gamma provides a priming, differentiating signal for potential functions, associated at that stage with very little ongoing new functions. Indeed under clean experimental conditions in vitro, IFN-gamma induces some increase in intra-cellular IL-1 activity, but no detectable enhancement of IL-1 activity in cell supernatants. Similarly, IFN-gamma induction of HLA class II antigens is incomplete when autocrine secretion of TNF is blocked by a neutralizing antibody, or in cells unable to produce TNF under the influence of IFN-gamma. However, IFN-gamma-treated cells appear to better perform IL-1 secretion and DR-gene expression when a second signal triggers such functions. Such triggering signals are mimicked by LPS or Poly IC in vitro, and are likely to be associated in vivo with phagocytosis of micro-organisms, or binding of specific immune complexes on membrane Fc-gamma receptors. This would lead to induction of monokine secretion, which in turn appear to be able to auto-activate monocytes in a loop of amplification, leading to a state of full cellular activation. Monokines indeed induce their own gene and those of a series of other mediators in monocytes (Philip and Epstein, 1986). We have shown that one of these monokines, namely TNF, amplifies IFN-gamma-induced HLA-class II expression in monocytes, thus completing through autocrine secretion the priming signal provided by IFN-gamma. This may facilitate further class II-restricted immune responses to micro-organisms.

Such amplification mechanisms, if they were not regulated in vivo, might have deleterious consequences for the body. This may well be what happens in histiocytosis X bone lesions, where the local accumulation of Langerhans cells able to secrete IL-1 is associated with massive osteolysis, a known property of IL-1 in vitro. Excessive secretion of monokines by systemically activated monocytes might result in severe general inflammatory and toxic reactions. It may be, however, that the "double-lock" mechanism of monocyte activation that we have discussed above does not normally permit full systemic activation of the monocytic system. Monocyte primed locally by IFN-gamma would then circulate without actually performing their potential functions. They would enter autocrine secretion steps and full activation state only when and where needed, presumably in sites of tissue lesions and local immune responses.

It is also possible to envisage that, in inflammatory lesions, production of TNF by activated macrophages enhances IFN-gamma-induced HLA class II gene expression in cell types other than monocyte. It may be that the "hyperexpression" of class II antigens that we have recently analysed in synovial cells of joints from rheumatoid patients (Teyton et al, 1987) is due to the local release, among other immunoregulatory cytokines, of TNF able to facilitate class II antigen induction by IFN-gamma secreted in low amounts by infiltrating T lymphocytes.

Finally, the observations discussed herein open up the possibility that IFN-gamma increases the antigen-presenting functions of human APC through its ability to enhance class II antigen expression. There is clear evidence that the density of class II antigens is important in antigen presentation. Thus the magnitude of a T cell response has been shown to be a function of the concentration of both antigen and class II molecules (Matis et al, 1983). Furthermore, the antigen-presenting capacity of Ia+ L cell transfectants is correlated with the amounts of Ia on the transfectants (Lechler et al, 1985). This is likely to be true for classical antigen-presenting cells, since enhanced antigen-presenting capacity of cultured Langerhans cells has been shown to correlate with markedly increased expression of Ia antigens (Shimada et al, 1987). It is likely that increased IL-1 production and class II antigen expression in APC participate in the reported adjuvanticity of IFN-gamma administration in vivo together with specific antigens in mice (Nakamura et al, 1984) and man (Grob et al, 1984).

Aknowledgement

We gratefully acknowledge support by NATO grant No. 85/0487.

REFERENCES

Arenzana-Seisdedos, F., Virelizier, J.L. and Fiers, W., 1985, Interferons as macrophage-activating factors. III. Preferential effects of interferon gamma on the interleukin 1 secretory potential of fresh or aged human monocytes, J.Immunol., 134:2444.
Arenzana-Seisdedos, F., Barbey, S., Virelizier, J.L., Kornprobst, M. and Nezelof, C., 1986, Histocytosis X: Purified (T6+) cells from bone granuloma produce interleukin 1 and prostaglandin E2 in culture, J.Clin.Invest., 77:326.
Arenzana-Seisdedos, F., Mogensen, S.C., Vuillier, F., Fiers, W. and Virelizier, J.L., 1988, Autocrine secretion of tumor necrosis factor under the influence of interferon-gamma amplifies HLA-DR gene induction in human monocytes, Proc.Natl.Acad.Sci.USA, 85:000.
Bjerck, S., Elgo, J., Braathen, L. and Thorsby, E., 1984, Enriched epidermal Langerhans cells are potent antigen-presenting cells for T cells, J.Invest.Dermatol., 77:286.
Burchett, S.K., Weaver, W.M., Westall, J.A., Larsen, A., Kronheim, S. and Wilson, C.B., 1988, Regulation of tumor necrosis factor-cachectin and IL-1 secretion in human mononuclear phagocytes, J.Immunol., 140:3473.
Chang, R.J. and Lee, S.H., 1986, Effects of interferon-gamma and tumor necrosis factor on the expression of an Ia antigen on a murine macrophage cell line, J.Immunol., 137:2853.
Collart, M.A., Belin, D., Vassalli, J.D., de Kossodo, S. and Vassalli, P., 1986, Gamma interferon enhances macrophage transcription of the tumor necrosis factor-cachectin, interleukin 1 and urokinase genes, which are controlled by short-lives repressors, J.Exp.Med., 164:2113.
Collins, T., Lapierre, L.A., Fiers, W., Strominger, J.L. and Pober, J.S.,

1986, Recombinant human tumor necrosis factor increase mRNA levels and surface expression of HLA-A-B antigens in vascular endothelial cells and dermal fibroblasts in vitro, Proc.Natl.Acad.Sci.USA, 83:446.

Germain, R.N., 1986, The ins and outs of antigen processing and presentation, Nature, 322:687.

Grob, P.J., Joller-Jemelka, H.I., Binswanger, V., Zaruba, K., Descoendres, C. and Fernex, M., 1984, Interferon as an adjuvant for hepatitis B vaccination in non- and low responder populations, Eur.J.Clin.Microbiol., 3:195.

Inaba, K. and Steinman, R.M., 1984, Resting and sensitized T lymphocytes exhibit distinct stimulatory (antigen-presenting cells) requirements for growth and lymphokine release, J.Exp.Med., 160:1717.

Kurt-Jones, E.A., Virgin, H.W. and Unanu, E.R., 1985, Relationship of macrophage Ia and membrane IL-1 expression to antigen presentation, J.Immunol., 135:3652.

Lechler, R.I., Norcross, M.A. and Germain, R.N., 1985, Qualitative and quantitative studies of antigen-presenting cell function by using I-A expressing L cells, J.Immunol., 135:2914.

Matis, L.A., Glimcher, L.H., Paul, W.E. and Schwartz, R.H., 1983, Magnitude of response of histocompatibility-restricted T cell clones is a function of the concentration of antigen and Ia molecules, Proc.Natl.Acad.Sci.USA, 80:6019.

Nakumura, M., Mauser, T., Pearson, G.D.N., Daley, M.J. and Gefter, M.L., 1984, Effect of IFN-gamma on the immune response in vivo and on gene expression in vitro, Nature, 307:381.

Philip, R. and Epstein, L., 1986, Tumor necrosis factor as immunomodulator and mediator of monocyte cytotoxicity induced by itself, interferon and interleukin 1, Nature, 323:86.

Rosa, F. and Fellous, M., 1984, The effect of gamma-interferon on MHC antigens, Immunol.Today, 5:261.

Shimada, S., Canghman, W., Sharrow, S.O., Stephany, D. and Katz, S.I., 1987, Enhanced antigen-presenting capacity of cultured Langerhans cells is associated with markedly increased expression of Ia antigen, J.Immunol., 139:2551.

Sztein, M.B., Steeg, P.S., Johnson, H.M. and Oppenheim, J.J., 1984, Regulation of human peripheral blood monocyte DR antigen expression in vitro by lymphokines and recombinant interferons, J. Clin.Invest., 73:556.

Teyton, L., Lotteau, V., Turmel, P., Arenzana-Seisdedos, F., Virelizier, J.L., Pujol, J.P., Loyau, G., Piattier-Tonneau, D., Auffray, C. and Charron, D., 1987, HLA-DR, DQ and DP antigen expression in rheumatoid synovial cells: a biochemical and quantitative study, J.Immunol., 138:1730.

Unanue, E.R., Beller, D.I., Lu, C.Y. and Allen, P.M., 1984, Antigen presentation: Comments on its regulation and mechanism, J.Immunol., 132:1.

Virelizier, J.L., Perez, N., Arenzana-Seisdedos, F. and Devos, R., 1984, Pure interferon-gamma enhances class II HLA antigens on human monocyte cell lines, Eur.J.Immunol., 14:106.

Walker, E.B., Maino, V., Sanchez-Lanier, M., Warner, N. and Stewart, C., 1984, Murine gamma interferon activates the release of a macrophage-derived Ia inducing factor that transfers Ia inductive capacity, J.Exp.Med., 159:1532.

CHEMICAL EVENTS IN IMMUNE INDUCTION: EVIDENCE FOR A COVALENT INTER-

CELLULAR REACTION ESSENTIAL IN ANTIGEN-SPECIFIC T-CELL ACTIVATION

J. Rhodes

Wellcome Research Laboratories, Langley Court
Beckenham, Kent BR3 3BS, UK

INTRODUCTION

Thymus-derived helper (T_H) lymphocytes are activated following recognition of foreign antigens in the context of class II major histocompatibility (MHC) gene products expressed by syngeneic antigen presenting cells (APC) (see chapter by Grey in this volume). Current evidence indicates that processed antigen and class II (1a) molecules interact to form a stable complex whose orientation effectively selects a particular determinant for recognition. The presentation of antigen requires prolonged physical interaction between APC and T_H-Cell, and studies at the macromolecular level have begun to define ancillary molecules that facilitate this process (Weiss and Imboden, 1987; Springer et al, 1987) (see Fig. 1). The present study focuses on events at the chemical level in the inductive interaction between class II positive APC and T-cell. Remarkably, the evidence reviewed here shows that a covalent reaction occurs between ligands on the APC surface and the T-cell surface and that the formation of these reversible covalent bonds is an essential process in antigen-induced T-cell activation.

In 1972 Novogrodsky and Katchalski of the Weizmann Institute, Rehovot, Israel, showed that oxidation of mononuclear cells with sodium periodate could produce a vigorous proliferative response. They found that the effect was due to the generation of an aldehyde group on cell-surface sialic acid (Novogrodsky and Katchalski, 1972). This was the first demonstration of a chemically defined method for activating T-cells and it remains the only method. The phenomenon was taken up by Rosenthal and colleagues who showed an obligatory requirement for accessory cells (macrophages). These authors concluded that the generation of aldehydes on sialic acid by periodate results in a covalent reaction between lymphocyte and accessory cell (Greineder and Rosenthal, 1975). The same effect was produced by generating aldehydes on the galactose or n-acetylgalactosaminyl residues exposed after neuraminidase treatment by means of galactose oxidase, and the reaction was fully reciprocal: aldehydes on either accessory cell or lymphocyte being equally effective. Although the response was Ia-dependent it was not MHC restricted, and allogeneic as well as syngeneic macrophages could serve as accessory cells. The helper T-cell specificity and the requirement for 1a positive accessory cells in various species has also been shown by others (Biniaminov et al, 1974; Thurman et al, 1974). Phillips and co-workers (1980) speculated that the aldehyde group somehow formed a neoepitope recognized by T-cells. Subsequent work, however, has confirmed that the T-cell

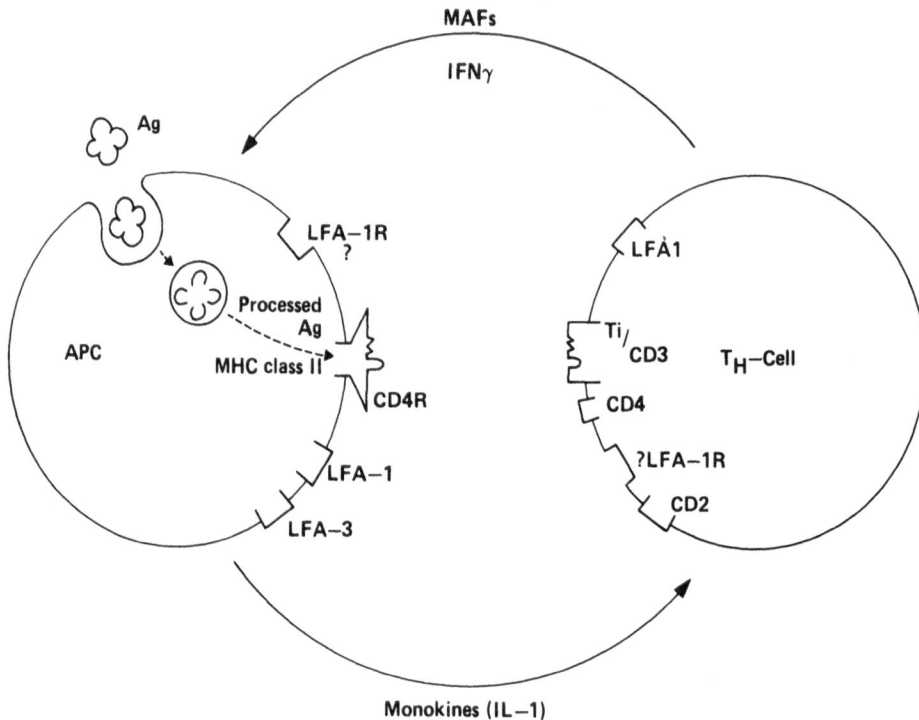

Fig. 1. Schematic representation of the interaction between antigen
presenting cell and T-cell.

response depends upon a covalent reaction between T-cell and accessory cell
involving Schiff base formation (Fleischer, 1983). This reaction involves a
carbonyl group and an amino group and has the general form shown in Fig. 2.
The carbonyl donor can be an aldehyde (R'=H) or a ketone and the reaction is
reversible. The Schiff base reaction can be rendered irreversible, however,
by reduction with the weak, highly selective reducing agent sodium cyanoborohydride (Fig. 3). This reagent is specific for Schiff bases at pH
values greater than 5 (Jentoft and Dearborn, 1979; Gray, 1978).

Recent work on oxidative mitogenesis has explored the requirement for
Ia and interleukin 1 (Smith et al, 1985), the possible macromolecular sites
involved (Suthanthrinan, 1988) and the use of covalent coupling of allogeneic accessory cells to T-cell clones as a means of non-restricted, antigen-
independent propagation (Fleischer, 1988). Our own studies of T-cell activation have been concerned with antigen-specific stimulation. In the course
of studies on the response of freshly prepared peripheral blood human T-
cells to recall antigens we found antigen presentation by autologous accessory cells (monocytes) to be exquisitely sensitive to inhibition by soluble
polymeric aldehydes (Rhodes and Tite, 1988). Brief (30 second) treatment of
antigen-pulsed presenting cells with low doses of glutaraldehyde or para-
formaldehyde results in complete inhibition of the T-cell proliferative
response. This gentle treatment does not affect antigen processing or
membrane turnover, and treated cells recover antigen-presenting ability
after approximately 24 h provided that the availability of antigen is not
limiting. The treatment abolished the expression of HLA-DR detectable in a
rosette assay although DR determinants remain accessible to soluble antibodies in FACS analysis. The chemical reactivity of glutaraldehyde is well-
known: its carbonyl groups react with free lysyl€-amino groups to form
Schiff bases (Peters and Richards, 1977). We therefore suggested that low-
dose aldehyde may disrupt the tertiary structure of the class II molecule by

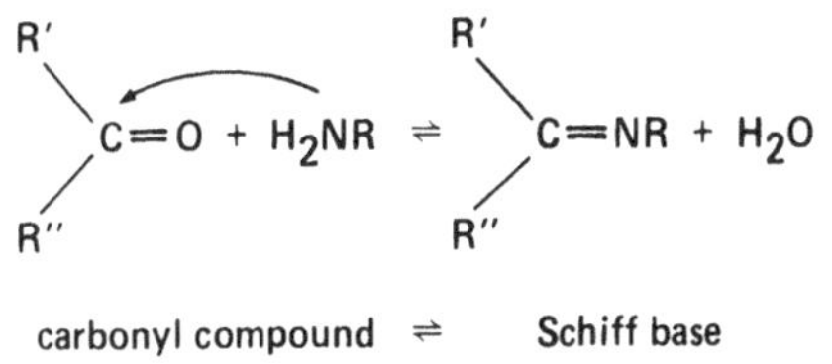

$$R'\diagdown$$
$$C=O + H_2NR \rightleftharpoons \diagup_{R''}^{R'} C=NR + H_2O$$

carbonyl compound $\rightleftharpoons$ Schiff base

Fig. 2. General formula describing Schiff base formation.

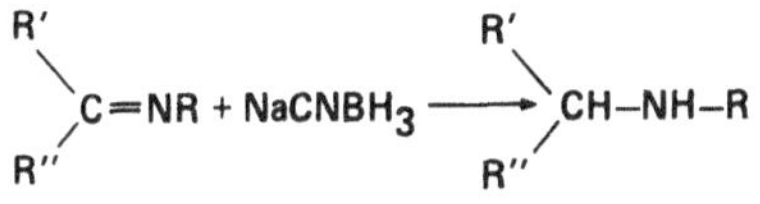

$$\diagup_{R''}^{R'} C=NR + NaCNBH_3 \longrightarrow \diagup_{R''}^{R'} CH-NH-R$$

Fig. 3. Reduction of Schiff base by sodium cyanoborohydride.

cross-linking susceptible lysines (Rhodes and Tite, 1988).

Subsequently, however, dimeric and monomeric aldehydes which do not cross-link glycoproteins were found to have the same effect and it was at this point that an alternative explanation suggested itself. The fact that aldehydes on T-cell surfaces promote inductive interaction with accessory cells through covalent Schiff base formation while pre-treatment of APC with soluble aldehydes prevents antigen-specific inductive interactions suggest that Schiff base formation on APC by soluble aldehydes may pre-empt the physiological formation of Schiff bases between T-cell and APC essential for antigen-induced T-cell activation (Fig. 4).

Such a mechanism is unprecedented in physiologic interactions between cells but there are good precedents in other physiologic systems. The process of molecular recognition depends on information residing in the three-dimensional conformation of macromolecules which dictates their stereo-specific reactivity. Both conformation and stereospecific reactions are stabilized by weak but numerous non-covalent bonds. Nevertheless, some enzymes form transient covalent bonds with their substrates and frequently these are Schiff bases formed between free amino groups on the substrate and carbonyl groups on the enzyme. Aldolase is an example; the co-enzyme pyridoxal phosphate is another (Fersht, 1984).

PREDICTIONS OF THE INTER-CELLULAR SCHIFF BASE MODEL

If the formation of covalent Schiff bases between APC and T-cell is an essential process in immune induction then we should observe the following: (1) Periodate (T-cell surface aldehyde) induced T-cell activation should be inhibited by treating the accessory cells with soluble aldehydes in the same way as antigen-induced T-cell activation; (2) Antigen-induced T-cell activation should be inhibited not only by aldehydes but by other, non-cross-linking, donors of carbonyl groups; (3) If Schiff base formation occurs reciprocally between cells (as it does in experimental oxidative mitogenesis), then pre-treatment of T-cells with aldehydes might also inhibit antigen-induced activation but only transiently (when antigen is not limiting) because membrane turnover will renew free ε-amino groups; (4) Exogenous amino groups in the form of lysine and other amino acids should inhibit antigen-induced T-cell activation, and periodate-induced T-cell activation should be inhibited in the same way; (5) The highly selective reducing agent sodium cyanoborohydride which reacts only with Schiff bases at pH>5 should produce substantial effects on antigen-induced T-cell activation by rendering the covalent bond between T-cell and access-

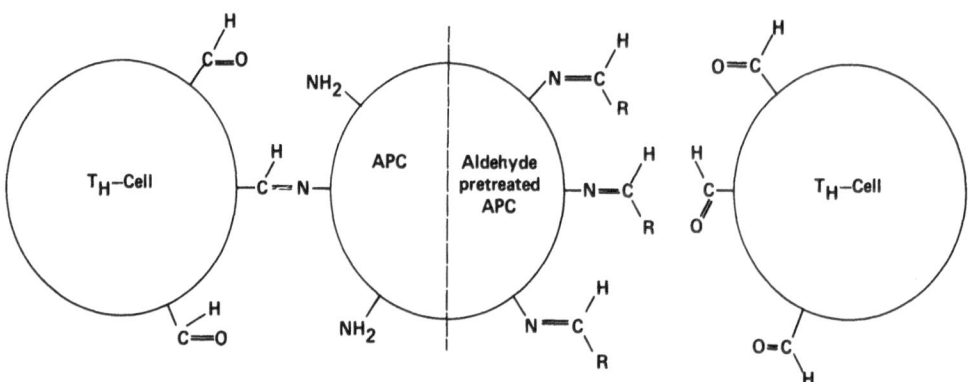

Fig. 4.   Schematic representation of intercellular Schiff base formation and
its prevention by aldehyde pre-treatment of APC.

ory cell irreversible. Whatever the effect, periodate-induced T-cell activation should be affected in precisely the same way; and finally (6) it
should be possible to demonstrate Schiff base formation occurring between
cells directly and definitively by specific radiolabelling with sodium
cyanoborotritiide (see Fig. 3).

EXPERIMENTAL EVIDENCE FOR AN INTER-CELLULAR COVALENT REACTION

When aldehydes are generated on cell-surface sialic acid by means of
periodate, vigorous proliferative responses are obtained whether the aldehydes are on T-cell or accessory cell. Treating both cell types produces an
additive effect. In order to test whether the periodate response behaves in
the same way as the antigen-specific T-cell response with regard to the
requirement for free ε -amino groups on the accessory cells (prediction 1)
human blood adherent mononuclear cells (monocytes) were treated for 30 s
with low-dose glutaraldehyde prior to addition of periodate-treated T-cells.
This treatment consistently results in substantial inhibition of the periodate response (about 60%) fulfilling the first prediction and indicating
that free ε -amino groups provide the acceptor ligand in T-cell aldehyde
mitogenesis. Complete inhibition is not to be expected here because of the
regeneration of free ε -amino groups that occurs after low-dose glutaraldehyde treatment. Responses to accessory cells pulsed with PHA or Con A are
unaffected by glutaraldehyde treatment (Table 1). In order to test the
second prediction it is necessary to add a non-crosslinking, non-aldehyde
carbonyl donor to the antigen-presentation system. When the seven-carbon
ketoacid 4, 6-dioxoheptanoic acid is added to cultures containing lymphocytes and antigen- or mitogen-pulsed presenting cells, antigen (but not
mitogen) responses are inhibited (Table 1). This compound is a carbonyl
donor known to form Schiff bases with free amino groups (Manabe et al,
1985).

The third prediction is testable by treating lymphocytes with low dose
glutaraldehyde for 30 s. After halting the reaction with excess lysine and
washing the cells, their response to antigen-pulsed accessory cells is
measured. This treatment produces transient inhibition, delaying the proliferative response by about 24 h. This is to be expected because of the
regeneration of free ε-amino groups which will reconstitute the response.
The complete inhibition that occurs when antigen-pulsed APC are treated with
glutaraldehyde is due to the limited availability of antigen which becomes
committed to T-cell contact elements. A second antigen pulse following
glutaraldehyde treatment restores responsiveness (Rhodes and Tite, 1988).

Table 1. Experimental Evidence for Inter-cellular Covalent Reaction

PREDICTION	RESULT
(1) Periodate (T-cell surface aldehyde) responses should be inhibited by aldehyde treatment of accessory cells in the same way as Ag-induced responses.	About 60% inhibition of the periodate response consistently occurs.
(2) Non-cross linking donors of carbonyl groups other than aldehydes should also inhibit T-cell responses to Ag.	4-6 dioxoheptanoic acid inhibits antigen but not mitogen responses.
(3) If Schiff base formation occurs reciprocally as in oxidative mitogenesis, then aldehyde pre-treatment of lymphocytes should also be inhibitory.	Aldehyde pre-treatment of lymphocytes (0.0025% glut. for 30s) delays the proliferative response to antigen by 24h.
(4) Exogenesis amino groups in the form of lysine and other amino acids should inhibit antigen-induced responses and periodate-induced responses.	Complete inhibition of Ag and periodate responses occurs in the presence of excess lysine. The same concentrations dramatically enhance PHA responses.
(5) $NaCNBH_3$ should produce the same effect in the periodate system as in the antigen-induced response.	Both are inhibited, with the same dose-response relation. PHA responses are prolonged.
(6) Schiff-base formation should be directly detectable by labelling with $NaCNB[^3H]_3$ in mixed lymphocyte-accessory cell cultures.	Significant radiolabelling occurs only in mixed lymphocyte-monocyte cultures. When the two cell types are separated, labelling returns to the lower levels seen in pure monocyte and pure lymphocyte cultures.

The addition of lysine at concentrations 50–100 fold higher than that of normal tissue-culture medium consistently produces dose-dependent inhibition of both antigen and periodate induced T-cell responses fulfilling the fourth prediction. Thus, by providing an excess of either reactive ligand (aldehyde or amino group) in the Schiff base reaction antigen presentation is inhibited. The effect of lysine on T-cell responses to PHA continuously present in culture is dramatic. The proliferative response, at first delayed, then develops to a peak some three times greater than with PHA alone. This clearly rules out any non-specific toxic effect of lysine but it also raises the intriguing question of how lysine is potentiating the PHA response. One potential explanation is than an excess of lysyl amino groups induces the reactive carbonyl donor on T-cells thus facilitating and prolonging the interaction between lymphocyte and accessory cell in the presence of non-limiting amounts of mitogen.

Sodium cyanoborohydride is a weak reducing agent which, unlike sodium borohydride, is specific for Schiff bases at neutral pH. It does not reduce aldehydes or ketones. When this reagent is added to lymphocyte cultures stimulated with antigen-pulsed accessory cells a dose-dependent inhibition of the response is observed. Periodate-induced responses (which as we have seen depend upon the formation of Schiff base bonds between T-cell and accessory cell) are inhibited in precisely the same way over the same dose-range, fulfilling the fifth prediction. At the same concentrations, cyanoborohydride has no effect on the spontaneous proliferation of transformed T-cell lines while the response to PHA over a comparable time course is enhanced. In fact, responses to PHA with the mitogen continuously present are markedly prolonged by cyanoborohydride so that the peak of proliferation occurs on day 4 rather than day 3.

Specific reduction by cyanoborohydride is definitive for Schiff bases at neutral pH (Jentoft and Dearborn, 1979). It is therefore of interest to know whether Schiff base formation between accessory cell and T-cell can be directly detected by means of cyanoborotritiide as anticipated by prediction 6. This is indeed the case; Schiff base formation defined by specific $NaCNB(^3H)_3$ labelling occurs spontaneously in mixed cultures of monocytes and lymphocytes, but not in cultures of either all-type alone. When, however, monocytes and lymphocytes are cultured together and then reseparated into adherent and non-adherent populations, Schiff base formation returns to the initial low values indicating that intra-cellular bond formation does not account for the Schiff base formation occurring in mixed cultures (Table 1).

CONCLUSIONS

All the evidence obtained in vitro at the chemical level in a human system indicates that the formation of covalent Schiff bases between ligands on APC and T-cell is essential for antigen-induced T-cell activation. In the light of these results, experimental oxidative mitogenesis is seen as an experimental amplification of normal physiologic mechanism.

The spontaneous, antigen-independent nature of Schiff base formation detectable between class II positive accessory cells and lymphocytes is consistent with descriptions of the antigen independent reversible binding of lymphocytes to macrophages first described by Lipsky and Rosenthal (1973). Subsequent studies have shown that dendritic cells are particularly active with respect to spontaneous physical associations with lymphocytes (Inaba and Steinman, 1986; Greene and Jotte, 1985; Martz, 1987). Such spontaneous inter-cellular associations are thought to facilitate the subsequent antigen-specific selection and activation of T-cells. The prevailing view of T-cell activation is that all stimuli activate a common pathway (the T-cell receptor does not transduce a unique signal; Alcover et al, 1987; Linch et al, 1987). Spontaneous covalent associations may therefore drive the autologous mixed leukocyte reaction and provide the essential basis for further activation, either selectively by antigen, or by mitogens or antibodies to signal-transducing determinants insofar as the latter stimuli are accessory-cell dependent. In Schiff base formation between cells the identity of the carbonyl donor is as yet unknown. It is not NANA or galactose and is perhaps more likely to be an inducible molecule similar to pyridoxal phosphate. The observations described here open a new level of investigation in studies of immune induction, namely, events occurring at the chemical level. It is the first time that evidence for a covalent reaction mediating a physiological interaction between cells has been described, and this provides the basis for a rational chemical approach to future studies of immune induction.

REFERENCES

Alcover, A., Ramali, D., Richardson, N.E., Chang, H-C. and Reinherz, E.L., 1987, Functional and molecular aspects of human T lymphocyte activation via T3-Ti and T11 pathways, Immunol.Revs., 95:31.

Biniaminov, M., Ramot, B. and Novogrodsky, A., 1974, Effect of macrophages on periodate-induced transformation of normal and chronic lymphatic leukaemia lymphocytes, Clin.Exp.Immunol., 16:235.

Fersht, A., 1984, "Enzyme Structure and Mechanism", W.H. Freeman and Co., New York.

Fleischer, B., 1983, Activation of human T lymphocytes. I. Requirements for mitogen-induced proliferation of antigen-specific T lymphocyte clones, Eur.J.Immunol., 13:970.

Fleischer, B., 1988, Non-specific propagation of human antigen-dependent T lymphocyte clones, J.Immunol.Methods, 109:215.

Gray, G.R., 1978, Antibodies to carbohydrates: Preparation of antigens by coupling carbohydrates to proteins by reductive animation with cyanoborohydride, in: "Methods in Enzymology" Volume L, Part C, V. Ginsburg, ed., Academic Press Inc., London.

Greene, J. and Jotte, R., 1985, Interactions between T helper cells and dendritic cells during the rat mixed leukocyte reaction, J.Exp.Med., 162:1546.

Greineder, D.K. and Rosenthal, A.S., 1975, The requirement for macrophage-lymphocyte interaction in T lymphocyte proliferation induced by generation of aldehydes on cell membranes, J.Immunol., 115:932.

Inaba, K. and Steinman, R.M., 1986, Accessory cell - T lymphocyte interactions: Antigen-dependent and independent clustering, J.Exp.Med., 163:247.

Jentoft, N. and Dearborn, D.G., 1979, Labelling of proteins by reductive methylation using sodium cyanoborohydride, J.Biol.Chem., 254:4359.

Linch, D.C., Wallace, D.L. and O'Flynn, K., 1987, Signal transduction in him T lymphocytes, Immunol.Revs., 95:154.

Lipsky, P.E. and Rosenthal, A.S., 1973, Macrophage-lymphocyte interaction. I. Characteristics of the antigen-independent-binding of guinea pig thymocytes and lymphocytes to syngeneic macrophages, J.Exp.Med., 138:900.

Manabe, S., Sassa, S. and Kappas, A., 1985, Hereditary tyrosinemia: Formation of succinylacetone-amino acid adducts, J.Exp.Med., 162:1060.

Martz, E., 1987, LFA-1 and other accessory molecules functioning in adhesions of T and B lymphocytes, Human Immunol., 18:3.

Novogrodsky, A. and Katchalski, E., 1972, Membrane site modified on induction of the transformation of lymphocytes by periodate, Proc.Nat. Acad.Sci.USA., 69:3207.

Peters, K. and Richards, F.M., 1977, Chemical cross-linking: Reagents and problems in studies of membrane structure, Ann.Rev.Biochem., 46:523.

Phillips, M.L., Parker, J.W., Frelinger, J.A. and O'Brien, R.L., 1980, Characterization of responding cells in oxidative mitogen stimulation. II. Identification of an Ia-bearing adherent accessory cell, J.Immunol., 124:2700.

Rhodes, J. and Tite, J., 1988, Functional abolition of monocyte HLA-DR by aldehyde treatment. A novel approach to studies of class II restriction elements in antigen presentation, J.Immunol., 140:3344.

Smith, L.A., Cohen, D.A., Lachman, L.B. and Kaplan, A.M., 1985, Sodium periodate-induced T-cell mitogenesis: an analysis of the requirement for Ia and IL-1, J.Immunol., 135:1137.

Springer, T.A., Dustin, M.L., Kishimoto, T.K. and Marlin, S.D., 1987, The lymphocyte function-associated LFA-1, CD2 and LFA-3 molecules: cell adhesion receptors of the immune system, Ann.Rev.Immunol., 5:223.

Suthanthiran, M., 1988, Oxidative mitogenesis: Participation of CD2 antigen in the generation and/or transduction of obligatory accessory signals, Cellular Immunol., 114:117.

Thurman, G.B., Giovanella, B. and Goldstein, A.L., 1974, Evidence for the
    T-cell specificity of sodium periodate-induced lymphocyte blasto-
    genesis, J.Immunol., 113:810.
Weiss, A. and Imboden, J.B., 1987, Cell surface molecules and early events
    involved in human T lymphocyte activation, Advan.Immunol., 41:1.

ALUMINIUM SALTS: PERSPECTIVES IN THEIR USE AS ADJUVANTS

R. Bomford

Wellcome Biotechnology, Langley Court
Beckenham, Kent, BR3 3BS, UK

INTRODUCTION

The adjuvant activity of aluminium salts was discovered more than 60 years ago (Glenny et al, 1926). Diphtheria toxoid was being purified by the addition of alum ($KAlSO_4$) followed by NaOH, which leads to the precipitation of a gel of $Al(OH)_3$ according to the equation in Table 1. The toxoid adsorbs to the gel by ionic interaction, and Glenny found that the complex was more immunogenic than the free toxoid. $Al(OH)_3$, and similar mineral gels (Table 1) are still the only adjuvants licensed for human use although, as other contributions to this volume should make plain, they are likely to be joined by others in the not too distant future. The wide application of aluminium salts in human vaccines has been reviewed elsewhere (Aprile and Wardlaw, 1966; Edelman, 1980).

Although the alum precipitation procedure is still followed in the preparation of some vaccines, it is now more usual to start with a preformed $Al(OH)_3$ or $AlPO_4$ gel which is mixed with the antigen under optimal conditions for adsorption. An aqueous solution of $Al(OH)_3$ has a pH of around 6.0, and the gel is positively charged. Most proteins will adsorb at this pH, although when working with a new protein it is wise to check the adsorption at 0.5 pH intervals over the acceptable physiological range. The latter may be restricted due to the instability of the antigen. For instance, foot-and-mouth disease virus loses immunogenicity at a pH lower than 7.0. Another factor affecting the degree of adsorption is the concentration of salts and buffering ions. Monovalent ions at isotonic concentrations are not inhibitory, but divalent anions do interfere with adsorption. The adjuvant effect depends on the ratio of antigen to gel (Hennessen, 1965; Schmidt, 1967; Joo, 1973). Increasing the quantity of gel at first improves the immune response, but too much gel causes a decline. The point at which all the antigen is adsorbed is on the ascending slope of the curve.

The conditions governing the optimal preparation of complexes of diphtheria toxoid and $AlPO_4$ have been extensively reviewed by Holt (1950).

Complexes of $Al(OH)_3$ and antigen are very stable, and indeed their immunogenicity can improve with storage (Holt, 1950). This may be connected with the progressively greater difficulty with time of desorbing the antigen from the gel by high salt concentrations, indicating a change from an ionic binding to a hydrophobic interaction. Freezing of $Al(OH)_3$-

Table 1. Varieties of Mineral Gel Adjuvants

1. Alum, $KAl(SO_4)_2$
   Forms a precipitate of $Al(OH)_3$ when NaOH is added:
   $$2KAl(SO_4)_2 + 6NaOH \quad 2Al(OH)_3 + 3Na_2SO_4 + K_2SO_4$$

2. $Al(OH)_3$
   Available under a variety of tradenames (Alhydrogel, Amphojel, etc.)

3. $AlPO_4$

4. $Ca_3(PO_4)_2$

5. Maalox, a mixture of $Al(OH)_3$ and $Mg(OH)_2$

6. $Be(OH)_2$

---

adjuvanted vaccines is not recommended, but lyophilization is potentially possible (Rethy et al, 1985).

A number of other mineral gels apart from $Al(OH)_3$ have been used as adjuvants (Table 1). Both aluminium and calcium phosphates have been included in human vaccines. $Be(OH)_2$ has only been exploited as an experimental adjuvant. The interesting observation has recently been made that it is possible to obtain an immunostimulatory effect by complexing the beryllium ion to the antigen, rather than by forming an insoluble complex with $Be(OH)_2$ (Hall, 1988).

THE ADJUVANT EFFECT OF ALUMINIUM GELS

$Al(OH)_3$ generates a sustained primary antibody response to proteins, and is superior to some other adjuvants in boosting the response to intrinsically weak antigens. This can be illustrated by two examples. First, $Al(OH)_3$ was compared to saponin as an adjuvant for the hapten dinitrophenol (DNP) coupled to the protein carriers bovine serum albumin (BSA), fowl gamma globulin (FGG) or keyhole limpet haemocyanin (KLH). In the mouse BSA is a weak antigen and carrier for haptens, whereas FGG and KLH are stronger. $Al(OH)_3$ was a better adjuvant than saponin for DNP coupled to BSA, but the reverse was the case for the FGG and KLH conjugates (Bomford, 1984). These results might have been explained by the selective physical association of the antigens with the adjuvants, but this possibility was ruled out by the second example, in which $Al(OH)_3$ and saponin were compared as adjuvants for human growth hormone (hGH). The immune response to hGH in mice is genetically controlled at the H-2 locus. CBA mice are low responders in that they mount a poor response to hGH when it is administered in Freund's complete adjuvant (FCA), and Balb/c mice are high responders. In other words, hGH is a weak antigen in CBA mice and a strong one in BALB/c mice. The relative efficacy of $Al(OH)_3$ and saponin as adjuvants for hGH in the two strains of mice conformed to what might have been expected from the results with DNP conjugates; $Al(OH)_3$ performed best in the low-responder CBA mice, and saponin in the high-responder Balb/c mice (Bomford et al, 1985).

These results on the relative potency of $Al(OH)_3$ and saponin for weak and strong antigens are of general significance in two respects. First, it is clearly an advantage for $Al(OH)_3$ that it can boost the response to weak antigens, as this will maximise the chances that it will be an effective adjuvant for a wide range of vaccines in individuals who differ in their genetic responsiveness to the antigens. Secondly, the ability of Al(OH)$_3$ to reverse the genetic non-responsiveness of CBA mice to hGH offers a clue to at least part of its mechanism of action as an adjuvant. The triggering of T lymphocytes requires that they should recognise fragments of antigen complexed to class II MHC antigen on the surface of antigen-presenting cells (APC), and genetic non-responsiveness is believed to occur when there is a lack of the appropriate class II antigen to pick up the antigen fragments, which are generated by proteolytic degradation ("processing") in the APC. Al(OH)$_3$ could be affecting the processing of hGH so that new fragments capable of binding to the CBA class II MHC molecules are created.

Al(OH)$_3$ differs from other adjuvants not only in the range of antigens with which it is effective, but also in the isotype profile of the antibodies which are produced, which could influence the degree of protection afforded by a vaccine (see chapter by Allison in this volume). This aspect has been studied in the mouse, where Al(OH)$_3$ is much less effective than FCA in potentiating IgG2 (Warner et al, 1968; Bomford, 1980a), but much more so for IgE (Hamaoka et al, 1973). This latter property of stimulating anaphylactic antibody might seem to be a disadvantage in a vaccine adjuvant, except when IgE contributes to protective immunity as is the case for schistosomes. In a study of vaccination against Schistosoma mansoni in the mouse Al(OH)$_3$ but not FCA provided protection, and this was correlated with anti-schistosomal IgE antibody (Horowitz et al, 1982).

The effect of Al(OH)$_3$ on the IgE response in humans does not appear to have been investigated. Human anti-tetanus toxoid antibodies are distributed in all IgG subclasses (IgG1, IgG2, IgG3 and IgG4) (Shakib and Stanworth, 1980a, b). Further information on the isotype profile of the antibody response after immunization with vaccines containing Al(OH)$_3$ or any new adjuvants that may be tested would be highly desirable.

Al(OH)$_3$ is a poor adjuvant for cell-mediated immunity (CMI) as assessed by delayed-type hypersensitivity (DTH) in the guinea-pig (Bomford, 1980b).

THE MECHANISM OF ACTION OF ALUMINIUM SALTS

The original hypothesis of the adjuvant action of aluminium gels was put forward by Glenny himself, who proposed that the gel forms a depot at the site of injection from which antigen is slowly released to the immune system so that a sustained antibody response is generated after a single injection of antigen (Glenny et al, 1931). This hypothesis was subsequently challenged on the grounds that excision of the complex of antigen and aluminium gel from a subcutaneous injection site quite shortly after injection does not diminish the adjuvant effect (Holt, 1950). Although this experiment demolishes the depot hypothesis in its simplest form, it does not rule out a contribution to the adjuvant effect from antigen persisting at some other location, perhaps in the draining lymph node. It is worth pointing out that a controlled slow release system which would release a pulse of antigen at a predetermined interval, thus simulating a second injection of antigen, would be very valuable for vaccines, but has not yet been developed for this purpose.

Another possible mechanism of action of aluminium gels is antigen targeting, or the improvement of contact between antigen and cells of the immune system. Antigen adsorbed onto Al(OH)$_3$ is more readily taken up by human monocytes than free antigen (Mannhalter et al, 1985), and this could result in altered processing of antigen as was discussed above. Human monocytes that have been exposed to Al(OH)$_3$ secrete interleukin 1 (IL-1) (Mannhalter et al, 1985). Many other adjuvants such as MDP or LPS have the same effect, and may also induce the release of other mediators which could have a role in the triggering of T lymphocytes.

Finally, aluminium salts activate complement (C') by the alternate pathway (Ramanathan et al, 1979), another property that is shared with many other adjuvants such as some pluronics (see chapter by Hunter in this volume). This could have a variety of immunostimulatory consequences by generating chemotactic factors, activating macrophages, or targeting antigen to the C'-receptor-bearing dendritic cells of the lymphoid follicles which present antigen to memory B lymphocytes (Klaus and Humphrey, 1977).

In summary, $Al(OH)_3$ possesses a range of properties that could potentiate immune responses by altering the distribution and processing of antigen, and by inducing the release of active mediators such as C' components and IL-1. It is doubtless this combination of characteristics that endows $Al(OH)_3$ with its reliable adjuvanticity for a wide range of antigens.

POSSIBLE SHORTCOMINGS OF ALUMINIUM SALTS

Cell-mediated Immunity

It is likely that the protective effect of most of the currently used human vaccines against both viruses and bacteria is mediated by antibody, which neutralises viruses and opsonises bacteria or neutralises their toxins. Possibly the only human vaccine where CMI definitely plays a role is that against tuberculosis (the BCG vaccine) and this is a living attenuated vaccine and does not require an adjuvant. Thus up to now the inability of $Al(OH)_3$ to stimulate CMI, if indeed this applies to all types of CMI and to man as well as guinea-pigs, has not been a handicap, although it could become so for the new vaccines under development against parasites and viruses such as retroviruses. The situation with regard to adjuvants for anti-parasite vaccines has been reviewed elsewhere (Bomford, 1988).

When considering whether adjuvants can stimulate CMI, it is as well to have in mind a clear definition of the phenomenon. It is here considered to be a state of immunity that is mediated by T lymphocytes and does not involve antibody. CMI can be further divided into two categories. First, there are cytotoxic T cells, which are mostly of the CD8 rather than the CD4 phenotype, and which are generated most efficiently by immunization with living pathogens (Bevan, 1987). As yet it has proved difficult to boost this type of immunity with adjuvants. Secondly, there is the immunity mediated by CD4 cells, of which there are multiple subsets and which release a range of biologically-active factors which kill pathogens directly, or affect the migration or state of activation of macrophages. DTH is an _in vivo_ manifestation of this type of immunity, although the subsets of T cells which play a part have not been completely defined. The proliferative response of T cells to antigen _in vitro_, which is sometimes taken as an index of CMI, probably involves CD4 cells of many different subsets, including those which function as helper T cells for the antibody response and not all of which will take part in CMI as defined above. In a recent paper in which $Al(OH)_3$ was used as an adjuvant for the gp 120 of the HIVI virus in chimpanzees a claim for the stimulation of CMI was made on the grounds of lymphocyte proliferation alone (Berman et al, 1988). This could be correct, but only an analysis of the mediator production by the proliferating T cells could have provided a decisive answer and data of this sort will be needed in future studies of the effects of $Al(OH)_3$ and other adjuvants on CMI in man.

Sub-optimal Stimulation of the Antibody Response

As discussed above, $Al(OH)_3$ is a relatively good adjuvant for weak antigens in the mouse, but conversely it is out-performed by saponin or FCA for strong antigens (Bomford, 1984; Bomford et al, 1985). There are also an increasing number of examples where $Al(OH)_3$ is not the optimal adjuvant for microbial antigens. For instance, the antibody response of guinea-pigs to a synthetic peptide from the foot-and-mouth disease virus was boosted by liposomes but not by $Al(OH)_3$ (Francis et al, 1985, 1987), and the adjuvant effect of $Al(OH)_3$ for a genetically-engineered malaria antigen in rabbits was improved when the antigen was incorporated into liposomes containing lipid A (Richards et al, 1988; chapter by Alving in this volume).

There are a number of reasons why $Al(OH)_3$ might be a sub-optimal adjuvant for some antigens. First, the adsorption may be poor due to the small size or charge of the peptide. Secondly, $Al(OH)_3$ may cause antigens to be proteolytically degraded with great efficiency. This could be beneficial for weak antigens, possessing few T cell epitopes perhaps located in the interior of the molecule, but it might not be helpful for peptides of low molecular weight requiring little or no processing whose immunogenicity would be destroyed by proteolytic cleavage. Finally, the adsorption of antigens to $Al(OH)_3$ could lead to some denaturation, with the loss of conformational antigenic determinants.

CONCLUSIONS

In view of their efficacy, cheapness and acceptability to regulatory authorities, the mineral gels may well continue to be used as adjuvants for many of the traditional human vaccines for some time to come. $Al(OH)_3$ is of necessity being used for the first clinical trials of the new generation of genetically-engineered human vaccines such as those against hepatitis B (Scolnick et al, 1984) and malaria (Ballou et al, 1988; Herrington et al, 1987; Etlinger et al, 1988; Patarroyo et al, 1988). In the former case the level of immunity was satisfactory and $Al(OH)_3$-adjuvanted hepatitis vaccines have reached the market place. The malaria vaccines provided some immunity, but insufficient for a useful vaccine. It remains to be seen whether the problem can be solved by improving the antigen, or whether a new adjuvant, perhaps chosen from amongst those described in this volume, will be needed.

REFERENCES

Aprile, M.A. and Wardlaw, A.C., 1966, Aluminium compounds as adjuvants for vaccines and toxoids in man; A review, Can.J.Public Health, 57:343.
Ballou, W.R., Hoffman, S.L., Sherwood, J.A., Hollingdale, M.R., Neva, F.A., Hockmeyer, W.T., Gordon, D.M., Schneider, I., Wirtz, R.A., Young, J.F., Wasserman, G.F., Reeve, P., Diggs, C.L. and Chulay, J.D., 1988, Safety and efficacy of a recombinant DNA Plasmodium falciparum sporozoite vaccine, Lancet, i:1277.
Berman, P.W., Groopman, J.E., Gregory, T., Clapham, P.R., Weiss, R.A., Ferriani, R., Riddle, L., Shimasaki, C., Lucas, C., Lasky, L.A. and Eichberg, J.W., 1988, Human immunodeficiency virus type 1 challenge of chimpanzees immunized with recombinant envelope glycoprotein gp 120, Proc.Natl.Acad.Sci.USA, 85:5200.
Bevan, M.J., 1987, Class discrimination in the world of immunology, Nature, 325:192.

Bomford, R., 1980a, The comparative selectivity of adjuvants for humoral and cell-mediated immunity. 1. Effect on the antibody response to bovine serum albumin and sheep red blood cells of Freund's incomplete and complete adjuvants, alhydrogel, Corynebacterium parvum, Bordetella pertussis, muramyl dipeptide and saponin, Clin Exp. Immunol., 39:426.

Bomford, R., 1980b, The comparative selectivity of adjuvants for humoral and cell-mediated immunity. II. Effect on delayed-type hypersensitivity in the mouse and guinea-pig, and cell-mediated immunity to tumour antigens in the mouse of Freund's incomplete and complete adjuvants, alhydrogel, Corynebacterium parvum, Bordetella pertussis, muramyl dipeptide and saponin, Clin.Exp.Immunol., 39:435.

Bomford, R., 1984, Relative adjuvant efficacy of Al(OH)$_3$ and saponin is related to the immunogenicity of the antigen, Int.Archs.Allergy Appl.Immun., 75:280.

Bomford, R., Aston, R. and Ivanyi, J., 1985, Reversal of H-2-restricted hyporesponsiveness to human growth hormone by the use of aluminium hydroxide as adjuvant, Immunogenetics, 21:505.

Bomford, R., 1988, Adjuvants for antiparasite vaccines, Parasitol.Today, in press.

Edelman, R., 1980, Vaccine adjuvants, Rev.Infect.Dis., 2:370.

Etlinger, H.M., Felix, A.M., Gillesen, D., Heimer, E.P., Just, M., Pink, J.R.L., Sinigaglia, F., Sturchler, D., Takacs, B., Trzeciak, A. and Matile, H., 1988, Assessment in humans of a synthetic peptide-based vaccine against the sporozoite stage of the human malaria parasite, Plasmodium falciparum, J.Immunol., 140:626.

Francis, M.J., Fry, C.M., Rowlands, D.J., Brown, F., Bittle, J.L., Houghten, R.A. and Lerner, R.A., 1985, Immunological priming with synthetic peptides of foot-and-mouth disease virus, J.gen.Virol., 66:2347.

Francis, M.J., Fry, C.M., Rowlands, D.J., Bittle, J.L., Houghten, R.A., Lerner, R.A. and Brown, F., 1987, Immune response to uncoupled peptides of foot-and-mouth disease virus, Immunology, 61:1.

Glenny, A.T., Pope, C.G., Waddington, H. and Wallace, U., 1926, The antigenic value of the toxin-antitoxin precipitates of Ramon, J.Pathol.Bacteriol., 29:31.

Glenny, A.T., Buttle, G.A.H. and Stevens, M.F., 131, Rate of disappearance of diphtheria toxoid injected into rabbits and guinea-pigs: Toxoid precipitated with alum, J.Pathol.Bacteriol., 34:267.

Hall, J.G., 1988, Studies on the adjuvant action of beryllium. IV. The preparation of beryllium containing macromolecules that induce immunoblast responses in vivo, Immunology, 64:345.

Hamaoka, T., Katz, D.H. and Benacerraf, B., 1973, Hapten-specific IgE antibody responses in mice. II. Cooperative interactions between adoptively transferred T and B lymphocytes in the development of IgE response, J.exp.Med., 138:538.

Hennessen, W., 1965, The mode of action of mineral adjuvants, Progr. Immunobiol.Standard, 2:71.

Herrington, D.A., Clyde, D.F., Losonsky, G., Cortesia, M., Murphy, J.R., Davis, J., Baquar, S., Felix, A.M., Heimer, E.P., Gillessen, D., Nardin, E., Nussenzweig, R.S., Nussenzweig, V., Hollingdale, M.R. and Levine, M.M., 1987, Safety and immunogenicity in man of a synthetic peptide malaria vaccine against Plasmodium falciparum sporozoites, Nature, 328:257.

Holt, L.B., 1950, "Developments in Diphtheria Prophylaxis", Heinemann, London.

Horowitz, S., Smolarsky, M. and Arnon, R., 1982, Protection against Schistosoma mansoni achieved by immunization with sonicated parasite, Eur.J.Immunol., 12:327.

Joo, I., 1973, Mineral carriers as adjuvants, Symp.Series Immunobiol. Standard, 22:123.

Klaus, G.G.B. and Humphrey, J.H., 1977, The generation of memory cells. I. The role of C3 in the generation of B memory cells, Immunology, 33:31.

Mannhalter, J.W., Neychev, H.O., Zlabinger, G.J., Ahmad, R. and Eibl, M.M., 1985, Modulation of the human immune response by the non-toxic and non-pyrogenic adjuvant aluminium hydroxide: Effect on antigen uptake and antigen presentation, Clin.exp.Immunol., 61:143.

Patarroyo, M.E., Amador, R., Clavijo, P., Moreno, A., Guzman, F., Romero, P., Tascon, R., Franco, A., Murillo, L.A., Ponton, G. and Trujillo, G., 1988, A synthetic vaccine protects humans against challenge with asexual blood stages of Plasmodium falciparum malaria, Nature, 332:158.

Ramanathan, V.D., Badenoch-Jones, P. and Turk, J.L., 1979, Complement activation by aluminium and zirconium compounds, Immunology, 37:881.

Rethy, L., Solyom, F., Bacskai, L., Geresi, M., Gerhardt, Z., Koves, B., Kriston, K., Magyar, T., Masek, I., Nagy, B. and Nemesi, M., 1985, Design and control of new type vaccines. Efficacy testing of adsorbed and freeze-dried toxoid-virus-bacterium combined vaccines, Ann. Immunol.Hung., 25:49.

Richards, R.L., Hayre, M.D., Hockmeyer, W.T. and Alving, C.R., 1988, Liposomes, lipid A and aluminium hydroxide enhance the immune response to a synthetic malaria sporozoite antigen, Infect.Immun., 56:682.

Schmidt, G., 1967, The adjuvant effect of aluminium hydroxide in influenza vaccine, Symp.Series Immunbiol.Standard, 6:275.

Scolnick, E.M., McLean, A.A., West, D.J., McAleer, W.J., Miller, W. and Bunyak, E.B., 1984, Clinical evaluation in healthy adults of a hepatitis B vaccine made by recombinant DNA, J.Am.Med.Assoc., 251: 2812.

Shakib, F. and Stanworth, D.R., 1980a, Human IgG subclasses in health and disease (A review) Part I, La Ricerca Clin.Lab., 10:463.

Shakib, F. and Stanworth, D.R., 1980b, Human IgG subclasses in health and disease (A review) Part II, La Recerca Clin.Lab., 10:561.

Warner, N.L., Vaz, N.M. and Ovary, Z., 1968, Immunoglobulin classes in antibody responses in mice. I. Analysis by biological properties, Immunology, 14:725.

SAPONINS AS IMMUNOADJUVANTS

R. Bomford

Wellcome Biotechnology, Langley Court
Beckenham, Kent BR3 3BS, UK

Saponins are very widely distributed in plants (Price et al, 1987), bu the one that is used as an adjuvant is extracted from the bark of the South American tree Quillaia saponaria (Dalsgaard, 1978). It is a triterpene glycoside, consisting of the hydrophobic triterpene polycyclic ring, to which is attached two chains of sugars (Fig. 1), which are hydrophilic. This gives the molecule surface-active properties, and saponin is used as a foaming agent in soft drinks. Its potential as an adjuvant was discovered more than 50 years ago (Thibault and Richou, 1936), and it has since become widely used in veterinary vaccines, particularly those against foot-and-mouth disease virus (Dalsgaard, 1978). Recently new impetus has been given to research on saponin by the discovery that it can form small ordered structures, ISCOMS or immunostimulating complexes, with phospholipid, cholesterol, and proteins with a suitable hydrophobic tail such as those extracted from viral envelopes (Morein, 1988 and chapter by Morein in this volume). This chapter reviews what is known of the adjuvant properties of saponin and its mechanism of action.

THE ADJUVANT ACTIVITY OF SAPONIN

Using model antigens in the mouse, saponin was found to be a powerful adjuvant for strong antigens, but was relatively inefficacious for weak antigens (Bomford, 1984, Bomford et al, 1985, chapter by Bomford on Al(OH)$_3$ in this volume). It has regularly performed very well in comparative adjuvant studies of experimental vaccines against parasitic protozoa such as malaria (Desowitz, 1975, Mitchell et al, 1979, McColm et al, 1982). In a study on Trypanosoma cruzi, the causative agent of Chagas' disease in South America, the isotype profile of the antibody response was studied (Scott et al, 1984). In contrast to Al(OH)$_3$, saponin boosted the IgG2 as well as the IgG1 response, and also cell-mediated immunity (CMI) measured by delayed-type hypersensitivity. Protective immunity against T. cruzi was correlated with CMI.

Recently the interesting observation has been made that the immuno-stimulating effects of saponin can be detected after after oral admini-stration (Chavali and Campbell, 1987).

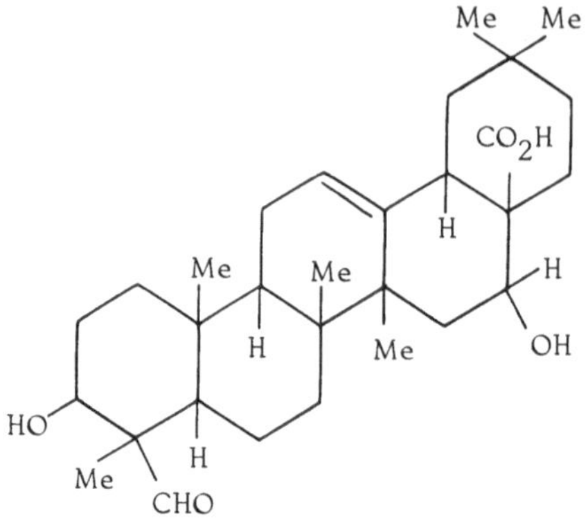

Fig. 1.   The triterpene ring from the saponin of Quillaia
saponaria. The sugar chains, which are not shown,
are attached to the two hydroxyl groups.

MECHANISM OF ACTION OF SAPONIN

The injection of free saponin in an adjuvant-active dose in the mouse
causes a local inflammatory response and the retention of antigen (keyhole
limpet haemoaganin, KLH) at the site of injection (Scott et al, 1985). Both
of these phenomena, but not the adjuvant effect, are blocked by including
liposomes containing cholesterol in the injection mixture. Saponin binds to
cholesterol in cell membranes or liposomes resulting in circular lesions
which in the electron microscope resemble those induced by the cytolytic
components of complement (Dourmashkin et al, 1962). This experiment shows
that the retention of antigen at the site of injection is a consequence of
the inflammatory reaction, and is not responsible for the adjuvant effect.

The hypothesis that saponin assists in targeting antigen to antigen-
presenting cells (APC) is attractive, especially when the saponin is formulated as ISCOMS, but the evidence that this is happening is up to now
rather indirect. The potentiation of the antibody response to sheep red
blood cells in mice requires the presence of T lymphocytes (Bomford, 1982),
which at least shows that saponin cannot bypass the need for T cell help for
T-dependent antigens, and is consistent with the view that saponin, like
many other adjuvants, is acting at the level of the macrophage or APC.
However there is evidence that saponin works at other points in the immune
response than the APC-T cell interaction, since it can act as an adjuvant
for T-independent antigens (Flebbe and Braley-Mullen, 1986). It is reasonable to propose that ISCOMS should be an excellent carrier for the presentation of a matrix of undernatured antigen molecules to the immunoglobulin
receptors of B lymphocytes.

The question of whether the binding of saponin to cholesterol in the
cell membrane plays a role in adjuvant activity has yet to be resolved.
There are two pieces of evidence in favour of this view. First, the adjuvant action of saponin for SRBC in the mouse is blocked by liposomes containing cholesterol, and this occurs whether they are added to the saponin
before mixing with the SRBC, thus preventing haemolysis, or whether they are
added to premixed saponin and SRBC (Bomford, 1980). These data are in contrast to those with KLH mentioned above (Scott et al, 1985), where the adjuvant action of saponin was not inhibited by cholesterol in liposomes, and
the reason for this has not been established, but could be connected with
some factor such as the relative dose of cholesterol. The second piece of
indirect evidence pointing to a role for cholesterol-binding in adjuvanticity is the adjuvant action of another group of compounds, the polyene
antibiotics such as nystatin and amphotericin B, which, although chemically

dissimilar to saponin, also bind cholesterol (Blanke et al, 1977; Bomford, 1980). However, just to complicate the story, the glycoside digitonin, which is rather similar in structure to saponin and also binds cholesterol, lacks adjuvant activity for SRBC or KLH in the mouse (Bomford, 1980; Scott et al, 1985). The present situation appears to be that cholesterol-binding is not sufficient for adjuvanticity, and it remains to be determined if it is necessary.

CONCLUSIONS

The adjuvant activity of saponin is superior to that of $Al(OH)_3$ for boosting the antibody response to strong antigens in the mouse, and protective immunity and CMI to model vaccines against parasitic protozoa. There is thus plenty of incentive to extend its use from veterinary to human vaccines, provided toxicity can be avoided. Although there is a dearth of published data, it is well-known to those who work with saponin that the toxicity of currently available preparations in the mouse takes the form of acute local inflammation peaking a few hours after injection, followed by malaise or death which does not always manifest itself immediately, but appears by 48 hours after treatment. The doses of saponin at which the latter effects occur are in the range of about 50-200 µg/mouse (not very much above the adjuvant-active dose in the free form) depending on the route of injection, intraperitoneal being more dangerous than subcutaneous. Although the toxicity of saponin in the mouse cannot be ignored, it does seem that the therapeutic ratio is unusually unfavourable in this species, and that relatively lower doses of saponin can be used in larger species, which accounts for its successful application in foot-and-mouth disease vaccines for decades. Most importantly, the ISCOM technology allows a massive reduction in the dose of saponin, sufficient for it to be included in equine and feline vaccines, where an absence of toxicity is paramount (Morein, 1988, and chapter in this volume). There is also a chance that the therapeutic ratio will be improved by the fractionation of saponin. The first step in this direction was taken with the Quil A preparation (Dalsgaard, 1978) which consists solely of triterpene glycosides, but of several different sorts, probably differing in their sugar side chains. It will be very valuable to have more highly fractionated samples available for evaluation. Meanwhile, until some definitive toxicology has been undertaken, the question of the human application of saponin remains open.

REFERENCES

Blanke, T.J., Little, J.R., Shirley, S.F. and Lynch, R.G., 1977, Augmentation of murine immune responses by amphotericin B, Cell Immunol., 33:180.

Bomford, R., 1980, Saponin and other haemolysins (vitamin A, aliphatic amines, polyene antibiotics) as adjuvants for SRBC in the mouse. Evidence for a role for cholesterol-binding in saponin adjuvanticity, Int.Archs.Allergy appl.Immun., 63:170.

Bomford, R., 1982, Studies on the cellular site of action of the adjuvant activity of saponin for sheep erythrocytes, Int.Archs.Allergy appl. Immun., 67:127.

Bomford, R., 1984, Relative adjuvant efficacy of $Al(OH)_3$ and saponin is related to the immunogenicity of the antigen, Int.Archs.Allergy appl.Immun., 75:280.

Bomford, R., Aston, R. and Ivanyi, J., 1985, Reversal of H-2-restricted hyporesponsiveness to human growth hormone by the use of aluminium hydroxide as adjuvant, Immunogenetics, 21:505.

Chavali, S.R. and Campbell, J.B., 1987, Immunomodulatory effects of orally-administered saponins and nonspecific resistance against rabies infection, <u>Int.Archs.Allergy appl.Immun.</u>, 84:129.

Dalsgaard, K., 1978, A study of the isolation and characterization of the saponin Quil A. Evaluation of its adjuvant activity, with a special reference to the application in the vaccination of cattle against foot-and-mouth disease, <u>Acta Vet.Scan.</u>, Suppl., 69:1.

Desowitz, R.S., 1975, Plasmodium berghei: Immunogenic enhancement of antigen by adjuvant addition, <u>Exp.Parasitol.</u>, 38:6.

Dourmashkin, R.R., Dougherty, R.M. and Harris, R.J.C., 1962, Electron microscope observations on Rous sarcoma virus and cell membranes, <u>Nature</u>, 194:1116.

Flebbe, L.M. and Braley-Mullen, H., 1986, Immunopotentiation by SGP and Quil A. II. Identification of responding cell populations, <u>Cell Immunol.</u>, 99:128.

McColm, A.A., Bomford, R. and Dalton, L., 1982, A comparison of saponin with other adjuvants for the potentiation of protective immunity by a killed Plasmodium yoelii vaccine in the mouse, <u>Parasite Immunol.</u>, 4:337.

Mitchell, G.H., Richards, W.H.G., Coller, A., Kietrich, F.M. and Dukor, P., 1979, Nor-MDP, saponin, corynebacteria and pertussis organisms as immunological adjuvants in experimental malaria vaccination of macaques, <u>Bull.WHO</u>, 57(Suppl. 1):189.

Morein, B., 1988, The ISCOM antigen-presenting system, <u>Nature</u>, 332:287.

Price, K.R., Johnson, I.T., and Fenwick, G.R., 1987, The chemistry and biological significance of saponins in foods and feeding stuffs, <u>CRC Crit.Rev.Food Sci.Nutr.</u>, 26:27.

Scott, M.T., Bahr, G., Moddaber, F., Afchain, D. and Chedid, L., 1984, Adjuvant requirements for protective immunization of mice using a <u>Trypanosoma cruzi</u> 90K cell surface glycoprotein, <u>Int.Archs.Allergy appl.Immun.</u>, 74:373.

Scott, M.T., Goss-Sampson, M. and Bomford, R., 1985, Adjuvant activity of saponin: Antigen localisation studies, <u>Int.Archs.Allergy appl.Immun.</u>, 77:409.

Thibault, P. and Richou, R., 1936, Sur l'accroissement de l'immunité antitoxique sous l'influence de l'addition de diverses substances a l'antigene (anatoxines diphthérique et tetanique), <u>C.R. Seances Soc. Biol.</u>, 121:718.

THE IMMUNOADJUVANT DIMETHYLDIOCTADECYLAMMONIUM BROMIDE

H. Snippe and C.A. Kraaijeveld

Laboratory of Microbiology
Faculty of Medicine
Utrecht University, Utrecht, The Netherlands

IMMUNOLOGICAL STUDIES

History

The first reports on DDA as an immunoadjuvant have been made by Gall (1966, 1967).

Interaction of DDA with Antigens and Cells

DDA is a positively charged lipophilic quaternary amine (MW 631) which acts as a cationic surfactant (Fig. 1). Relatively little is known about the binding of adjuvants (e.g. DDA) to antigens. Dailey and Hunter (1974) suggested that DDA attached to the antigen by electrostatic bonds. Recently Baechtel and Prager (1982) investigated the interaction of DDA with soluble protein and cellular antigens using radiolabeled (methyl ^{14}C) DDA. Dose-dependent stable complexes appeared to be formed and little influence of ionic strength on DDA binding to ovalbumin was found. Further, binding of DDA to erythrocytes was independent of the number of negatively charged sialic acid residues on the cells. As DDA bound hardly to intracellular structures, it was concluded that binding occurs primarily to the cell membrane, presumably by hydrophilic interactions. The idea of membrane perturbation by DDA is supported by the observed leak of macromolecules (lactate dehydrogenase) from cells treated with DDA (Baechtel and Prager, 1982), however, complete lysis of these cells did not occur (Bomford, 1980).

It has been suggested that addition of DDA to antigens results in an altered trapping (Coon and Hunter, 1973; Dailey and Hunter, 1974; Alving, 1977), degradation (Dailey et al, 1974; Baechtel and Prager, 1982; Hilgers et al, 1984, 1988) and presentation (Dailey et al, 1977; Prager, 1985) in the host. This may be due to modification of the antigen by DDA, which becomes probably more lipophilic by the attachment of DDA. Covalent binding of lipids to protein antigens gave similar results (Coon and Hunter, 1975). Binding characterizations of DDA to sheep red blood cells (Baechtel and Prager, 1982) suggest that DDA can also easily bind to cells and might influence the immune response more directly by interaction with host cells. The necessity of DDA to form complexes with the antigen is demonstrated unambiguously by the monovalent lipophilic antigen A-PE which is 3-(p-azobenzenearsonate)-N-acetyl derivative coupled to phosphatidylethanolamine. A-PE induces cellular responses only when DDA is added (van Houte

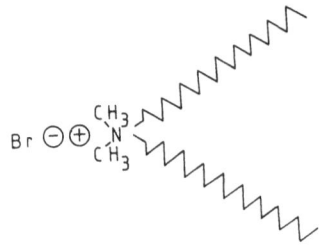

Fig. 1.   Dimethyldioctadecylammonium bromide.

et al, 1981a, b). This antigen forms stable insoluble complexes with DDA which are responsible for the induction of the cellular immune response. When the binding of DDA to A-PE is hindered by the addition of a blocking agent no cellular immune response is induced (Hilgers et al, 1986c).

DDA appeared to be rather toxic for macrophages in vitro (30 μg/ml, Bloksma et al, 1983) which prevented exact measurement on cells in vitro for a number of parameters. It was also noted that DDA was toxic for other cells in vitro, e.g. L-cells. These cells lost their viability at a concentration of 125 μg DDA/ml (Kraaijeveld and Snippe, unpublished observations). In vivo experiments revealed that 24 h after intraperitoneal administration of DDA without antigen, in mice a local accumulation (eight-fold increase of cell number) of both mononuclear and polymorphonuclear cells was induced. The macrophages in the peritoneal exudate fluid displayed normal to enhanced spreading (Gordon et al, 1980; Bloksma et al, 1983) but inhibited phagosome-lysosome fusion (Bloksma et al, 1983). The cells contained increased amounts of lysosomal enzymes (acid phosphatase) while membrane bound enzymes (alkaline phosphodiesterase I, leucine-amino-peptidase) were low in activity compared to resident macrophages (Snippe and Hilgers, unpublished observations). The phagocytic activity of these cells differed according to the particle tested, being normal for yeast cells (de Weger et al, 1982; Bloksma et al, 1983) and malignant cells (Prager et al, 1985), enhanced for sheep red blood cells (Gordon et al, 1980) and yeast cells (Hilgers et al, 1985) and diminished for Listeria monocytogenes (Gonggrijp et al, 1985).

Antibody Formation

Gall (1966, 1967) reported an enhanced antibody formation upon two-fold subcutaneous immunizations of guinea pigs on day 0 and day 28 with diphtheria toxoid and DDA. Similar results were obtained in chickens after two intramuscular injections, 6 weeks apart, with inactivated Infectious Bronchitis virus and DDA (Rijke and Lüttichen, 1983). On the other hand a single intraperitoneal injection of either sheep red blood cells, hapten-carrier complexes, Semliki Forest virus or Encephalomyocarditis virus mixed with DDA did result in an enhanced specific antiody formation in mice which depended on dose and time of administration (Snippe et al, 1981; Hilgers et al, 1984, 1985, 1986a, b; Kraaijeveld et al, 1983). Recently an enhanced antibody formation by DDA to bacterial and viral membrane proteins has been described (Teerlink et al, 1987). Incorporation of outer membrane proteins of Neisseria gonorrhoae in liposomes and addition of DDA to this complex was also effective in induction of specific antibodies (Kersten et al, 1988).

A single subcutaneous injection of antigen and DDA in mice failed to induce detectable antibody formation but instead sensitized these mice for delayed-type hypersensitivity (DH). The eliciting injection for DH with antigen alone in the footpads of these mice, however, resulted in the appearance of circulating antibodies to the antigen after subsidence of the

Table 1. Conditions Reported for Induction of DH Against Various Antigens with the Help of DDA in Different Animal Species

Species	Antigen	Dose range tested	Optimal dose	DDA Dose range tested	Optimal dose	Interval between sensitization and challenge Tested	Optimum	Reference
Mouse	SRBC[1]	2x10^5-2x10^9 cells	2x10^7	125 µg		5 days		Gordon, 1980
		2x10^8 cells		50 µg		5-14 days	5 days	Chiba, 1978
		4x10^7 cells		100 µg		1-15 days	5 days	Snippe, 1977, 1981
Mouse	dnp-BSA[2]	1-100 µg	25 µg	100 µg		1-18 days	5 days	Snippe, 1977
	dnp-BSA[3]	3-100 µg	10 µg	100 µg		5 days		Snippe, 1978, 1980
	A PE-lipo somes[4]	0.005-25 nmol	5 nmol	100 µg		1-20 days		Van Houte, 1981a, b
Mouse	L. mono-cytogenes[5]	10^4-10^8 cells	10^5 cells	100 µg		5-17 days	7 days	Van der Meer, 1977; Willers, 1979, 1982
	L. mono-cytogenes	10 µg fraction I (protein)	10 µg fraction I (protein)	300 µg		7 days		Antonissen, 1986
Mouse	SFV[6]	10^2-10^5 HAU[9]	10^3 HAU	100 µg		4-40 days	6 days	Kraaijeveld, 1980
Mouse	EMCV[7]	5-5x10^4 HAU	5x10^2 HAU	25-400 µg	200 µg	3-10 days	6 days	Kraaijeveld, 1983
Mouse	IBV[8]	10^7 EID50[10]		100-500 µg	100 µg			Rijke, 1983
Rat	SRBC	2.5x10^6		100 µg		2 weeks		Brocades Zaalb., 1980
Guinea pig	dnp-BSA[2]	100 µg		100 µg		3 weeks		Dailey, 1974
Guinea pig	dnp-BSA[2]	1-100 µg	100 µg	10-600 µg	200 µg	1-10 week	3 weeks	Snippe, 1982b
Rabbit	dnp-BSA[2]	1-200 µg	200 µg	10-2000 µg	600 µg	5 days-	1 week	Snippe, 1982a
Mouse	Lysozyme	1-1000 µg	30 µg	100 µg		3 weeks	8 days	Kraaijeveld, 1986

1 sheep red blood cells
2 dinitrophenylated bovine serum albumine
3 as 2, but a tripeptide-enlarged hapten dnp was used
4 3-(p-azobenzenearsonate)-N-acetyl-L-tyrosylglycylglycine coupled to phosphatidylethanolamine and incorporated into liposomal membranes
5 heat killed Listeria monocytogenes
6 purified inactivated Semliki Forest virus
7 purified inactivated Encephalomyocarditis virus
8 Infectious Bronchitis virus
9 haemagglutination units
10 50% egg infectious dosis

DH reaction (Chiba and Egashira, 1978; Snippe et al, 1980). Similar observations on antibody formation after induction of DH were made in rabbits and guinea pigs (Snippe et al, 1982a, b). It is suggested that priming with DDA and antigen results in an expanded pool of T-helper cells, which promote antibody formation by B-cells upon reexposure (elicitation) to the same antigen without DDA (Snippe et al, 1980; van Houte et al, 1981a).

Cell-mediated Immune Reactions

DDA was shown to favour induction of DH to a number of antigens in various animal species after subcutaneous or intramuscular immunization (Table 1). DH reactions were assessed by delayed skin reactions to a local eliciting injection of antigen (erythema and induration of abdominal skin; swelling of ear or footpad). In some experiments (Dailey and Hunter, 1974; Dailey et al, 1977; Snippe et al, 1977, 1982) DH was histologically verified by the predominant mononuclear cell infiltrate consisting of lymphocytes and histiocytes/macrophages. Sporadically basophilic granulocytes were observed (Snippe et al, 1982; van Houte et al, 1981b).

The time required to induce optimal DH after immunization with antigen and DDA appeared to be rather short when compared with another adjuvant, Freund's complete adjuvant (FCA). In mice optimal DH responses were elicited 5-7 days after immunization. In other animal species different intervals are reported: rats 2 weeks, rabbits 1 week, guinea pigs 3 to 6 weeks. An example of how the footpad swelling develops in the foot of an immunized animal is presented in Fig. 2. Although the animal species tested differ considerably in total body weight, the optimal adjuvant dose of DDA was always between 100 and 600 ug in each species. This observation might indicate a local rather than a systemic mode of action of the agent DDA.

As mentioned above the adjuvant activity of DDA is not species restricted. Great intraspecies variations of inducibility of DH reactions, however, have been observed using various inbred mouse strains, but a correlation between H-2 haplotype and intensity of DH could not be demonstrated (Snippe et al, 1980).

A frequently used in vitro parameter of induction of cell mediated immune reactions is blast transformation and enhanced DNA synthesis of lymphoid cells upon exposure to the antigen. Splenocytes from mice immunized subcutaneously with DDA and antigens like sheep erythrocytes (Snippe et al, 1977; Chiba and Egashira, 1978), hapten-carrier complexes (Snippe et al, 1977), Infectious Bronchitis virus (Rijke and Luttichen, 1983) or Herpes Simplex virus type 1 (Smith and Ziola, 1985; Ziola et al, 1987) were found to react this way, whereas splenocytes of normal mice immunized with the antigen alone were inactive in this respect.

In conclusion, the results presented above indicate that DDA strongly enhances cell mediated immune reactions to a number of investigated antigens.

VACCINES

Non-specific Resistance to Micro-organisms

Many immunomodifying agents are known inducers of non-specific resistance against pathogenic micro-organisms and tumors (Borek, 1977; Chedid et al, 1978; Hadden, 1981). DDA did not enhance the non-specific bacterial resistance in mice and guinea pigs to challenge with either viable Listeria monocytogenes, Salmonella enteritidis or Brucella abortus 45/20 (Table 2). This is consistent with the observation that the total activity of the

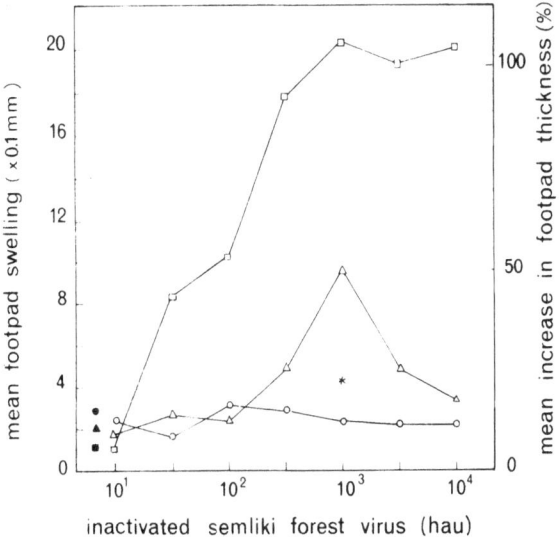

Fig. 2. Induction of strong delayed type hypersensitivity by use of DDA.

Groups of 10 week old female Balb/c mice (n=5) were immunized with graded doses of ultraviolet inactivated Semliki Forest virus (SFV) mixed with DDA (100 ug per animal). Control groups received either DDA or SFV alone. Dilutions of SFV, expressed as haemagglutinating units (HAU), were injected intracutaneously in four divided doses (4 x 0.05 ml) in the neighbourhood of peripheral lymph nodes in axillae and groins. At day 6 of immunization all mice were chall-enged in the left hind footpad with 500 HAU of SFV (0.05 ml). The increase in footpad thickness was measured with an electronic foot-pad meter at selected time intervals. The open (connected) symbols represent the means of swelling of groups of immune mice at 3 h (o), 8 h (△) and 24 h (□) respectively. The black symbols (●,▲, ■) represent the mean footpad swelling of non-immune control mice at the indicated time points. Also given is the mean footpad swelling at 24 h (*) of mice immunized with 1000 HAU of SFV without addition of DDA.

mononuclear phagocytic system in vivo, as measured by carbon clearance cap-acity, was stimulated by DDA within 24 h after intraperitoneal injection, but returned to normal levels afterwards (Willers et al, 1979; Bloksma et al, 1983). On the other hand non-specific protection against Semliki Forest virus and Encephalomyocarditis virus was enhanced 24 h after intraperitoneal injection of DDA (Table 2). As intraperitoneal injection of 1500 μg DDA in mice resulted in the induction of interferon in both serum and peritoneal fluid, protection is probably due to induction of interferon (vide infra, Kraaijeveld et al, 1982). The interferon induced was stable at pH 2 (24 h at 4°C) which indicates alpha or beta specificity (formerly type I inter-feron). The serum levels of interferon in Balb/c mice after injection were relatively low compared to a strong interferon inducer such as synthetic double stranded polynucleotide polyinosinic-polycytidylic acid (poly I:C, Kraaijeveld et al, 1986). On the other hand, reasonable interferon titres were demonstrable in the peritoneal fluid between 10 and 14 h after DDA injection. Peritoneal exudate cells collected from mice, which were treated intraperitoneally before with DDA, produced interferon in vitro (Kraaijeveld et al, 1982, 1986).

Table 2. Capacity of DDA to Induce Non-specific Resistance Against Various Pathogens in Mice[1]

DDA	Challenge		Pro-tec-tion	Ref.
Dose range tested	Pathogen	Dose		
30–100 μg	L. monocytogenes	50 LD_{50}	no	Gonggrijp, 1985
				V.d. Bosch,1986
				V.d. Meer, 1977
30–100 μg	S. enteritidis	100 cells	no	V.d. Meer, 1977
5000 μg	B. abortus 45/20	4×10^6 cells	no	Woodward, 1980
100–1500 μg	SFV	7 LD_{50}	yes[2]	Kraaijeveld,1983
100–1500 μg	EMC	6 LD_{50}	yes[2]	Kraaijeveld,1983

1 Brucella abortus 45/20 was tested in guinea pigs.
2 Maximal protection 1 day after administration of 1500 μg DDA intraperitoneally.

Specific Resistance to Micro-organisms

DDA has been successfully used to enhance specific resistance to several pathogens. Table 3 summarizes the data available. Intraperitoneal injection of heat-killed Listeria monocytogenes and DDA in mice resulted in specific acquired cellular resistance to L. monocytogenes (van der Meer et al, 1977; Antonissen, 1986a, b). Optimal conditions for the induction of acquired cellular resistance were simultaneous intraperitoneal injection of 300 μg DDA and 10^7 or 10^8 heat-killed listeria. The optimal interval (time period) between immunization and challenge was 7 days. The interval during which the mice were protected could be extended considerably by booster injection with heat-killed listeria (10^8) and DDA (Willers et al, 1982). No protection was found against challenge with a lethal dose of Salmonella enteritidis, suggesting that the protection was specific. Protective immunity to L. monocytogenes could also be obtained by immunization of mice with a ribonuclease sensitive fraction derived from the cell content in combination with DDA (Willers et al, 1979; van den Bosch, 1986; Antonissen, 1987). Similar results are reported for Pseudomonas aeruginosa (Gonggrijp et al, 1980).

An enhanced delayed type hypersensitivity to Semliki Forest virus in the absence (undetectable) of specific (neutralizing) antibodies could be induced with UV inactivated virus and DDA in mice (Kraaijeveld et al, 1980). An enhanced DH to SFV was associated with an enhanced survival rate after an otherwise lethal challenge at day 7 and day 42 (Kraaijeveld et al, 1980). Intracutaneous immunization of mice with low doses of Encephalomyocarditis virus (EMCV) and DDA resulted in a state of resistance to infection with virulent EMC virus. Transfer experiments with either peritoneal exudate cells or serum suggested that protection seemed more related to the presence of specific T-cells than the prevalence of specific antibodies at the time of challenge (van der Meer et al, 1977; Kraaijeveld et al, 1984).

Enhanced protective immunity to Plasmodium berghei could be induced with the help of selenium and DDA (Desowitz, 1980). Swiss-Webster mice were immunized intraperitoneally with 500 ug protein (derived from late schizont or merozoite stage) and 200 μg DDA. The selenium was administered in drinking water (2.5 μg/ml). After two immunizations (day 0, day 12) the mice were challenged with 5×10^4 infected mouse erythrocytes at day 26. The group's mortality rate, average peak parasitemia and average cumulative parasitemia were used as criteria for estimating the level of protective immunity. Selenium treatment potentiated the immunostimulating effect of

Table 3. Conditions Reported for Induction of Specific Resistance Against Various Pathogens by DDA in Mice[1]

Pathogen	Immunization Dose range tested	Range affording protection	DDA dose	Challenge dose	Enhancement of survival at day	Ref
L. monocytogenes[2]	10^5-10^8 cells	10^6-10^8 cells	20-1000 µg	50 LD50	5, 7 and 10	Gonggrijp, 1985; Willers, '79, '82; v.d. Meer, 1980; Antonissen, 1986
B. abortus 45/20[1]	300 µg whole cells	none	5000 µg	4×10^4 cells		Woodward, 1980
SFV[3]	300-1000 HAU	300-1000 HAU	100 µg	10 LD50	7, 42 and 63	Kraaijeveld, '80
EMC[4]	150-5000 HAU	150-5000 HAU	200 µg	10 LD50	6 (NS)	Kraaijeveld, '83
P. berghei[5]	500 µg protein[5]	500 µg protein[5]	200 µg	5×10^4 infected erythrocytes	14	Desowitz, 1980
P. yoelli[6]	10^8 FFFP[6]	none	100 µg	10^7 blood stage parasites		McColm, 1982

1 Brucella abortus 45/20 was tested in guinea pigs
2 heat killed Listeria monocytogenes
3 purified inactivated Semliki Forest virus
4 purified inactivated Encephalomyocarditis virus
5 formaldehyde treated Plasmodium berghei parasites in late schizont or merozoite stage
6 formalin fixes free parasites (FFFP) as in 5.

DDA. The survival rate for this group was 90% and the parasitemia in these animals rarely exceeded 10%.

DDA was unable to potentiate acquired cellular resistance to Brucella abortus 45/20 in guinea pigs. These animals were immunized subcutaneously with 300 μg killed whole cell vaccines and 5000 μg DDA and challenged six weeks later intramuscularly with viable B. abortus 2308. Neither non-specific nor specific resistance to B. abortus was enhanced as was concluded from infection rate, mean splenic B. abortus colony forming units and mean splenic weights after challenge with viable B. abortus (Woodward et al, 1980). Also, DDA appeared to be ineffective in potentiating protective activity of a killed Plasmodium yoelli (YM strain) vaccine in outbred CD-1 mice, whereas other adjuvants such as saponin and the polyene antibiotics nystatin and amphotericin B were active (McColm et al, 1982).

MEDICAL APPLICATION

Immunoprophylaxis against Tumours

DDA has been used successfully to increase the immunogenicity of lymphoma cells in mice (Prager and Gordon, 1978, 1979, 1980). Treating iodoacet-amide-modified viable lymphoma cells with DDA increased their immunogenicity as evidenced by the increased capacity of syngeneic, vaccinated hosts to reject subsequent implants of the same lymphoma. Enhancement of the immune response was dependent on DDA dosage (250 μg) and was most striking when DDA was directly complexed to iodoacetamide-modified lymphoma cells (10^7 cells). Several syngeneic mouse leukemia-lymphoma systems were studied: 6C3HED in C3H mice, P 1798 in Balb/c mice, L1210 in DBA/2 mice and the Moloney virus induced YAC leukemia in A/Sn mice. Specific resistance to syngeneic tumour implants was recorded after only one or repeated immunization. In nine experiments 23% of 91 animals vaccinated in the absence of DDA and 71% of 129 animals vaccinated in the presence of DDA survived (p<0.01). In the absence of antigen, DDA was ineffective in immunoprophylaxis. Note: the tumour cells used for immunization were not X-irradiated.

DDA has also been used in a clinical trial on human renal cell carcinoma (Prager et al, 1980, 1981, 1982). Analogous tumour tissue was dispersed mechanically or enzymatically. A single cell suspension (containing approximately 25 mg tumour protein) was complexed with 125 μg DDA and given intradermally divided among 10-12 injection sites selected so as to get reasonably rapid entry into draining lymph nodes. In this trial three out of 31 evaluable patients cleared all metastatic lesions and became without evidence of disease. After 101, 167 and 179 weeks these patients were still disease-free (Prager et al, 1982). In each of these three cases the patient was treated with material from his own tumour, and all had lung metastases. Immunization of one patient with another patient's tumour failed to give any response.

DDA as Adjuvant for Autoblast Tolerization

DDA has been safely used as an adjuvant for immunization of human subjects with autologous lymphoblasts (Chambers et al, 1980). Peripheral blood lymphocytes of nine patients (all children with various immune deficiency disorders) and of 19 parents or immediate relatives were stimulated in vitro with irradiated allogeneic lymphocytes for six days. Responder lymphoblasts were purified, irradiated, mixed with DDA and injected intradermally at four sites in the lymphocyte donor concerned. Induction of tolerization was assessed in mixed lymphocyte cultures. Twenty-seven tolerizations were attempted with a success rate of 30%.

Although the mode of action of DDA is far from clear, based on the data reviewed in this paper, it can be best classified as an immuno adjuvant (Hadden, 1981). According to the classification of immunomodulators proposed in that paper one should compare DDA with an agent like muramyl dipeptide (MDP, Chedid et al, 1978), another immunoadjuvant. Such agents are most effective when administered in combination with the antigen. In this way a specific immune reaction against the antigen can be initiated, regulated (cell-mediated immune response vs humoral response) or tolerance can be prevented or terminated. So far no studies have been initiated to compare DDA and MDP in their efficacy to stimulate immune reactions.

The data presented in this paper indicate that optimal specific immune reactions are obtained when DDA is added to an antigen (vaccine). The vaccine should include micro-organisms, tumour cells or even subcellular fractions derived thereof (Willers et al, 1979; Gonggrijp et al, 1980). Depending on the dose of antigen and route of administration, antibody formation and cellular immune reactions can be favoured selectively. Care must be taken to use viable antigens such as tumour cells because DDA was found to promote growth of UV-attenuated Meth A fibrosarcoma cells on mixed intraperitoneal administration (Bloksma and Snippe, unpublished observations). Intratumoural administration of DDA may be also dissuaded for this reason.

CONCLUSIONS

In our view the immunoadjuvant DDA has a number of advantages over other clinically applied adjuvants. Prime examples are: the synthetic origin, the non-immunogenicity, the suspensibility and activity in aqueous solutions and the absence of deleterious reactivities at the site of injection. DDA should be used in vaccines containing non-proliferating antigenic material and in case of tumours, intralesial injection should be avoided.

Acknowledgement

The authors are indebted to Dr. N. Bloksma, for constructive criticism in the preparation of this manuscript.

REFERENCES

Alving, C.A., 1977, Immune reactions of lipids and lipid model membranes, in "The Antigens", M. Sela, ed., vol. IV, Academic Press, New York.

Antonissen, A.C., Lemmens, P.J.M.R., Van den Bosch, J.F. and Van Boven, C.P.A., 1986a, Dissociation between enhanced resistance and delayed hypersensitivity induced with subcellular preparations from Listeria monocytogenes and the adjuvant dimethyl dioctadecyl amonium bromide, Ant.van Leeuwenhoek, 52:75.

Antonissen, A.C., Lemmens, P.J.M.R., Van den Bosch, J.F. and Van Boven, C.P.A., 1986b, Purification of a delayed hypersensitivity-inducing protein from Listeria monocytogenes. FEMS Micr.Lett., 34:91.

Antonissen, A.C., Lemmens, P.J. and Van den Bosch, J.F., 1987, Transfer of enhanced resistance against Listeria monocytogenes induced with ribosomal RNA and DDA, Immunol.Lett., 14:21.

Baechtel, F.S. and Prager, M.D., 1982, Interaction of antigens with dimethyl dioctadecyl ammonium bromide, a chemically defined biological response modifier, Cancer Res., 42:4959.

Bloksma, N.M., de Reuver, M.J. and Willers, J.M.N., 1983, Influence on

macrophage functions as a possible basis of immunomodification by
microbial agents, tilorene and dimethyl dioctadecyl ammonium bromide,
Ant.van Leeuwenhoek, 49:13.

Bomford, R., 1980, Saponin and other haemolysins (vitamin A, aliphatic
amines, polyene antibiotics) as adjuvants for SRBC in the mouse.
Evidence for a role for cholesterol-binding in saponin, Int.Arch.
Allergy, 67:139.

Borek, F., 1977, Adjuvants, in: "The Antigens", M. Sela, ed., vol. IV, ch.
6, Academic Press, New York.

Van den Bosch, J.F., Kanis, I.Y.R., Antonissen, A.C.J.M., Buurman, W.A. and
Van Boven, C.P.A., 1986, T-cell independent macrophage activation in
mice induced with rRNA from Listeria monocytogenes and dimethyl dioctadecyl ammonium bromide, Infec.and Imm., 53:611.

Brocades Zaalberg, O. and Gerbrandy, J.L.T., 1980, Evaluation of immunologic methods for toxicity studies in animals, Bulletin 1980-5,
Medical Biological Laboratory TNO, Rijswijk, The Netherlands.

Chambers, J.D., Thomas, C.R. and Hobbs, J.R., 1980, Induction of specific
transplantation tolerance in man by autoblast immunization, Blut.,
41:229.

Chedid, L., Audibert, F. and Johnson, A.G., 1978, Biological activities of
muramyl dipeptide, a synthetic glycopeptide analogous to bacterial
immunoregulating agents, in: "Progress in Allergy", P. Kallos, B.H.
Waksman and A.L. de Weck, eds., Karger, Basel.

Chiba, J. and Egashira, Y., 1978, Adjuvant effect of cationic surface active
lipid, dimethyl dioctadecyl ammonium bromide, on the delayed-type
hypersensitivity to sheep red blood cells in mice, Jap.J.Med.Sci.
Biol., 37:362.

Coon, J. and Hunter, R., 1973, Selective induction of delayed hypersensitivity by a lipid conjugated protein antigen which is localized in
thymus dependent lymphoid tissue, J.Immunol., 110:183.

Coon, J. and Hunter, R., 1975, Properties of conjugated protein immunogens
which selectively stimulate delayed-type hypersensitivity, J.Immunol.
114:1518.

Dailey, M.O. and Hunter, R.L., 1974, The role of lipid in the induction of
hapten-specific delayed hypersensitivity and contact sensitivity,
J.Immunol., 112:1526.

Dailey, M.O., Post, W. and Hunter, R.L., 1977, Induction of cell mediated
immunity to chemically modified antigens in guinea pigs. II. The
interaction between lipid-conjugated antigens, macrophages and T
lymphocytes, J.Immunol., 118:963.

Desowitz, R.S. and Barnwell, J.W., 1980, Effect of selenium and dimethyl
dioctadecyl ammonium bromide on the vaccine-induced immunity of
Swiss-Webster mice against malaria (Plasmodium berghei), Infect.
Immun., 27:87.

Gall, D., 1966, The adjuvant activity of aliphatic nitrogenous bases,
Immunology, 11:369.

Gall, D., 1967, Observations on the properties of adjuvants, Int. Symp. on
Adjuvants of Immunity, Utrecht, 1966, Symp. Series Immunobiol.
Standard vol. 6, Karger, Basel.

Gonggrijp, R., Mullers, W.J.H.A., Lemmens, P.J.M.R. and van Boven, C.P.A.,
1980, Ribonuclease-sensitive ribosomal vaccine of Pseudomonas aeruginosa, Immunology, 27:204.

Gonggrijp, R., Mullers, W.J.H.A., Dullens, H.F.J. and van Boven, C.P.A.,
1985, Antibacterial resistance, macrophages influx and activation induced by bacterial rRNA with DDA, Inf. and Imm., 50:728.

Gordon, W.C., Prager, M.D. and Carroll, M.C., 1980, The enhancement of humoral and cellular immune responses by dimethyl dioctadecyl ammonium
bromide, Cell.Immunol., 49:329.

Hadden, J.W., 1981, The immunopharmacology of immunotherapy: an update, in:
"Advances of Immunopharmacology", J. Hadden et al, eds., Pergamon
Press.

de Heer, E., Kersten, M.C., van der Meer, C., Linnemans, W.A. and Willers, J.M.N., 1980, Electron microscopic observations on the interaction of Listeria monocytogenes and peritoneal macrophages of normal mice, Lab.Invest., 43:449.

Hilgers, L.A.Th., Snippe, H., Jansze, M. and Willers, J.M.N., 1984, Immunomodulating properties of two synthetic adjuvants: dependence upon type of antigen, dose and time of administration, Cell.Immunol., 86:393.

Hilgers, L.A.Th., Snippe, H., Jansze, M. and Willers, J.M.N., 1985, Effect of in vivo administration of different adjuvants on the in vitro candidacidal activity of mouse peritoneal cells, Cell.Immunol., 90:14.

Hilgers, L.A.Th., Snippe, H., Jansze, M. and Willers, J.M.N., 1985, Combination of two synthetic adjuvants: Synergistic effects of a surfactant and a polyanion on the humoral immune response, Cell. Immunol., 92:203.

Hilgers, L.A.Th., Snippe, H., Jansze, M. and Willers, J.M.N., 1986a, Route dependent immunomodulation: local stimulation by a surfactant and systemics stimulation by a polyanion, Int.Arch.Allergy Appl.Immunol., 79:388.

Hilgers, L.A.Th., Snippe, H., Jansze, M. and Willers, J.M.N., 1986b, Synergistic effects of synthetic adjuvants on the humoral immune response, Int.Arch.Allergy Appl.Immunol., 79:392.

Hilgers, L.A.Th., Snippe, H., Jansze, M. and Willers, J.M.N., 1986c, Suppression of the cellular adjuvanticity of lipophilic amines by a polyanion, Int.Arch.Allergy Appl.Immunol., 80:320.

Hilgers, L.A.Th., Zigterman, G.J.W.J. and Snippe, H., 1988, Immunomodulating properties of amphiphilic agents, in: "Autoimmunity and Toxicology, Immune disregulation induced by drugs and chemicals", M.E. Kammuler, N. Bloksma and W. Seinen, eds., Elsevier Science Publ., Amsterdam.

Hofhuis, F.M.A., van der Meer, C., Kersten, M.C.M., Rutten, V.P.M.G. and Willers, J.M.N., 1981, Effects of dimethyl dioctadecyl ammonium bromide on phagocytosis and digestion of Listeria monocytogenes by mouse peritoneal macrophages, Immunology, 43:425.

van Houte, A.J., Snippe, H., Peulen, G.T.M. and Willers, J.M.N., 1981a, Characterization of immunogenic properties of haptenated liposomal model membranes in mice. II. Induction of delayed-type hypersensitivity, Immunology, 42:165.

van Houte, A.J., Snippe, H., Peulen, G.T.M. and Willers, J.M.N., 1981b, Characterization of immunogenic properties of haptenated liposomal model membranes in mice. III. Specificity of delayed-type hypersensitivity and antibody formation, Immunology, 42:233.

van Houte, A.J., Snippe, H. and Willers, J.M.N., 1981a, Characterization of immunogenic properties of haptenated liposomal model membranes in mice. IV. Introduction of IgM memory, Immunology, 43:627.

van Houte, A.J., Snippe, H., Schmitz, M.G.J. and Willers, J.M.N., 1981b, Characterization of immunogenic properties of haptenated liposomal model membranes in mice. V. Effect of membrane composition on humoral and cellular immunogenicity, Immunology, 44:561.

Kersten, G.F., Teerlink, T., Derks, H.J., Verkley, A.J., Van Wezel, T.L., Crommelin, D.J. and Beuvery, E.C., 1988, Incorporation of the major outer membrane protein of Neisseria gonorrhoeae in saponin-lipid complexes (iscoms): chemical analysis, some structural features and comparison of their immunogenicity with three other antigen delivery systems, Infect.Immun., 56:432.

Kraaijeveld, C.A., Snippe, H., Harmsen, M. and Khader Boutahar-Trouw, B., 1980, Dimethyl dioctadecyl ammonium bromide as an adjuvant for delayed type hypersensitivity and cellular immunity against Semliki Forest virus in mice, Arch.Virol., 65:211.

Kraaijeveld, C.A., Snippe, H., Harmsen, T. and Benaissa-Trouw, B., 1982,

Enhancement of delayed type hypersensitivity and induction of interferon by the lipophilic agent DDA and CP-20,961, Cell.Immunol., 74:277.

Kraaijeveld, C.A., La Riviere, G., Benaissa-Trouw, B.J., Jansen, J., Harmsen, T., and Snippe, H., 1983, Effect of the adjuvant dimethyl dioctadecyl ammonium bromide on the humoral and cellular immune responses to encephalomyocarditis virus, Antiviral Res., 3:137.

Kraaijeveld, C.A., Benaissa-Trouw, B., Harmsen, M. and Snippe, H., 1984, Delayed-type hypersensitivity against Semliki Forest virus in mice: local transfer of delayed-type hypersensitivity with thioglycollate-induced peritoneal exudate cells, Int.Arch.Allergy Appl.Immunol., 73:342.

Kraaijeveld, C.A., Kamphuis, W., Benaissa-Trouw, B.J., van Haarlem, H., Harmsen, M. and Snippe, H., 1986, Potentiation of the cellular immune response by adjuvants: a limited role for adjuvant induced interferon, Int.Arch.Allergy Appl.Immunol., 8:148.

Kraaijeveld, C.A., Kamphuis, W., Benaissa-Trouw, B.J., Harmsen, M. and Snippe, H., 1986, Modulation of adjuvant enhanced delayed-type hypersensitivity by the interferon inducers Poly IC and New Castle Disease virus, Int.Arch.Allergy Appl.Immunol., 79:86.

McColm, A.A., Bomford, R. and Dalton, L., 1982, A comparison of saponin with other adjuvants for the potentiation of protective immunity by a killed Plasmodium yoelii vaccine in the mouse, Parasite Immunol., 4:337.

van der Meer, C., Hofhuis, F.M.A. and Willers, J.M.N., 1977, Delayed-type hypersensitivity and acquired cellular resistance in mice immunized with Listeria monocytogenes and adjuvants, Immunology, 37:77.

Prager, M.D. and Gordon, W.C., 1978, Enhanced response to chemo-immuno-therapy and immuno-prophylaxis with the sue of tumor-associated antigens with a lipophilic agent, Cancer Res., 38:2052.

Prager, M.D. and Gordon, W.C., 1979, Immunoprophylaxis and therapy with lipid conjugated lymphoma cells, GANN Monograph on Cancer Research, 23:143.

Prager, M.D. and Gordon, W.C., 1980, Specific immunoprophylaxis in experimental tumour-host systems, CMA Journal, 122:780.

Prager, M.D., Peters, P.C., Baechtel, F.S. and Brown, G., 1980, Specific immunotherapy of metastatic human renal cell carcinoma: preliminary results, Proc.Am.Ass.Cancer Res., Abstract No. 856:213.

Prager, M.D., Baechtel, F.S., Peters, P.C., Brown, G.L. and Greene, C.L., 1981, Specific immunotherapy of human metastatic renal cell carcinoma, Proc.Am.Ass.Cancer Res., Abstract No. 647:163.

Prager, M.D., Baechtel, F.S., Peters, P.C., Brown, G.L. and Greene, C.L., 1982, Anti-tumor effects of dimethyl dioctadecyl ammonium bromide (DDA), a chemically defined biologic response modifier, Proc.Am.Ass. Cancer Res., Abstract No., 1780:314.

Prager, M.D., Kanar, M.C., Farmer, J.L. and Vanderzee, J., 1985, Effect of DDA induced macrophages on malignant cell proliferation, Cancer Lett., 27:225.

Prager, M.D., 1985, Dimethyl dioctadecyl ammonium bromide (DDA) as an immunologic adjuvant, Kodak Laboratory Chemicals, 56:1.

Rijke, E.O. and Luttichen, D.L., 1983, Personal communication, Intervet International B.V., Boxmeer, The Netherlands.

Smith, R.H. and Ziola, B., 1985, CY and DDA immunopotentiate the DTH response to inactivated enveloped viruses, Immunology, 58:245.

Snippe, H., Belder, M. and Willers, J.M.N., 1977, Dimethyl dioctadecyl ammonium bromide as adjuvant for delayed hypersensitivity in mice, Immunology, 33:931.

Snippe, H., Johannessen, L., Inman, J.K. and Merchant, B., 1978, Specificity of murine delayed-type hypersensitivity to conjugates of large or small haptens on protein carriers bearing lipid groups, Immunology, 34:947.

Snippe, H., Willers, J.M.N., Inman, J.K. and Merchant, B., 1980, The specificity of antibody formation in mice following immunization with hapten-carrier complexes mixed with the surfactant dimethyl dioctadecyl ammonium bromide, Immunology, 39:361.

Snippe. H., Johannessen, L., Lizzio, E. and Merchant, B., 1980, Variable expression of delayed hypersensitivity in different mouse strains using dimethyl dioctadecyl ammonium bromide as an adjuvant, Immunology, 39:399.

Snippe, H., de Reuver, M.J., Willers, J.M.N., Strickland, F. and Hunter, R., 1981, Adjuvant effect of nonionic surface active polyols in both humoral and cellular immunity, Int.Arch.Allergy Appl.Immunol., 65: 390.

Snippe, H., de Reuver, M.J., Beunder, J.W., van der Meer, J.B., van Wichen, D.F. and Willers, J.M.N., 1982a, Delayed-type hypersensitivity in rabbits: comparison of the adjuvants dimethyl dioctadecyl ammonium bromide and Freund's complete adjuvant, Int.Arch.Allergy Appl. Immunol. 67:139.

Snippe, H., de Reuver, M.J., Kamperdijk, E.W.A., van den Berg, M. and Willers, J.M.N., 1982b, Adjuvanticity of dimethyl dioctadecyl ammonium bromide in guinea pigs. I. Skin test reactions, Int.Arch.Allergy Appl.Immunol., 68:201.

Teerlink, T., Beuvery, E.C., Evenberg, D. and Van Wezel, T.L., 1987, Synergistic effect of detergents and aluminium phosphate on the humoral immune response to bacterial and viral membrane proteins, Vaccine, 5:307.

de Weger, R.A., Pels, E. and den Otter, W., 1982, The induction of lymphocytes with the capacity to render macrophages cytotoxic in an allogenic murine system, Immunology, 47:541.

Willers, J.M.N., Bloksma, N., van der Meer, C., Snippe, H., van Dijk, H., de Reuver, M.J. and Hofhuis, F.M.A., 1979, Regulation of the immune response by macrophages, Antonie van Leeuwenhoek, 45:41.

Willers, J.M.N., Hofhuis, F.M.A. and van der Meer, C., 1982, Prolongation of acquired cellular resistance to Listeria monocytogenes, Immunology 46:787.

Woodward, L.F., Toone, N.M. and McLaughlin, C.A., 1980, Comparison of muramyl dipeptide, trehalose dimycolate and dimethyl dioctadecyl ammonium bromide as adjuvants in Brucella abortus 45/20 vaccines, Infect. Immun., 30:409.

Ziola, B., Smith, R.H. and Quattiere, L.F., 1987, In vitro proliferation of lymphocytes from cyclophosphamide-pretreated mice immunized with antigen mixed with dimethyl dioctadecyl ammonium bromide, J.Immunol. Meth., 97:159.

BACTERIAL ENDOTOXINS: RELATIONSHIPS BETWEEN CHEMICAL STRUCTURE AND BIOLOGICAL ACTIVITY

Ernst Th. Rietschel, Lore Brade, Ulrich Schade
Ulrich Seydel, Ulrich Zähringer, Harald Loppnow
Hans-Dieter Flad and Helmut Brade

Forschungsinstitut Borstel, Institut für Experimentelle
Biologie und Medizin, Parkallee 22
D-2061 Borstel, FRG

Bacterial endotoxins induce in higher organisms a great number of different acute pathophysiological effects such as fever, hypotension, leukopenia followed by leukocytosis, disseminated intravascular coagulation and, in higher doses, irreversible shock (Galanos et al, 1977). In view of the obvious similarity between these biological effects and the symptoms of gram-negative bacteremia, endotoxins (released from multiplying or disintegrating bacteria), these appear to be important causative agents of certain manifestations of sepsis. Because of their pathogenic role, endotoxins have (for several decades) received worldwide scientific attention, research having been and being performed with the aim to immunologically or pharmacologically control endotoxicosis.

In more recent years, however, it has been recognized that endotoxins, in addition to their toxic properties, are endowed with a wide spectrum of biological activities which may be considered beneficial for the host. Such activities include induction of nonspecific resistance against bacterial and viral infections, resistance to irradiation, tumor necrosis and the modulation of the host's immune system. Notably, the potent immunostimulating properties of endotoxins have raised the possibility of a clinical application of endotoxin which, however, was hitherto hampered by the intrinsic toxic properties of this molecule.

Many attempts have been made to prepare, by chemical or enzymatic modification, endotoxin derivatives, analogues or partial structures exhibiting low or no toxicity but still expressing beneficial activities. During recent years a number of such products have been described (compare, for example, McIntire et al, 1976, Nowotny, 1983, Takayama et al, 1984). Due to the intrinsic heterogeneity of endotoxins such preparations, however, were also heterogeneous and, therefore, it was difficult to correlate biological activity with a defined structure. An alternative strategy was to chemically synthesize homogeneous and chemically well defined endotoxin-related structures. In order to pursue this strategy it was necessary to characterize those regions of the endotoxin molecule which harbour biological activity and to elucidate the chemical structure of these regions. In the following, our present knowledge on the structure of endotoxin and some selected biological effects is summarized (for comprehensive reviews see R. Proctor (ed.) Handbook of Endotoxin, 1984, 1985, 1986).

The investigations of a number of laboratories demonstrated that endotoxins of all gram-negative bacteria studied so far conform to a common
structural principle (Fig. 1): They consist of a hydrophilic heteropoly-
saccharide component and a covalently linked hydrophobic lipid portion,
termed lipid A (Lüderitz et al, 1982). Hence, chemically, endotoxins are
lipopolysaccharides (LPS).

The polysaccharide component of enterobacterial lipopolysaccharides
consists of two regions which differ in their genetic determination, biosynthesis and architecture. These regions are the O-specific chain and the
core oligosaccharide. A variety of non-enterobacterial wild-type strains of
phototrophic and some human pathogenic gram-negative bacteria including
Neisseria, Acinetobacter, Bordetella, Bacteroides and Haemophilus form lipopolysaccharides which consist only of the core and lipid A region, thus
lacking the O-specific chain (for literature see Hitchcock et al, 1986).

O-Specific Chain

The O-specific chain is a polymer of repeating oligosaccharide units
which contain up to six sugar residues. A large diversity of the constituent components of repeating units has been revealed within different gram-
negative bacteria. The nature, ring form, type of linkage, and type of substitution of the individual monosaccharide residues as well as their sequence within a repeating unit is characteristic and unique for a given lipopolysaccharide and the parental bacterial strain i.e. a bacterial species.
Thus, the O-specific chain is species-specific (Lüderitz et al, 1982).
Because of the diversity of constituents and their linkages, an enormous
number of structures of O-specific chains is conceivable and also verified
in nature. Therefore, an immense structural variability is revealed if the
O-specific chains of distinct bacterial origin are compared.

Core Oligosaccharide

The core region of enterobacterial lipopolysaccharides consists of a
hetero-oligosaccharide which can be formally subdivided into the O-chain-
proximal outer core and the lipid A-proximal inner core. The outer core
contains the common sugars D-glucose, D-galactose, and N-acetyl-D-glucosamine, whereas the inner core region is composed of the unusual sugars hep-
tose, mainly in the L-glycero-D-manno (L,D-heptose) and the D-glycero-D-
manno configuration, and 2-keto-3-deoxyoctonic acid (KDO, systematically
termed 3-deoxy-D-manno-2-octulosonic acid, dOc1A). These residues are, in
general, substituted by charged groups such as phosphate, pyrophosphate,
phosphorylethanolamine and pyrophosphorylethanolamine, often in nonstoichiometric amounts. Therefore, the inner core region exhibits microheterogen-
eity and a considerable accumulation of charged residues.

The structural variability of the core within different bacterial
species is limited. Thus, in the genus Salmonella only one core type (Ra
core) exists for all serotypes, and in Escherichia coli so far five core
types (R1, R2, R3, R4 and K-12) have been described for more than a hundred
different serotypes (Jann et al, 1984). The structural variability of core
types relates primarily to the outer region, while the KDO-containing inner
core appears to be structurally more uniform. In the chemical analysis of
the KDO-containing inner core, enterobacterial rough-(R)-mutants which synthesize lipopolysaccharides lacking the O-specific chain and parts of the
core proved to be most valuable. Using the lipopolysaccharide of a S.
minnesota Rd_1P^- mutant (strain R7), the structure of the enterobacterial
inner core region could be established as shown in Fig. 2 (Tacken et al,
1986). Accordingly, one KDO residue is present in the main core oligosacch-

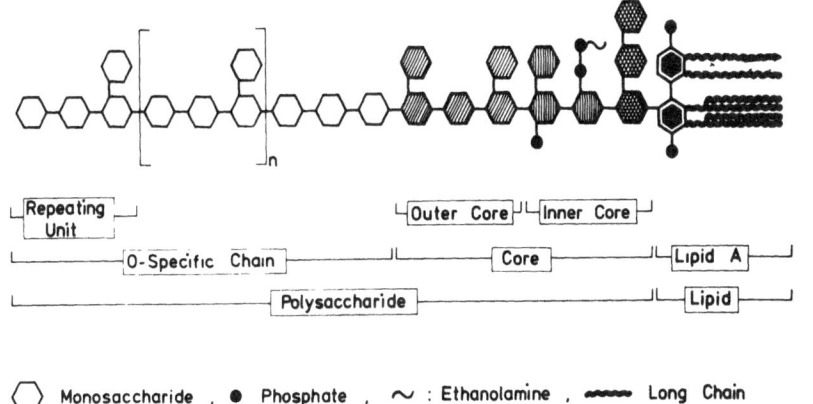

Fig. 1. Architecture of an enterobacterial lipopolysaccharide (Rietschel et al, 1987).

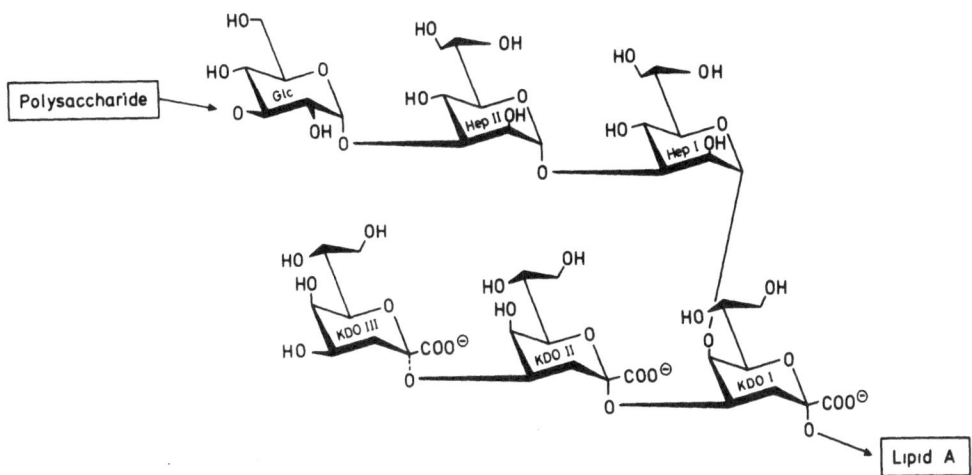

Fig. 2. Chemical structure of the core oligosaccharide in the lipopoly-saccharide of the S. minnesota chemotype RcP⁻ (strain R5). KDO I and KDO II are the only core constituents in Re-type lipopoly-saccharide. KDO III is only found in mutants other than Re where it is not always present in stoichiometric amounts (as indicated by the dotted line). KDO-bound phosphorylethanolamine is not shown (Brade et al, 1988).

aride chain (KDO I) being substituted in position 4 by α-linked KDO II, which in turn may carry, at position 4, nonstoichiometric amounts of a further α-linked KDO residue (KDO III). All KDO groups are present as pyranosides. In its position 5, the KDO I residue carries α-bound L,D-heptose (Hep I) to which a second L,D-heptose residue (Hep II) is α-linked (position 3). In lipopolysaccharides of less defective mutants and wild-type bacteria, the saccharide chain (outer core) extends from the hydroxyl group in position 3 of Hep II and in Fig. 2 the terminal D-glucose residue of the outer core is shown (structure of the S. minnesota RcP⁻core). KDO I is a α-ketosidically linked to the primary hydroxyl group of the non-reducing D-glucosamine residue (GlcN II) of the lipid A backbone (see Fig. 3).

The analysis of the inner core architecture of other bacterial groups was, and still is, very difficult (Brade et al, 1988). This is due to the problems encountered with the analytical chemistry of KDO and the nonavailability of enzymes cleaving the ketosidic linkage of KDO. KDO is a polyfunctional sugar acid with eight carbon atoms harbouring carboxyl, keto, deoxy and hydroxyl groups. This accumulation of different functional groups renders KDO extremely sensitive to the action of acid and other chemical reagents. Thus, under the experimental conditions of its isolation, purification or derivatization, KDO undergoes numerous side reactions yielding many artefacts.

On the other hand, KDO may carry other residues, a fact which further hampers the elucidation of the inner core structure. In principle, each hydroxyl group of KDO may be substituted. A special case was encountered in the lipopolysaccharide of Acinetobacter calcoaceticus strain NCTC 10305 (Kawahara et al, 1987). Here, KDO I is positionally replaced by an octulosonic acid, which has the D-glycero-D-talo configuration and, therefore, is isosteric to KDO. Some lipopolysaccharides (e.g. of Vibrio cholerae) had previously been claimed to lack KDO, and in V. cholerae KDO was postulated to be positionally replaced by D-fructose. It was recently found, however, that in V. cholerae KDO-phosphate (detected as KDO-5-phosphate) is present (Brade, 1985) and that D-fructose, rather, occupies a branch position (Kaca et al, 1986). Also the lipopolysaccharide of Bordetella pertussis, Bacteroides strains, V. parahaemolyticus, Rhodopseudomonas sphaeroides, and Haemophilus influenzae contain phosphorylated KDO (for literature see Caroff et al, 1987; Helander et al, 1988; Kondo et al, 1988; Salimath et al, 1984; Zamze et al, 1987). In this context the lipopolysaccharide of a Haemophilus influenzae deep rough mutant (strain I-69 Rd$^-$/b$^+$) which was genetically constructed by Moxon et al (Moxon, 1985) is of interest. As our chemical analysis shows, the lipopolysaccharide contains only one KDO residue which is phosphorylated and α-linked to lipid A (Helander et al, 1988). The phosphate group is mainly (about 75%) present at position 4 of KDO. This lipopolysaccharide is remarkable in that its core is represented by a single KDO (phosphate) residue. Since this H. influenzae strain is able to multiply, it follows that one (phosphorylated) KDO group in the lipopolysaccharide suffices for the growth and division of a gram-negative bacterial cell.

According to present knowledge all lipopolysaccharides, independent of their bacterial origin, contain at least one pyranosidic or furanosidic KDO residue (or a derivative thereof) with a free carboxyl group occupying an internal position in the inner core region (KDO I in Fig. 2). KDO or a derivative, therefore, represents a common and obligatory constituent of lipopolysaccharides. In all cases studied, this KDO group is α-ketosidically bound to the primary hydroxyl group of the distal glucosamine unit (GlcN II) of the lipid A disaccharide backbone (see Fig. 3). It is this KDO residue which carries the polysaccharide chain and, thus, mediates the link between the polysaccharide and lipid A components in lipopolysaccharides.

Lipid A

Lipid A represents the covalently linked lipid component of lipopolysaccharides. In Enterobacteriaceae, polysaccharide-free lipid A does not exist, a fact which is related to the biosynthesis of lipid A. Thus KDO is transferred to a lipid A precursor molecule (tetraacyl precursor Ia) before the completion of the lipid A structure by addition of nonhydroxylated fatty acids (Raetz, 1987). Enzymes which cleave the polysaccharide-lipid A bond are not known and hence, polysaccharide-deprived free lipid A can only be prepared by acid catalyzed hydrolysis of lipopolysaccharide.

The primary structure of lipid A of enterobacterial and some nonenterobacterial lipid As has been elucidated (for literature see Rietschel et al,

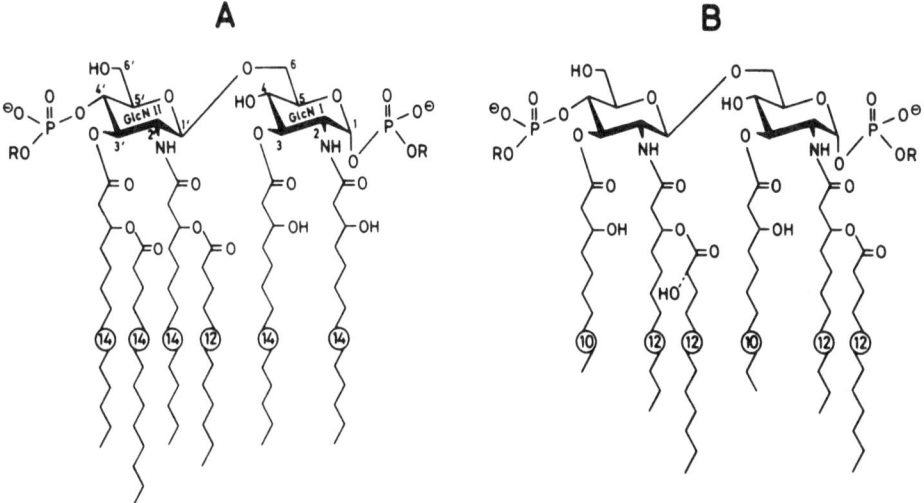

Fig. 3. Chemical structure of the lipid A component of (A) <u>Escherichia coli</u> and (B) <u>Chromobacterium violaceum</u> (Rietschel et al, 1988).

1984; Rietschel et al, 1988). In Fig. 3, two examples of lipid A structures are shown, i.e. those of <u>E. coli</u> and <u>Chromobacterium violaceum</u> with molecular weights of 1796 and 1655 Da (or 1671 Da), respectively. In both cases lipid A is composed of a β-<u>D</u>-glucosaminyl-(1-6)-α-<u>D</u>-glucosamine disaccharide which carries two phosphoryl groups: one in position 4' (of the distal glucosaminyl residue, GlcN II) and one in position 1 (of the reducing glucosaminyl residue, GlcN I). This hydrophilic lipid A backbone is, in both cases, acylated by four residues of (R)-3-hydroxy fatty acids at positions 2, 3, 2', and 3'. As a further common feature both lipid As contain two free hydroxyl groups at positions 4 and 6'. The latter primary hydroxyl group is only free in polysaccharide-deprived lipid A (free lipid A) as obtained on acid treatment of lipopolysaccharide. It is this hydroxyl group which serves as the attachment site of KDO and, thus, the polysaccharide component.

Differences between the two lipid As shown in Fig. 3 are noted with regard to the chain length of hydroxylated and nonhydroxylated fatty acids and the location of acyl groups. In <u>E. coli</u>, (R)-3-hydroxytetradecanoic acid [14:0(3-OH)] is present in ester and amide linkage. The hydroxyl groups of the two 14:0(3-OH) residues bound to GlcN II at positions 2' and 3' carry dodecanoic (12:0) and tetradecanoic acid (14:0), respectively. The two 14:0(3-OH) residues bound to GlcN I are not acylated. In <u>C. violaceum</u> lipid A, two moles of (R)-3-hydroxydodecanoic acid [12:0(3-OH)] are amide-bound and two moles of (R)-3-hydroxydecanoic acid [10:0(3-OH)] are ester-linked to the lipid A backbone. The latter are not substituted at their 3-hydroxyl group while the amide-bound 12:0(3-OH) residues at position 2 and 2' each carry 12:0. The 12:0(3-OH) residue at position 2' may carry (S)-2-hydroxydodecanoic acid instead of 12:0. Thus, two molecular species of lipid A are revealed differing in the substitution of the amino group of GlcN II by either 3-(dodecanoyloxy)dodecanoic acid or 3-(2-hydroxydode-canoyloxy)-dodecanoic acid. <u>E. coli</u> type lipid A with an asymmetrical acylation pattern has been detected in many enterobacterial lipopolysaccharides and, for example, in <u>Haemophilus influenzae</u> (Helander et al, 1988), while the <u>C. violaceum</u> type structure was shown to be present in <u>Neisseria gonorrhoeae</u> (Takayama et al, 1986), and our analyses suggest that it also occurs in <u>Pseudomonas</u>, <u>Xanthomonas</u> and <u>Bacteroides</u> strains (Wollenweber et al, 1984). As indicated in Fig. 3, the phosphoryl groups in positions 1 and 4' may be substituted (symbol R). Phosphate (<u>E. coli</u>), <u>D</u>-glucosamine and 4-

amino-4-deoxy-L-arabinopyranose (C. violaceum) have been identified as polar headgroups. Since their distribution over phosphoryl groups is presently under investigation and since they are not of major importance for endotoxic activity, these substituents are not shown in Fig. 3.

As these analyses show, endotoxically active lipid As of different bacterial origin are structurally closely related. Characteristic and common to them is the presence of a bisphosphorylated β(1-6)-linked D-glucosamine disaccharide. This structure has so far not been identified in other natural compounds and, hence, it is unique to lipid A. The lipid A backbone carries, in general, approximately four mole equivalents of (R)-3-hydroxy fatty acids (carbon numbers 10-18), two of which occupy amino functions and two of which are linked to backbone hydroxyl groups (position 3 and 3'). Both amide- and ester-bound (R)-3-hydroxy fatty acids are, in part, acylated at their 3-hydroxyl group. Such (R)-3-acyloxyacyl residues were found in evolutionarily distinct groups of gram-negative bacteria and they are also characteristic of lipid A (Wollenweber et al, 1984). It is noteworthy that cyclopropane and unsaturated fatty acids are, in general, absent from lipid As. The two phosphate groups of the lipid A backbone may be substituted (as in E. coli or C. violaceum) by nonacylated and, in most cases, amino function-carrying residues. Some examples are known, however, in which one or both backbone phosphate groups are not substituted.

It should be mentioned that some components of lipid A (fatty acids, phosphate substituents) are not always present in stoichiometric amounts. Hence, in a lipid A preparation several molecules may be present that possess an identical backbone but differ in the substitution pattern of the backbone. This structural diversity is at least partially responsible for the well known intrinsic heterogeneity of lipid A and lipopolysaccharide (Nowotny, 1984).

Despite these variations, the lipid A component is the least variable region of biologically active lipopolysaccharides. Lipid A is an obligatory constituent of lipopolysaccharides and its structure appears to be highly conserved.

Certain gram-negative microbes, notably photosynthetic bacteria, produce lipopolysaccharides which are less or not endotoxically active. (Thus, not all lipopolysaccharides are endotoxins.) Their lipid A component has been shown to differ structurally from those lipid As described above (for literature see Weckesser and Mayer, 1988).

Synthetic Lipid A

Based on the results of these analyses, lipid A has been chemically synthesized by Shiba, Kusumoto and colleagues (Imoto et al, 1987). The first fully synthetic lipid A molecule (preparation 506 or LA-15-PP) corresponds in structure to E. coli lipid A (Figs. 3 and 4). Later, other lipid As and Lipid A partial structures were prepared which all contain a β(1-6)-linked D-glucosamine disaccharide but which differ in the acylation and phosphorylation pattern (Fig. 4). These preparations include the heptaacyl species of S. minnesota lipid A (compound 516 or LA-16-PP), the tetraacyl precursor Ia (406 or LA-14-PP), the pentaacyl precursor Ib (LA-20-PP), an isomer of precursor Ib (LA-21-PP), hexaacyl E. coli lipid A (506 or LA-15-PP) as well as the 1-dephospho (compound 504 or LA-15-PH) and the 4'-dephospho (505 or LA-15-HP) partial structures of E. coli lipid A (Fig. 4). Also, lipid A disaccharide analogues with an acylation pattern which is distinct from that of bacterial lipid A have been chemically synthesized (for literature see Shiba and Kusumoto, 1984).

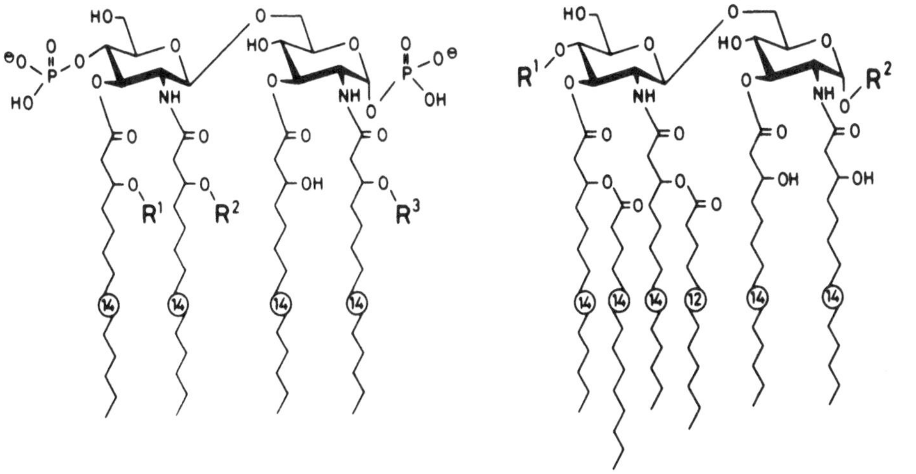

Designation of Preparation		Nature of		
Bacterial	Synthetic	R¹	R²	R³
Precursor Ia	LA-14-PP (406)	H	H	H
Precursor Ib	LA-20-PP	H	H	16 0
Isomer of Precursor Ib	LA-21-PP	H	16 0	H
E coli Lipid A	LA-15-PP (506)	14 0	12 0	H
S minnesota heptaacyl Lipid A	LA-16-PP (516)	14 0	12 0	16 0

Designation of Preparation		Nature of	
Bacterial	Synthetic	R¹	R²
Lipid A-HCl (Monophosphoryl Lipid A)	LA-15-PH (504)	PO(OH)₂	H
———	LA-15-HP (505)	H	PO(OH)₂
E coli Lipid A	LA-15-PP (506)	PO(OH)₂	PO(OH)₂

12 0, dodecanoic, 14 0, tetradecanoic, 16 0, hexadecanoic acid

Fig. 4. Chemical structure of natural and synthetic lipid As and disaccharide-containing partial structures. A: compounds differing in the number and location of phosphoryl groups (Rietschel et al, 1988).

Further, a great number of monosaccharide partial structures with a different acylation and phosphorylation pattern have been prepared by several groups (for literature see Haselberger et al, 1987; Kumazawa et al, 1988; Rietschel et al, 1987). These compounds include synthetic counterparts of bacterial products such as lipid X and lipid Y, as well as other partial structures and analogues corresponding to either the reducing or the distal glucosamine unit of lipid A.

RELATIONSHIP BETWEEN CHEMICAL STRUCTURE AND BIOLOGICAL ACTIVITY

Endotoxicity

For a long time it has been postulated, and experimental evidence has been provided, that the lipid A component represents the endotoxic principle, which is responsible for the induction of the manyfold pathophysiological effects of lipopolysaccharides (Westphal and Luderitz, 1954; Galanos et al, 1977). The successful chemical synthesis of lipid A and its partial structures therefore offered the possibility of confirming this concept as well as the structural proposal made for lipid A. Further, the biological analysis of synthetic, i.e. homogeneous, lipid A and its partial structures was expected to allow to establish relationships between the chemical and physical structure and endotoxic activity at a molecular level. Thus, the synthetic preparations were analyzed in typical _in vivo_ (eg. pyrogenicity,

Shwartzman reactivity, lethality) and in vitro endotoxin test systems and compared with the corresponding natural structures (for summaries of the many original papers compare Brade et al, 1988; Rietschel et al, 1987; Rietschel et al, 1988; Luderitz et al, 1988). The most important results of these investigations can be summarized as follows:

1. Bacterial and synthetic E. coli lipid A expressed, with identical effective doses, the same degree of in vivo and in vitro activity in all biological assays employed. This result shows that indeed lipid A represents the endotoxic center of lipopolysaccharides and that variable substituents of bacterial lipid A (e.g. residues R in Fig. 3) are not required for the expression of endotoxicity. The chemical and physical identity of bacterial and synthetic lipid A proves that the structural proposal made for lipid A, as derived from analytical work, is correct.

2. The in vivo activity (e.g. pyrogenicity) of the bisphosphorylated disaccharide-containing preparations differing in the acylation pattern (Fig. 4A) was, as compared to E. coli lipid A, similar (pentaacyl precursor Ib and its isomer) or significantly lower (tetraacyl precursor Ia, and heptaacyl S. minnesota lipid A). The latter two compounds were not capable of preparing for or of eliciting the dermal Shwartzman reaction. A synthetic analogue containing five fatty acids, but in an arrangement different to that of E. coli lipid A, exhibited only marginal endotoxicity. Therefore, the location of fatty acids, the number of acyl chains present (and possibly the acyl chain length and configuration of 3-hydroxy fatty acids) are of importance for lipid A endotoxic activity in vivo.

3. The 1- and 4'-monodephospho partial structures such as compounds 504 (LA-15-PH) and 505 (LA-15-HP) (Fig. 4B) were, in general, less active in vivo than the bisphosphorylated parent compounds as pyrogens, lethal toxins, and in preparing for or in eliciting the local Shwartzman reaction.

4. Glucosamine monosaccharide partial structures show activity in some in vitro systems. They lack, however, without exception, endotoxic in vivo activity such as pyrogenicity, Shwartzman reactivity and lethal toxicity.

Interleukin 1-Inducing Capacity

The immunostimulatory properties of endotoxins are assumed to be largely mediated by host-derived endogenous mediators such as Interleukin 1 (IL-1) Interleukin 6 (IL-6) and tumor necrosis factor (TNF). These mediators are produced and released by blood and tissue mononuclear cells on exposure to endotoxin. We have recently examined the structural prerequisites of endotoxins with respect to their capacity to induce IL-1 production in human peripheral monocytes (Loppnow et al, 1986; Loppnow et al, 1988). In these studies, various S- and R-form lipopolysaccharides, lipid As and lipid A partial structures (bacterial and synthetic) and core oligosaccharide partial structures were employed. Some of the results obtained with cells of different low and high responder donors are shown in Fig. 5.

In Fig. 5 the minimal concentration of lipopolysaccharide-related compounds required for the induction of a defined amount of IL-1 (1 unit/ml in the thymocyte comitogenic proliferation assay) are given. S and R-form lipopolysaccharides (with the exception of Bacteroides fragilis) were very potent IL-1 inducers (concentration range of 1 to 100 pg), while E. coli lipid A (506) and S. minnesota heptaacyl lipid A (516) as well as the monophosphoryl partial structures 505 and 504 were less active (concentration range 100 pg to 100 ng). The pentaacyl precursor Ib (LA-20-PP), its synthetic isomer (LA-21-PP), the tetraacyl precursor Ia (406), the bisacyl compound 606 and various monophosphoryl partial structures exhibited

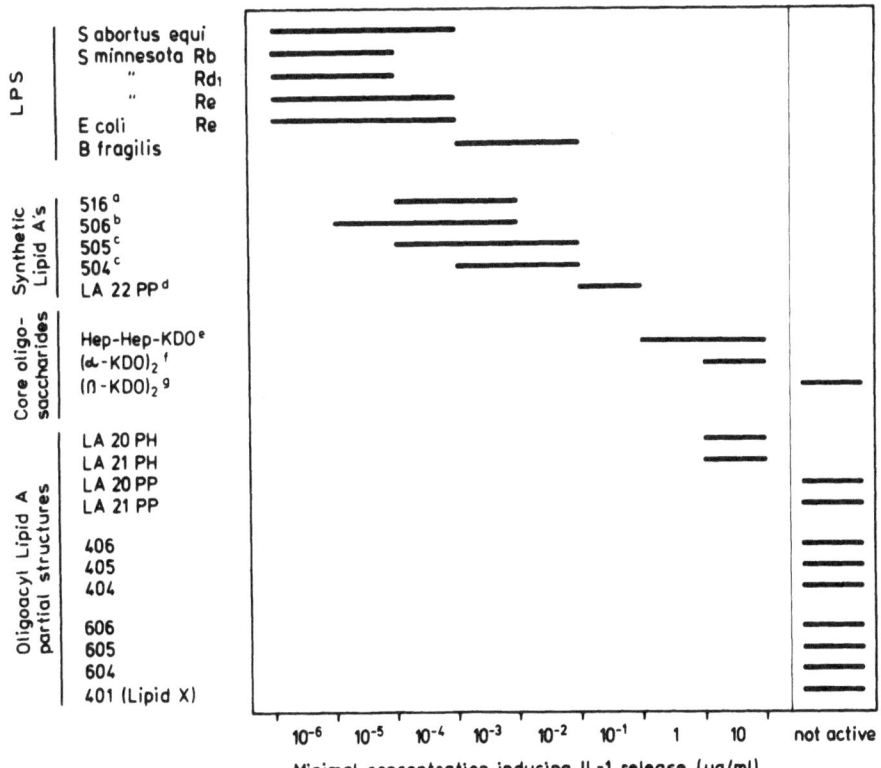

Fig. 5. Interleukin-1-inducing capacity of lipopolysaccharide, lipid As and partial structures. Shown are the minimal concentrations required for the induction of ≥1 units/ml of released IL-1 (Loppnow et al, 1988). a, S. minnesota heptaacyl lipid A; b, E. coli hexaacyl lipid A; c, Monodephospho partial structures of E. coli lipid A; d, C. violaceum type lipid A; e, α-Hep-(1-3)-α-Hep-(1-5)-KDO; f, α-KDOp-(2-4)-α-KDOp-OMe; g, β-KDOp-(2-4)-β-KDOp-OMe.

very low or no activity. This was also true for the monosaccharide structure lipid X (preparation 401).

It has previously been claimed that isolated lipid A induced only the generation of intracellular IL-1 (Haeffner-Cavaillon, 1988). Since lipopolysaccharides induce both the synthesis and release of IL-1, it was postulated that the (KDO-containing) polysaccharide portion plays a role in the release of IL-1. In recent studies relating to this question using defined lipopolysaccharide and synthetic compounds we measured intracellular and extracellular IL-1 in human monocytes. The data shown in Table 1 prove that synthetic and bacterial lipid A, like lipopolysaccharide, is capable of inducing IL-1 production and release and, further, that compound 406 as expected, causes neither an extracellular release nor an intracellular accumulation of IL-1.

It is noteworthy (see Fig. 5) that the lipid A-free core-derived oligosaccharides Hep-α(1-3)-KDO- 2.4-KDO (bacterial) and the synthetic α-methylglycoside of KDO-α(2-4)-KDO exhibit (in higher doses; 1 to 10 µg) IL-1-inducing properties thus confirming previous findings (Lebbar et al, 1986). In contrast, the β-methylglycoside of the disaccharide KDO-β(2-4)-KDO was inactive in this respect. It therefore appears that KDO-containing oligosaccharides possess weak but significant IL-1 inducing capacity, the ex-

Table 1. Comparison of Released and Intracellular Interleukin 1
Activity after Treatment of Human Peripheral Monocytes
with Lipopolysaccharide or Partial Structures (Loppnow
et al, 1988)

Lipopolysaccharide or Partial Structure (10 pg/ml)	Interleukin 1 Activity	
	Intracellular	Released
	units/ml	
Lipopolysaccharide[a]	103	156
Lipid A (Compound 506)	48	55
Precursor Ia (Compound 406)	<1	<1
Medium Control	1	2

[a]LPS of S. abortus equi

pression of which depends on stereochemical factors and which, therefore, exhibits features of specificity.

In summary, our results show that:

1. lipid A represents that structure of endotoxins which is capable of inducing the synthesis and release of IL-1

2. the number of fatty acids of lipid A-related structures is of great importance, a minimum of six fatty acids being required for significant IL-1 inducing capacity

3. the phosphorylation pattern is of significance for IL-1 induction and that

4. glucosamine monosaccharide structures, independent of their acylation or phosphorylation pattern, are not active.

CONCLUSIONS

Research on endotoxin has, in the past, greatly contributed to the field of adjuvants and modulators of the immune system. Today we can define those structures within the complex endotoxin molecule which are capable of interacting with and activating humoral and cellular components of the immune system. These structures are mainly located in the most conservative region of endotoxins i.e. the lipid A component. The primary structure of lipid A has been elucidated, lipid A has been chemically synthesized and partial structural analogues have been prepared and analyzed in various biological systems. The results presented in this paper suggest that the structural requirements of lipid A for inducing endotoxic effects are identical or very similar to those for inducing IL-1 production. This is remarkable since it shows that an in vitro test detects the same biologically relevant lipid A determinants as an in vivo system (eg. pyrogenicity). The minimal structural requirements for endotoxicity and IL-1 inducing capacity can be described as follows:

Endotoxin activity is determined by a molecule containing two D-glucosamine residues (which are β(1-6)-interlinked), two phosphoryl groups and at least five, but not more than six fatty acids including one or two 3-acyloxyacyl groups in a defined location as it is present in E. coli lipid A (compound 506). Molecules lacking only one component, irrespective of its chemical nature, or molecules with a different distribution of components are less or not endotoxically active (Rietschel et al, 1987; Takahasi et al, 1987). This shows that slight modifications at any site of the E. coli lipid A architecture result in a significant reduction of biological activity, suggesting that endotoxicity is not dependent on one single lipid A constituent. It appears that this very structure leads to a unique supramolecular conformation of lipid A which allows the optimal expression of endotoxic activity. Being in this conformation, lipid A is bioavailable and capable of interacting, perhaps selectively, with humoral factors and or cellular and subcellular targets of the endotoxin-susceptible host. We currently favour the view that for the expression of endotoxic activity a particular supramolecular structure and (partial) melting of lipid A acyl chains at physiological temperature are prerequisites. It is not known at present which physical structure of lipid A is involved in endotoxic activity. It seems reasonable to assume that a higher fluidity of the hydrocarbon chains of lipid A should favour the interaction with the host cell membrane, the fluidity of which is higher at the physiological temperature than that of bound or free lipid A. This means that such biological effects which are provoked by direct incorporation of the lipid A portion into the host cell lipid matrix should proceed at a higher rate. This concept is supported by the fact that the biologically less active compound 516 exhibits a relatively high phase transition temperature while biologically active preparations have lower phase transition temperatures (for literature see Brandenburg and Seydel, 1988; Naumann et al, 1987). The biologically less active compound 516 which harbours seven fatty acids leading to an increase of the critical packing parameter may be assumed to adopt, in aqueous solution, to a larger extent more complex physical structures, including inverted phases. This is also suggested by our serological studies (Brade et al, 1987). On the other hand the tetraacyl preparation 406, which is endotoxically also less active than lipid A, should be preferentially present in the lamellar or even micellar state. It therefore appears that neither of these structures per se is responsible for triggering the initial steps of endotoxic events. Bacterial and synthetic hexaacyl E. coli lipid A is likely to adopt, under physiological conditions, both lamellar and inverted structures and it is possible that this ability to express both these supramolecular conformations is related to its potent endotoxic activity.

This view is speculative and we presently direct our efforts at a more precise characterization of lipid A conformation(s) important for its endotoxic properties. These studies are performed in the hope that knowledge of the physical structure(s) of endotoxically active lipid A will enable us to understand the initial steps of the lipid A-host interaction and, thus, the mechanisms involved in endotoxin-mediated adjuvanticity and immunostimulatory activity.

Acknowledgements

The financial support of the Deutsche Forschungsgemeinschaft (Br 731/4-1, Br 731/7, Scha 402/1-1, Scha 402/1-2), the Bundesministerium für Forschung und Technologie (HB, 01 ZR 8604, HDF, PTB 03 8667), the Kultusminister des Landes Schleswig-Holstein (EThR, 3156.46-7-2 and HDF, 3156.46-7-1), and the Fonds der Chemischen Industrie (EThR, HDF) is gratefully acknowledged. We also thank Mrs. M. Lohs, G. Stegelmann, and B.

Köhler for illustrations and photographic work, and Mrs. R. Mohr for typing this manuscript.

REFERENCES

Brade, H., 1985, Occurrence of 2-keto-3-deoxyoctonic acid 5-phosphate in lipopolysaccharides of Vibrio cholerae Ogawa and Inaba, J.Bacteriol., 161:795.

Brade, H., Brade, L. and Rietschel, E.Th., 1988, Structure-activity relationships of bacterial lipopolysaccharides (endotoxins), Zbl.Bakt. Hyg.A., 268:151.

Brade, L., Brandenburg, K., Kuhn, H-M., Kusumoto, S., Macher, I., Rietschel, E.Th. and Brade, H., 1987, The immunogenicity and antigenicity of lipid A are influenced by its physiochemical state and environment, Infect.Immun. 55:2636.

Brandenburg, K. and Seydel, U., 1988, Orientation measurements on membrane systems made from lipopolysaccharides and free lipid A by FT-IR spectroscopy, Eur.Biophys.J., 16:83.

Caroff, M., Lebbar, S. and Szabo, L., 1987, Detection of 3-deoxy-2-octulosonic acid in thiobarbiturate-negative endotoxins, Carb.Res., 161:04.

Galanos, C., Lüderitz, O., Rietschel, E.Th. and Westphal, O., 1977, Newer aspects of the chemistry and biology of bacterial lipopolysaccharides, with special reference to their lipid A component, in: "International review of Biochemistry, Biochemistry of lipids II", T.W. Goodwin, ed., University Park Press, Baltimore.

Haselberger, A., Hildebrandt, J., Lam, C., Liehl, E., Loibner, H., Macher, I., Rosenwirth, B., Schütze, E., Vyplel, H. and Unger, F.M., 1987, Immunopharmacology of lipopolysaccharides (endotoxins) from gram-negative bacteria, Triangle, Sandoz Journal of Medical Science, 26:33.

Haeffner-Cavaillon, N., Bacle, F., Caroff, M. and Cavaillon, J-M., 1988, Characteristics of lipopolysaccharide-induced interleukin-1 production by human monocytes. Clinical relevance in patients undergoing hemodialysis, Progr.Clin.Biol.Res., 272:89, Alan R. Liss Inc., New York.

Helander, I., Lindner, B., Brade, H., Altmann, K., Lindberg, A.A., Rietschel, E.Th. and Zähringer, U., 1988, Chemical structure of the lipopolysaccharide of Haemophilus influenzae strain I-69 Rd$^-$/b$^+$. Description of a novel deep rough chemotype, Eur.J.Biochem., in press.

Hitchcock, P., Leive, L., Mäkelä, P.H., Rietschel, E.Th., Strittmatter, W. and Morrison, D., 1986, Lipopolysaccharide nomenclature - past, present and future, J.Bacteriol., 166:699.

Imoto, M., Yoshimura, H., Shimamoto, T., Sakaguchi, N., Kusumoto, S. and Shiba, T., 1987, Total synthesis of E. coli lipid A, the endotoxically active principle of cell-surface lipopolysaccharide, Bull.Chem.Soc.Jpn., 60:2205.

Jann, K. and Jann, B., 1984, Structure and biosynthesis of O-antigens, in: "Handbook of Endotoxins", R. Proctor, ed., Vol. 1, "Chemistry of Endotoxin" E.Th. Rietschel, ed., Elsevier, Amsterdam.

Kaca, W., Brade, L., Rietschel, E.Th. and Brade H., 1986, The effect of removal of D-fructose on the antigenicity of the lipopolysaccharide from a rough mutant of Vibrio cholerae OGAWA, Carb.Res., 149:293.

Kawahara, K., Brade, H., Rietschel, E.Th. and Zähringer, U., 1987, Studies on the chemical structure of the core-lipid A region of the lipopolysaccharide of Acinetobacter calcoceticus NCTC 10305. Detection of a new 2-octulosonic acid interlinking the core oligosaccharide and lipid A component, Eur.J.Biochem., 163:489.

Kondo, S., Iguchi, T. and Hisatsune, K., 1988, Occurrence of thiobarbituric acid test-positive substances in lipopolysaccharides (LPS) of Vibrio-

naceae, in: "Adv. Res. Cholera and Related Diarrheas", Vol. 4, KTK
    Scientific Publishers, Tokyo.
Kumazawa, Y., Nakatsuka, M., Takimoto, H., Furuya, T., Nagumo, T., Yamamoto,
    A., Homma, J.Y., Inada, K., Yoshida, M., Kiso, M. and Hasegawa, A.,
    1988, Importance of fatty acid substituents of chemically synthesized
    lipid A-subunit analogs in the expression of immunopharmacological
    activity, Infect.Immun., 56:149.
Lebbar, S., Cavaillon, J-M., Caroff, M., Ledur, A., Brade, H., Sarfati, R.
    and Haeffner-Cavaillon, N., 1986, Molecular requirement for inter-
    leukin 1 induction by lipopolysaccharide-stimulated human monocytes:
    Involvement of the heptosyl-2-keto-3-deoxyoctulosonate region, Eur.J.
    Immunol., 16:87.
Loppnow, H., Brade, L., Brade, H., Rietschel, E.Th. and Flad, H-D., 1986,
    Induction of human interleukin 1 by bacterial and synthetic lipid A,
    Eur.J.Immunol., 16:1263.
Loppnow, H., Brade, H., Dinarello, C.A., Kusumoto, S., Rietschel, E.Th. and
    Flad, H-D., 1988, Interleukin 1 induction - capacity of defined lipo-
    polysaccharide partial structures, J.Immunol., Submitted.
Lüderitz, O., Freudenberg, M.A., Galanos, C., Lehmann, V., Rietschel, E.Th.
    and Shaw, D.H., 1982, Lipopolysaccharides of gram-negative bacteria,
    in: "Membrane Lipids of Procaryotes. Current Topics in Membranes and
    Transport", S. Razin and S. Rottem, eds., Academic Press, New York.
Lüderitz, O., Galanos, C., Mayer, H. and Rietschel, E.Th., 1988, Lipopoly-
    saccharides, the O-antigens and endotoxins of gram-negative bacteria:
    relationships of chemical structure and biological activity, in:
    "Vaccines: New Concepts and Developments", H. Kohler, P.T. LoVerde,
    eds., Longman Scientific and Technical.
McIntire, F., Hargie, M.P., Schenck, J.R., Finley, R.A., Sievert, H.W.,
    Rietschel, E.Th. and Rosenstreich, D.L., 1976, Biological properties
    of non-toxic derivatives of a lipopolysaccharide from E. coli K235,
    J.Immunol., 117:674.
Mayer, H. and Weckesser, J., 1988, Different lipid A types in lipopolysacc-
    harides of phototrophic and related non-phototrophic bacteria, FEMS
    Microbiol.Rev., 54:143.
Moxon, E.R., 1985, Antigen expression influencing tissue invasion of Haemo-
    philus influenzae type B, in: "Bayer-Symposium VIII. The Patho-
    genesis of Bacterial Infections", G.G. Jackson and H. Thomas, eds.,
    Springer Verlag, Heidelberg.
Naumann, D., Schultz, C., Born, J., Labischinski, H., Brandenburg, K., von
    Busse, G., Brade, H. and Seydel, U., 1987, Investigations on the
    polymorphism of lipid A from lipopolysaccharides of Escherichia coli
    and Salmonella minnesota by fourier-transform infrared spectroscopy,
    Eur.J.Biochem., 164:159.
Nowotny, A., ed., 1983, "The Beneficial Effects of Endotoxin", Plenum, New
    York.
Nowotny, A., 1984, Heterogeneity of endotoxins, in: "Handbook of Endotoxin",
    R. Proctor, ed., Vol. 1, Chemistry of Endotoxin, E.Th. Rietschel,
    ed., Elsevier, Amsterdam.
Proctor, R.A., ed., "Handbook of Endotoxins", Vol. 1, "Chemistry of Endo-
    toxin", 1984; Vol. 2, "Pathophysiology of Endotoxin", 1985; Vol. 3,
    "Cellular Biology of Endotoxins", 1985; Vol. 4, "Clinical Aspects of
    Endotoxin Shock", 1986, Elsevier, Amsterdam.
Raetz, C.R.H., 1987, Structure and biosynthesis of lipid A, in: "Escher-
    ichia coli and Salmonella typhimurium. Cellular and Molecular
    Biology", C. Neidhardt, J.L. Ingraham, K. Brooks Low, B. Magasanik,
    M. Schaechter and H.E. Umbarger, eds., Am.Soc.Microbiol., Washington,
    DC.
Rietschel, E.Th., Brade, H., Brade, L., Brandenburg, K., Schade, U.F.,
    Seydel, U., Zähringer, U., Galanos, C., Lüderitz O., Westphal, O.,
    Labischinski, O., Kusumoto, S. and Shiba, T., 1987, Lipid A, the
    endotoxic center of bacterial lipopolysaccharides: relation of chem-

ical structure to biological activity, Prog.Clin.Biol.Res., 231:25.

Rietschel, E.Th., Brade, L., Schade, U.F., Seydel, U., Zähringer, U. and Brade, H., 1988, Bacterial endotoxins: properties and structure of biologically active domains, in: "Surface Structures of Microorganisms and their Interaction with the Mammalian Host", E. Schrinner, M. Richmond, G. Seibert and U. Schwartz, eds., Verlag Chemie, Weinheim.

Rietschel, E.Th., Wollenweber, H.W., Brade, H., Zähringer, U., Lindner, B., Seydel, U., Bradaczek, H., Barnickel, G., Labischinski, H. and Giesbrecht, P., 1984, Structure and conformation of the lipid A component of lipopolysaccharides, in: "Handbook of Endotoxins" R. Proctor, ed., Vol. 1, "Chemistry of Endotoxin", E.Th. Rietschel, ed., Elsevier, Amsterdam.

Salimath, V., Tharanthan, R., Weckesser, J. and Mayer, H., 1984, The structure of the polysaccharide moiety of Rhodopseudomonas sphaeroides ATCC 17023 lipopolysaccharide, Eur.J.Biochem., 144:227.

Shiba, T. and Kusumoto, S., 1984, Chemical synthesis and biological activity of lipid A analogs, in: "Handbook of Endotoxin" R. Proctor, ed., Vol. 1, Chemistry of Endotoxin, E.Th. Rietschel, ed., Elsevier, Amsterdam.

Tacken, A., Rietschel, E.Th. and Brade, H., 1986, Methylation analysis of the heptose/3-deoxy-D-manno-2-octulosonic acid region (inner core) of lipopolysaccharide from Salmonella minnesota rough mutants, Carb. Res., 149:279.

Takahashi, I., Kotani, S., Takada, H., Tsujimoto, M., Ogowa, T., Shiba, T., Kusumoto, S., Yamamoto, M., Hasegawa, A., Kiso, M., Nishijima, M., Amano, F., Akamatsu, Y., Harada, K., Tanaka, S., Okamura, H., and Tamura, T., 1987, Requirement of a properly acylated β(1-6)-D-glucosamine disaccharide bisphosphate structure for efficient manifestation of full endotoxic and associated bioactivities of lipid A, Infect. Immun., 65:57.

Takayama, K., Quereshi, N., Hyver, K., Honovich, J., Cotter, R.J., Mascagni, P. and Schneider, H., 1986, Characterization of a structural series of lipid A obtained from the lipopolysaccharides of Neisseria gonorrhoea, J.Biol.Chem., 261:10624.

Takayama, K., Quereshi, N., Ribi, E. and Centrell, J.L., 1984, Separation and characterization of toxic and nontoxic forms of lipid A, Rev.Inf. Dis., 6:439.

Westphal, O. and Lüderitz, O., 1954, Chemische Erforschung von Lipopolysacchariden Gram-negativer Bakterien, Angew.Chem., 66:407.

Wollenweber, H-W., Seydel, U., Lindner, B., Lüderitz, O., Rietschel, E.Th., 1984, Nature and location of amide-bound (R)-3-acyloxyacyl groups in lipid A of lipopolysaccharides from various gram-negative bacteria, Eur.J.Biochem., 145:265.

Zamze, S.E., Ferguson, M.A.J., Moxon, E.R., Dwek, R.A. and Rademacher, T.W., 1987, Identification of phosphorylated 3-deoxy-manno-octulosonic acid as a component of Haemophilus influenzae lipopolysaccharide, Biochem. J., 245:583.

RECOMBINANT GAMMA INTERFERON AS AN IMMUNOLOGICAL ADJUVANT

J.H.L. Playfair, A.W. Heath and J.B. De Souza

Department of Immunology, University College and
Middlesex School of Medicine, Arthur Stanley House
40-50 Tottenham Street, London W1P 9PG, UK

INTRODUCTION

Given the substantial advances that have been made in the understanding
of how immune responses are induced and regulated, it should be possible to
design vaccine adjuvants deliberately rather than empirically. This approach has led to the demonstration that at least three of the cytokines
involved in interactions between immunological cells have significant adjuvant properties when given together with antigen.

Purified IL-1 has been shown to enhance secondary antibody responses to
BSA in mice (Staruch and Wood, 1983) and the recombinant molecule has the
same effect on the response to sheep RBC (Nencioni et al, 1987). Likewise,
recombinant IL-2 enhances the protection induced by vaccine against rabies
(Anderson et al, 1987) and Herpes Simplex virus (Weinberg et al, 1988),
though in this case the cytokine has to be given in repeated doses.

In our department we have been investigating the adjuvant properties of
recombinant gamma interferon (rIFNγ). We have mainly used a detergent-
soluble blood-stage malaria vaccine in mice, which only gives consistent
protection against the lethal infection when combined with a potent adjuvant
(Playfair and De Souza, 1986) and is therefore a good model for testing adjuvanticity. We found that 5000u of mouse rIFNγ (Genentech: kindly supplied
by Dr. G. Adolf, Boehringer, Vienna) gave a highly significant enhancement
of protection, both antibody and cell-mediated responses to the parasite
being augmented (Playfair and De Souza, 1987). The adjuvant activity of
rIFNγ has recently been confirmed with a cattle vaccine against vesicular
stomatitis virus (Anderson et al, 1988).

DOSAGE ROUTE AND TIMING

So far, we have found IFNγ to show adjuvanticity when given by the
intraperitoneal, subcutaneous, and intradermal routes, while the intramuscular route was less effective. Doses of 100-20,000 units are effective,
and there is some suggestion of a bimodal dose-response curve; we are investigating whether this represents enhancement of different types of response at different doses. The optimal timing (fortunately!) is when given
with the antigen or up to 6 hours before, but when given more than 6 hours

before, or more than 2 hours after the antigen, IFN$_\gamma$ paradoxically reduces the protective effect of the vaccine.

We have never found adjuvanticity when antigen and IFN$_\gamma$ were administered by different routes.

COMBINATIONS WITH OTHER ADJUVANTS

Since many antigens for human vaccination are given in alum-precipitated form, we tested IFN$_\gamma$ with our malaria vaccine, both soluble and alum-precipitated. Although alum itself improved the protective effect of the vaccine, there was still an enhancement when IFN$_\gamma$ was added.

Our attempts to combine IFN$_\gamma$ with other adjuvants and cytokines have produced variable results. We have shown no additive effects of IFN$_\gamma$ with either IL-1, IL-2 or TNF, in fact in all three cases some antagonism was apparent. IFN$_\gamma$ showed no synergy with MDP, although this was in aqueous solution and MDP itself was ineffective. When IFN$_\gamma$ was given with Saponin there was antagonism in some experiments and an additive effect in others. We are currently investigating these paradoxical effects more fully, as well as testing the efficacy of IFN$_\gamma$ with other cytokines.

MOUSE STRAIN AND AGE DIFFERENCES

Our original experiments were performed in (C57Bl x Balb/c) F1 mice, but we have since tested a variety of other strains. Of these, only Balb/c's failed to respond to IFN$_\gamma$ (and only weakly to saponin) while CBA, A/J, C57Bl and non-inbred (Tuck) mice responded well. Indeed, in the case of C57Bl and CBA mice, IFN$_\gamma$ appeared to be a stronger adjuvant than saponin.

Most vaccines are given to children and in the case of endemic diseases such as malaria, vaccination should ideally be at birth. However, there is then the complication that passively-acquired maternal antibody may inhibit proper priming of the newborn immune system (Harte & Playfair, 1982).

We therefore vaccinated young (3 wk) mice born to immune mothers with high titres of anti-malarial IgG antibody. Such offspring failed to respond to the vaccine unless adjuvant was used, and IFN$_\gamma$ appeared to be at least as good as, if not better than saponin. We are currently studying other models of vaccination in relatively immunodeficient mice, with a view to establishing whether IFN$_\gamma$ and/or other cytokines are likely to improve the results of vaccination in immunocompromised patients - with, of course, particular reference to AIDS.

THE MECHANISM OF ACTION

Our original concept was that IFN$_\gamma$ was probably working by ensuring maximum presentation of injected antigen by increased MHC Class II expression on the antigen presenting cells (Zlotnik et al, 1983). Several observations have made us doubt that this is the only effect of interferon contributing to its adjuvanticity. IFN$_\gamma$ was effective as an adjuvant by the i.d. route where presumably antigen presentation is by Langerhans cells, which constitutively express Class II (Skoglund et al, 1988). IFN$_\gamma$ was also effective as an adjuvant in A/J mice which have a defect in Class II up-regulation in response to IFN$_\gamma$ (Strassman et al, 1986). In experiments using Cr^{51} labelled lymphocytes we have shown that IFN$_\gamma$ will enhance homing to the injection site and to the local lymph nodes, which would be expected to facilitate contact of antigen and lymphocytes. Another possibility is

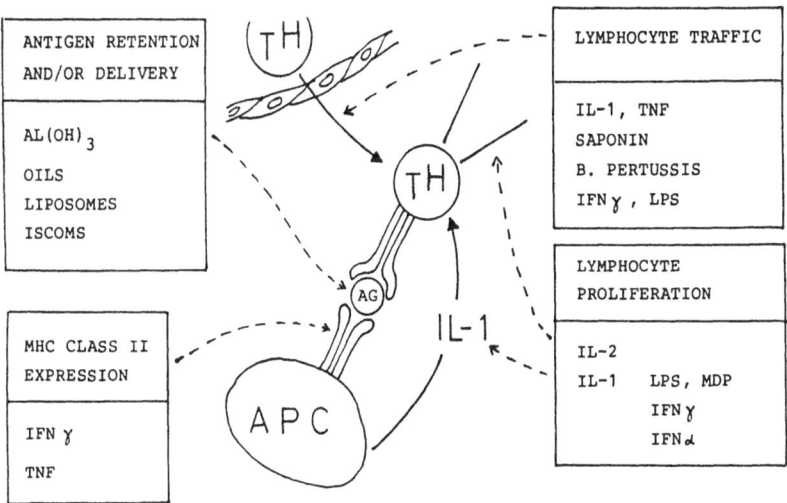

Fig. 1. Potential mechanisms of adjuvant activity.

that IFNγ is acting in concert with an LPS-like molecule in our lysate to induce IL-1 release from the antigen presenting cells (Gerrard et al, 1987), IL-1 being the true mediator of adjuvanticity. We are still investigating the roles that these possible modes of action might play. A general scheme illustrating the points at which IFNγ and other cytokines may operate is shown in Fig. 1.

OTHER ANTIGENS

We have shown IFNγ to be effective in enhancing the primary antibody responses to avidin and ovalbumin, and are now looking at secondary responses. IFNγ markedly enhanced the antibody response of low-affinity mice to DNP-BGG (M.W. Steward, personal communication). We are also looking at the use of IFNγ to enhance the response to inactivated influenza A virus.

Acknowledgement

This work was supported by a grant from the Wellcome Trust.

REFERENCES

Anderson, G., Urban, U., Fedorka-Cray, P., Newell, A., Nunberg, J. and Doyle, M., 1987, Interleukin 2 and protective immunity in Haemophilus pleuropneumoniae: preliminary studies, in: "Vaccines '87", R.M. Chanock, R.A. Lerner, F. Brown and H. Ginsberg, eds., Cold Spring Harbor Laboratory Symposia.

Anderson, K.P., Fennie, E.H., and Yilma, T., 1988. Enhancement of a secondary antibody response to Vesicular stomatatis virus "G" protein by IFNγ treatment at primary immunization, J.Immunol., 140:3599.

Gerrard, T.L., Siegel, J.P., Dyer, D.R. and Zoon, K., 1987, Differential effects of interferon-α and interferon-γ on IL-1 secretion by monocytes, J.Immunol. 138:2535.

Harte, P.G., De Souza, J.B. and Playfair, J.H.L., 1982, Failure of malaria vaccination in mice born to immune mothers, Clin.Exp.Imm., 49:509.

Nencioni, L., Villa, L., Tagliabue, A., Antoni, G., Presentini, R., Perin, F., Silvestri, S. and Boraschi, D., 1987, In vivo immunostimulatory activity of the 163-171 peptide of human recombinant IL1β, J.Immunol., 139:800.

Playfair, J.H.L. and De Souza, J.B., 1986, Vaccination of mice against malaria with soluble antigens. 1. The effect of detergent route, and adjuvant, <u>Parasite Immunol</u>. 8:406.

Playfair, J.H.L. and De Souza, J.B., 1987, Recombinant gamma interferon is a potent adjuvant for a murine malaria vaccine in mice, <u>Clin.exp. Immunol</u>., 67:5.

Skoglund, C., Scheynius, A., Holmdahl, R. and Van der Meide, P.H., 1988, Enhancement of DTH reaction and inhibition of the expression of class II transplantation antigens by <u>in vivo</u> treatment with antibodies against IFN-γ, <u>Clin.Exp.Imm</u>., <u>71:428.</u>

Staruch, M.J. and Wood, D.D., 1983, The adjuvanticity of interleukin 1 <u>in vivo</u>, J.Immunol., 130:2191.

Strassmann, G., Somers, S.D., Spencer, T.D., Adams D.O. and Hamilton, T.A., 1986, Biochemical models of interferon γ mediated macrophage activation; independent regulation of lymphocyte function associated antigen (LFA-1) and Ia antigen on murine peritoneal macrophages, <u>Cell. Immunol</u>., 97:110.

Weinberg, A. and Merigan, T.C., 1988, Recombinant interleukin 2 as an adjuvant for vaccine induced protection. Immunisation of Guineapigs with <u>Herpes simplex</u> virus subunit vaccines, J.Immunol., 140:294.

Zlotnik, A., Shimonkevitz, R.P., Gefter, M.L., Kappler, J. and Marrack, P.J., 1983, Characterisation of the γ-interferon mediated induction of antigen presenting ability in P388D1 cells, <u>J.Immunol</u>., 131:2814.

THE IMMUNOADJUVANT ACTION OF LIPOSOMES: ROLE OF STRUCTURAL CHARACTERISTICS

Gregory Gregoriadis, Lloyd Tan and Qifu Xiao

Medical Research Council Group, Academic Department of
Medicine, Royal Free Hospital School of Medicine, Pond Street
London NW3 2QG, UK

INTRODUCTION

Initial observations on the immunoadjuvant properties of liposomes
(Allison and Gregoriadis, 1974; Gregoriadis and Allison, 1974) have been now
confirmed and extended using a variety of antigens relevant to human and
veterinary immunization (for reviews see Gregoriadis, 1988). Antigens studied in conjunction with liposomes include Streptococcus pneumoniae serovar 3
(Snippe et al,1983), Salmonella typhimurium lipopolysaccharide (Desiderio
and Campbell, 1985), cholera toxin (Alving et al, 1980; Pierce and Sacci,
1984), adenovirus type 5 hexon (Kramp et al, 1982), Herpes Simplex virus
type 1 antigens (Naylor et al, 1982), hepatitis B virus surface antigen
(Manesis et al, 1979), Epstein-Barr virus gp 340 protein (Epstein et al,
1985), tetanus toxoid (Davis et al, 1986; Gregoriadis et al, 1987), synthetic peptides of foot-and-mouth disease virus (Francis et al, 1985),
poliovirus peptides (Xiao et al, 1989) and rat spermatozoal polypeptide
fraction (Mettler et al, 1983).

The unique variability of liposomes in terms of structural characteristics and mode of antigen accommodation has helped in rendering the immunoadjuvant action of the system versatile and provided a variety of possibilities in vaccine design (Gregoriadis, 1988). However, in spite of the considerable amount of work already carried out in this area, the nature or
mechanisms of liposomal immunoadjuvant action are essentially unknown
(Gregoriadis, 1985; Hedlund et al, 1984). Moreover, opposing views have
been aired regarding liposomal structural characteristics deemed optimal for
such action (Gregoriadis, 1986; Hedlund et al, 1984; Kinsky, 1978; Dancey et
al, 1978; Shek and Sabiston, 1982; van Rooijen and van Nieuwmegen, 1980).
In addition, there still exist problems with the procedures used for antigen
entrapment into liposomes which must be resolved if liposomal vaccines are
to compete with those presented in alum or other (proposed) adjuvants. Some
of the procedures, for instance, are complicated or yield low antigen entrapment values and are thus uneconomical. With other procedures, the
requirement of organic solvents, sonication or detergents (Gregoriadis,
1984) may lead to the masking or modification of antigenic sites. This
laboratory has recently investigated some of these questions and problems
and findings are discussed below.

We have recently reported a simple and mild method (Kirby and Gregoriadis, 1984; Gregoriadis, 1985; Gregoriadis et al, 1987) for the preparation of liposomes which is based on the fusion of preformed phospholipid vesicles by dehydration followed by rehydration in the presence of solute (e.g. antigen) destined for entrapment. Fusion leads to the formation of multilamellar liposomes (dehyration-rehydration vesicles, DRV), and in the process, up to 80% or more of the solute is entrapped. The procedure does not require organic solvents, sonication or detergents, is amenable to scale-up for industrial use (Kirby and Gregoriadis, 1984) and, therefore, particularly suitable for liposomal vaccines (Gregoriadis et al, 1987; Davis and Gregoriadis, 1987). DRV have therefore been adopted in our studies of liposomal immunoadjuvant action, mostly using aggregate-free immunopurified tetanus toxoid as a model antigen alone or together with other co-adjuvants (e.g. cytokines). Toxoid-containing DRV coated with a mannosylated ligand (expected to facilitate interaction with macrophages known to express the mannose receptor) were also used in attempts to obtain targeted adjuvanticity. In addition, the toxoid was covalently coupled to the surface of preformed multilamellar vesicles. This, in parallel with other studies (van Rooijen and van Nieuwmegen, 1980; Snyder and Vannier, 1984; Shek and Sabiston, 1982), provided us with an alternative means of antigen presentation in liposomes.

Antigen Entrapment in DRV Liposomes

In agreement with previous data obtained for a variety of solutes (e.g. drugs, proteins, etc.) (Kirby and Gregoriadis, 1984; Seltzer et al, 1988), entrapment of tetanus toxoid in DRV liposomes made of equimolar phospholipid and cholesterol and generated from preformed small unilamellar vesicles (SUV) was substantial (47-82%) depending on the phospholipid component and the surface charge (Gregoriadis et al, 1987). Work with other antigens, namely polio type 3-VP2 and 1-VP2 peptides (Xiao et al, 1989), reconstituted influenza virus envelopes (Tan et al, 1989), Leishmania major LV 39 antigens (Kahl et al, 1989), hepatitis B surface antigen (L. Tan and G. Gregoriadis, unpublished) has also shown high (about 40-50%) entrapment values. Recently, entrapment values of 67.0-77.4% for recombinant interleukin-2(rIL-2) (Tan and Gregoriadis, 1989) were achieved. DRV, as prepared following rehydration of the dehydrated mixture of SUV and solute, are heterogenous in size with an average diameter of 0.30 0.28 nm (Kirby and Gregoriadis, 1984). We have been recently investigating the possibility of reducing the size of these vesicles and at the same time achieving a more homogenous vesicle population without excessive loss of entrapped solute. Preliminary work (G. Gregoriadis and H. da Silva) indicates that after subjecting DRV (containing a model solute such as carboxyfluorescein or maltose) to one or two cycles of microfluidization (Seltzer et al, 1988), solute loss in the resulting smaller "DRV" is modest, especially if microfluidization occurs in the presence of the mother liquid (ie. unentrapped solute following DRV preparation). The size of such microfluidized DRV is at present under investigation.

Approach to Long-term Storage of Liposomes

Long-term storage of liposomal vaccines in a freeze-dried form is a desirable prerequisite for the commercialization of the system. To this end, we have carried out experiments with liposomal toxoid. DRV liposomes made of equimolar phospholipid and cholesterol with the toxoid passively entrapped or covalently linked (Gregoriadis et al, 1987) to their surface, were freeze-dried and then reconstituted in saline. In some experiments freeze-drying was carried out in the presence of trehalose, known to act as a cryoprotectant. Results (Gregoriadis et al, 1987) showed that after dehydration in the absence of trehalose, DRV liposomes composed of equimolar egg phosphatidylcholine (PC) and cholesterol retained on rehydration most (78.3%) of the entrapped toxoid. Retention of the toxoid by distearoyl phosphatidylcholine (DSPC) DRV was, however, quantitative (93.5%). DRV PC liposomes containing bovine serum albumin (BSA) also retained most of the protein (73.2%) when subjected to freeze-drying (Gregoriadis et al, 1987). Dehydration in the presence of trehalose further improved the retention of toxoid in DRV PC and DSPC liposomes. PC liposomes with covalently linked toxoid and dehydrated in the absence (or presence) of trehalose, also retained much of the toxoid with no obvious changes in antigen (surface) localization compared to untreated preparations (Gregoriadis et al, 1987).

IMMUNE RESPONSES FOLLOWING IMMUNIZATION WITH LIPOSOMAL TOXOID

In experiments in which Balb/c mice were injected twice intramuscularly with toxoid-containing liposomes, we observed (Davis et al, 1987) that adjuvant activity was reflected in various IgG immunoglobulin subclasses. Thus, when high doses (2 and 10 µg) of tetanus toxoid were used, IgG_1 response was similar for free and liposomal antigen. However, at a lower dose (0.1 µg), there was a significant increase in the level of anti-toxoid IgG_{2b} antibody at all liposomal antigen doses in comparison to free antigen. Adjuvant effect was less consistent in terms of IgG_{2a} and IgG_3 responses, but significant differences between the liposomal and free antigen were measured when the toxoid was injected in higher doses (Davis et al, 1987). It was also observed (Davis et al, 1987) that no liposomal adjuvant effect could be ascribed solely to events following the second injection. These results suggest, therefore, that liposomes increase the antibody response within an individual IgG subclass, independently of their effect on other IgG subclasses. It appears that with the antigen dose increasing, there is an increase both in the antibody response within an individual IgG subclass (whether the antigen is liposome-entrapped or not) and the number of IgG subclasses involved in the response.

EFFECT OF LIPOSOMAL STRUCTURAL CHARACTERISTICS ON ADJUVANTICITY

The uniquely versatile structure of liposomes and mode of antigen accommodation allows, as already stated, for a wide range of options in the design of effective vaccines. Some of the structural features of liposomes have been recently examined, in terms of their effect on adjuvanticity, using DRV liposomes with tetanus toxoid or other antigens. The choice of the dehyration-rehydration procedure for antigen entrapment into liposomes was based, as alluded to earlier, on findings (Kirby and Gregoriadis, 1984; Gregoriadis et al, 1987) of realistically and reproducibly high solute entrapment values. The method of diazotization was used for the coupling of antigen to the liposomal surface (Snyder and Vannier, 1984; Gregoriadis et al, 1987; Davis et al, 1986).

Relationship Between Liposomal Phospholipid to Tetanus Toxoid Mass Ratio and Antibody Response

It was reasonable to assume that liposomal adjuvanticity would be increased by ensuring a high concentration of antigen in individual vesicles and, therefore, presentation of sufficient quantities of antigen to antigen presenting cells. In three separate experiments (each with a variety of liposomal phospholipid to toxoid mass ratios, ranging from 2:1 to 5×10^5:1) however, the opposite became apparent, namely that the greater the phospholipid to toxoid mass ratio, the greater the (secondary) immune response (IgG_1 and IgG_{2b}) (Davis and Gregoriadis, 1987). Thus adjuvanticity increased considerably with increasing (up to about 2×10^3:1) ratios. At even greater ratios (e.g. 9×10^4:1 or above), however, adjuvanticity was reduced drastically. It thus appears that liposomal adjuvanticity in terms of

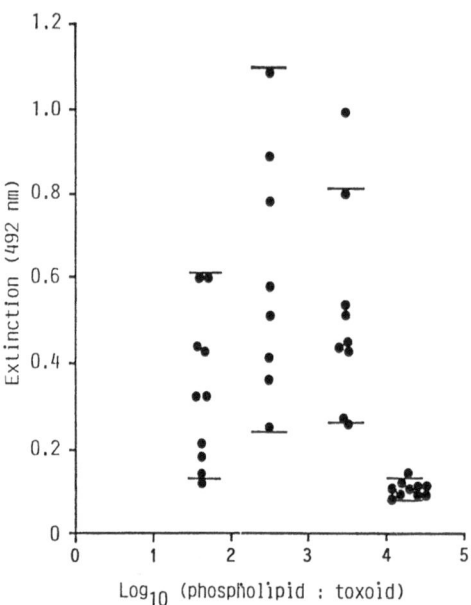

Fig. 1. Primary immune responses to tetanus toxoid-containing liposomes of varying phospholipid to toxoid mass ratios. Balb/c mice (groups of 8-10) were injected intramuscularly once with 0.2 µg tetanus toxoid entrapped in DRV liposomes composed of equimolar PC and cholesterol. The phospholipid to toxoid mass ratios were 37.4:1, 346:1, 2857:1 and 17,804:1. Animals were bled four weeks after injection and anti-toxoid IgG_1 measured by ELISA. Readings for individual mice are plotted against ratios with bars representing 95% confidence intervals of the median values. Differences in response between the highest ratio group and the other groups were significant according to the Kruskall Wallis test ($H=23.33$, $p<0.01$).

secondary response, is related (up to a certain level) to the lipid dose, presumably in conjunction with a slow rate of degradation and/or removal of a large liposomal lipid dose from the tissue on injection. Interestingly, a similar relationship of liposomal lipid to toxoid mass ratio and antibody (IgG_1) levels formed is observed when primary response to the liposomal toxoid is measured (Fig. 1) (D. Davis and G. Gregoriadis, unpublished). Such reduction in primary and secondary response at very high phospholipid to toxoid mass ratios could be tentatively attributed to an immunosuppressive effect of excessive phospholipid (Davis and Gregoriadis, 1987). Alternatively or concurrently, it is possible that at very high phospholipid to antigen mass ratios, the number of liposomal vesicles in the preparation is so great that only a small proportion of them contain antigen, probably at a concentration too low for it to be immunogenic.

Immune Response after Injection of Entrapped and Surface-linked Antigen

Measurement of immune response in Balb/c mice injected with entrapped and surface-linked tetanus toxoid, showed no significant difference in anti-toxoid IgG_1 and IgG_{2b} levels between the two liposomal preparations (Davis and Gregoriadis, 1987). The similarity in immune responses (IgG_1 and IgG_{2b}) between entrapped and surface-linked antigen was also found to apply to PC (and DSPC) liposomes of higher (up to about 4.7×10^4:1) phospholipid to toxoid mass ratios (Gregoriadis et al, 1988). This contrasts data by Shek and Sabiston (1982) who reported that liposomal surface-linked antigen (albumin) is not as efficient in stimulating indirect plaque-forming cells as the entrapped protein. Our results are also in disagreement with findings by van Rooijen and van Nieuwmegen (1980) that liposomes coated with the antigen (albumin) are better adjuvants than liposomes entrapping the antigen. In view of the effect of the lipid dose (Fig. 1) on liposomal adjuvanticity, it would appear essential that, in comparing antibody responses to antigen incorporated in different liposome preparations, the amounts of injected liposomal lipid and the antigen must be similar. However, such conditions have not been easy to fulfil since antigen entrapment values with the procedures used by these workers have been low and/or unpredictable. Furthermore, the significance of an optimal lipid to antigen mass ratio in achieving adjuvanticity had not been previously appreciated. Thus, the opposing conclusions reached by Shek and Sabiston (1982) and van Rooijen and van Nieuwmegen (1980) may be due to the fact that the lipid to protein ratios for the preparations coated with the antigen were four-fold and fifty-fold greater respectively than the same ratios for the entrapped antigen.

Relationship between the Nature of Liposomal Phospholipid and Antibody Responses to Entrapped Antigen

Secondary immune responses (IgG_1 and IgG_{2b}) in mice immunized with tetanus toxoid entrapped in DRV liposomes composed of various phospholipids and equimolar cholesterol were found (Fig. 2) similar when the gel-liquid crystalline transition temperature (Tc) of the phospholipids ranged between -32 and $41.5^\circ C$ but reduced significantly (both subclasses) when DSPC ($Tc=54^\circ C$) was the phospholipid component of DRV. The mass ratios of the phospholipid to protein in the preparations used were 21.2:1 (dilinoleoyl phosphatidylcholine; (DLPC), 24.2:1 (dioleyl phosphatidylcholine; DOPC), 23.0:1 (PC), 33.1:1 (dimyristoyl phosphatidylcholine; DMPC), 29.0:1 (dipalmitoyl phosphatidylcholine; DPPC) and 14.3:1 (DSPC). However, the significantly low or absence of response with the (water soluble) toxoid presented in DSPC liposomes contrasts responses obtained by others (Kinsky, 1978; Dancey et al, 1978; Bakouche and Gerlier, 1986) using membrane antigens. Thus, in work by Kinsky (1978) and Dancey et al (1978), liposomes incorporating a hapten-phospholipid complex (20:1 estimated lipid to complex mass ratio) promoted immune response to the hapten when beef sphingomyelin (Tc $37-39^\circ C$), but not egg phosphatidylcholine, was the phospholipid component. Similarly, a broad relationship between increasing values of phospholipid Tc and increasing antibody response to liposomal Gross virus cell surface antigen (estimated phospholipid to protein ratio 7.8:1 Gerlier et al, 1978), has been reported by Bakouche and Gerlier (1986). In this respect, it is of interest that recent findings from this laboratory (Tan et al, 1989) with reconstituted influenza virus envelopes (RIVE) incorporated in PC or DSPC DRV liposomes suggest (Table 1) that both primary and secondary immune responses (IgG_1) to the RIVE in Balb/c mice, are similar for the two lipid compositions. It thus seems that even for membrane antigens (e.g. RIVE), high melting phospholipids alone will not necessarily give a stronger response and that other parameters such as lipid to antigen mass ratios may play a role. For instance, in recent work (Davis and Gregoriadis, 1987; Gregoriadis et al, 1988) using a much greater DSPC to toxoid mass ratio (about $4.2 \times 10^3:1$), anti-toxoid antibody (IgG_1 and IgG_{2b}) responses to DSPC DRV were as high as those obtained with PC liposomes of a similar ratio. Further, a mixture of low ratio (12:1) PC DRV and "empty" (toxoid free) PC liposomes giving a high (about $2.7 \times 10^3:1$) overall ratio, also improved immune responses to levels approaching those obtained with toxoid entrapped in PC DRV of an identical (about $2.7 \times 10^3:1$) ratio. More recent work, however, has shown that using two different water soluble antigens, immune responses were unpredictable and did not follow the pattern seen with the toxoid. For instance, there was a similarity in primary and secondary immune response between PC and DSPC DRV with bovine serum albumin as the antigen (Table 2), whereas with poliovirus 3-VP2 peptide, (secondary) response was nil for PC DRV but substantial (and much greater than that seen with the free peptide) for DMPC and DSPC liposomes (Table 3). It would seem

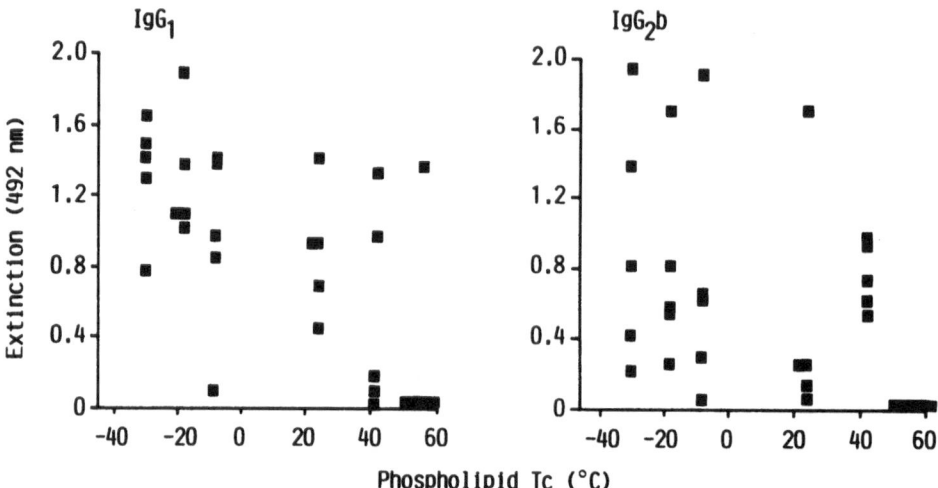

Fig. 2. Effect of the liposomal phospholipid Tc on liposome adjuvanticity. Balb/c mice in groups of five were injected twice with 0.25 μg of tetanus toxoid entrapped in DRV liposomes composed of equimolar phospholipid and cholesterol. Animals were bled 9-10 days after the booster injection and analysed for IgG_1 and IgG_{2b} by the ELISA immunosorbent assay. The phospholipid to toxoid mass ratios are given in the text. Readings for individual mice are plotted against the liquid-crystalline phase transition temperatures (Tc) (^{0}C in parentheses) of the DLPC (-32), DOPC (-20), PC (-10), DMPC (23), DPPC (41.5) and DSPC (54.0) components of DRV. Differences in response between DSPC DRV and the other DRV preparations were significant (IgG_1), H=15.37, p<0.01; (IgG_{2b}), H=11.23, p<0.05, as determined by the Kruskal-Wallis non-parametric test. (From Davis and Gregoriadis, 1987.)

then that tailoring of DRVs to optimise their action must be carried out individually for antigens. This may have favourable implications in the patenting of the system as a vaccine carrier.

Receptor Mediated (Targeted) Adjuvanticity of Liposomes

Conventional liposomes are engulfed by antigen presenting cells (such as macrophages) avidly and, in terms of adjuvanticity, modification of liposomes leading to selective uptake by the cells would seem superfluous. On the other hand, it has been shown (e.g. Perry and Ofek, 1984) that invading micro-organisms can interact through mannose-terminating ligands on their surface with mannose receptors on macrophages. As this event may relate to natural immunization, it was thought that liposomes bearing mannose-terminating ligands on their surface might show improved adjuvanticity (Gregoriadis et al, 1988; Garçon et al, 1988). Initial experiments were therefore carried out to confirm that toxoid-containing DRV which were coated (Garçon et al, 1986) with mannosylated albumin bind selectively to the relevant receptors on (mouse peritoneal) macrophages. As anticipated, such targeted liposomes interacted with the cells with greater affinity than when non-mannosylated DRV of the same composition and content were used (Gregoriadis et al, 1988; Garçon et al, 1988) (Fig. 3). In addition, the specificity of interaction with macrophages was confirmed, as in the presence of excess ⍺-methyl-D-mannoside, binding was reduced to levels seen with mannose-free DRV.

Table 1. Antibody (IgG$_1$) Responses to Free and Liposome-entrapped RIVE

	Primary response		Secondary response	
RIVE	0.1 µg	1.0 µg	0.1 µg	1.0 µg
Free	0.020	0.064	0.000	0.527
	0.040	0.025	0.000	0.387
	0.069	0.021	0.000	0.399
	0.019	0.046	0.004	0.112
	0.044	0.048	0.000	0.399
PC DRV	0.274	0.405	1.023	1.826
	0.049	0.238	0.782	1.819
	0.170	0.465	1.193	1.477
	0.349	0.479	1.537	1.596
	0.082	0.563	1.486	1.806
DSPC DRV	0.075	0.207	0.903	1.649
	0.160	0.662	1.485	1.708
	0.145	0.640	1.532	1.831
	0.131	0.506	1.049	1.871
	0.070	0.326	1.031	1.715

Results for primary and secondary responses (0.1 and 1.0 µg doses) are shown as ELISA readings at 492 nm for individual mice in each group. Median values are underlined. For other details, see the text. (From Tan et al, 1989).

Table 2. IgG$_1$ Response in Mice Immunized with Free and Liposome-entrapped Bovine Serum Albumin

Days after first injection	Free BSA	BSA in PC DRV[c]	BSA in DSPC DRV[c]
27	0.042	0.590	0.407
42[a]	0.072	1.957	1.712
69[b]	0.065	1.323	1.238

Balb/c mice (in groups of five) were injected intramuscularly on day 0 and day 28 with 1 µg BSA, free or entrapped in DRV liposomes composed of equimolar PC or DSPC and cholesterol (phospholipid to antigen mass ratio 50:1). Animals were bled at intervals and anti-BSA IgG$_1$ assayed in sera by ELISA (median readings shown). Comparison of ELISA readings was made by the method of Kruskall Wallis. The Mann-Whitney test was used to compare each of the liposomal BSA groups with the free BSA group (p<0.01). Differences between the two liposomal groups at all time intervals examined were not significant. [a]14 days after booster injection. [b]41 days after booster injection. [c]ELISA readings were still significantly (p<0.01) higher than readings for free BSA (0.467 and 0.605 median values for PC and DSPC DRV respectively) 240 days after the first injection. (From Gregoriadis and Panagiotidi, 1989.)

Experiments were subsequently carried out in Balb/c mice (Garçon et al, 1988; Gregoriadis et al, 1988) to study the effect, if any, of liposomal mannose residues on adjuvanticity. It is apparent from Fig. 4 that mannosylated DRV promote greater (eight-fold) (IgG_1) immune response than the two control preparations of uncoated or mannose-free albumin coated DRV. In further studies, mice were immunized with (a) DRV preparations coated with two different amounts of mannosylated ligand (different numbers of protein chains) which, however, had the same number (36) of mannose moles per mole albumin; (b) preparations coated with approximately the same amount of mannosylated albumin (same numbers of protein chains) which had either 8 or 36 moles mannose per mole albumin. Findings suggest (Gregoriadis et al, 1988; Garçon et al, 1988) that immune responses (IgG_1 and IgG_{2b}) to liposomal toxoid are dependent on the number of mannosylated albumin molecules available on the vesicle surface rather than the number of mannose residues present on the albumin. Assuming that targeted adjuvanticity (as determined here) is related to the extent (or firmness) of vesicle binding to macrophages, it may be that a certain number of mannose moles per mole albumin (8 in the present studies) mediate binding which cannot be improved upon by a larger number (36) of mannose moles. Alternatively, binding may be substantially augmented (or become firmer) by increasing the number of mannosylated ligand molecules on the surface of vesicles.

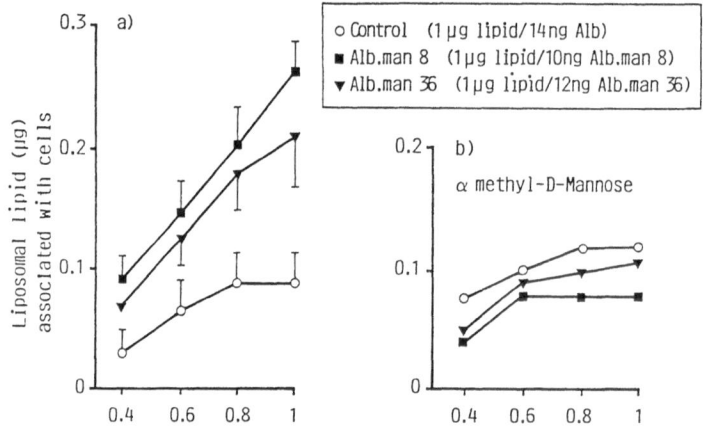

Fig. 3. Binding of mannosylated liposomes to macrophages. Peritoneal macrophages were harvested from male Balb/c mice weighing 20–25 g (Clinical Research Centre, Harrow, Middlesex) four days after i.p. injection with 1.0 ml of 10% thioglycolate per mouse. Cells were suspended in HBSS medium and grown at 37^0 in small petri-dishes (10^6 cells per dish) at 5% CO_2 and 95% humidity. After 3 hr, non-adherent cells were removed by washing the dishes twice with HBSS. At the end of their first day of growth, cells were incubated for 2 hr at 4^0 in the absence (a) or presence (b) of 100 mmoles methyl α-mannopyranoside with increasing amounts of radiolabelled, toxoid-containing DRV (10^4 d.p.m. [^3H]PC) coated with albumin only (O) or albumin coupled to 8 (■) or 36 (▼) moles mannose per mole protein. Albumin content of DRV was 10–12 ng/μg PC. Following incubation, cells were detached with a rubber policeman and assayed for lipid ^3H radioactivity. (From Garcon et al, 1988.)

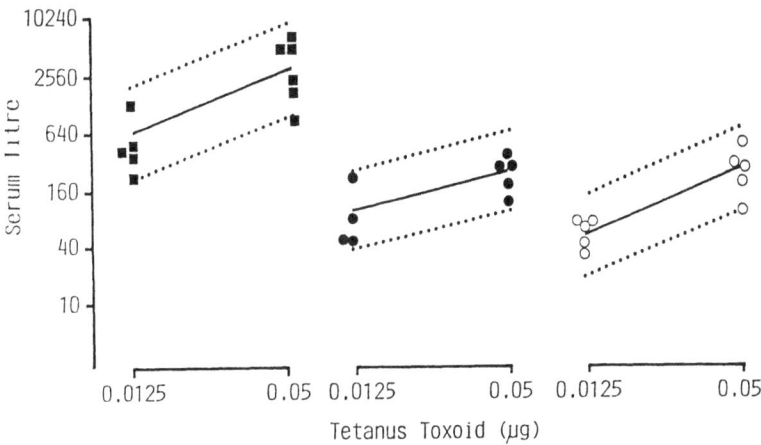

Fig. 4. Anti-tetanus toxoid IgG_1 titres in mice immunized with toxoid-
containing DRV coated with mannosylated albumin. Balb/c mice (in
groups of five) were immunized intramuscularly with tetanus toxoid
(0.0125 or 0.05 µg) entrapped DRV composed of equimolar PC and
cholesterol (90 µg lipid per 1 µg toxoid) (0) in similar DRV coated
with albumin (1 µg toxoid per 120 µg lipid per 2.4 µg albumin) (●)
or in similar DRV coated with mannosylated (36 mannose moles per
mole albumin (1 µg toxoid per 67 µg lipid per 1.36 µg albumin (■).
Three weeks after primary immunization, mice received identical
injections of the same preparations and were bled nine days later.
The least squares estimate of the regression line of IgG_1 antibody
titre against antigen dose was derived separately for the three
groups. Regression lines (o versus ● or ■) were significantly
different (p<0.05). Dotted lines denote 95% confidence interval of
the regression lines. (From Garçon et al, 1988.)

The Effect of Physiological and Non-physiological Mediators on the Primary Immune Response to Liposomal Tetanus Toxoid

The primary immune response to liposomal tetanus toxoid and possible
modulation of such response by interleukin-2 (IL-2), interferon-γ (IFN-γ),
the water soluble N-acetylmuramyl-L-threonyl-D-isoglutamine ([Thr^1]MDP) and
its liposoluble 6-0-stearoyl derivative (St.[Thr^1MDP]) were investigated.
These agents were administered in Balb/c mice via cholesterol-rich PC DRV
together with the toxoid using a variety of protocols (D. Davis and G.
Gregoriadis, in preparation). Findings suggest that primary immune response
(IgG_1) to the toxoid using optimal ratios (see Fig. 2), peaks at four weeks
and declines slowly thereafter. By 24 weeks ELISA readings attain levels
seen two weeks after injection. Non-purified recombinant mouse IL-2 (350
units) IFN-γ (585 or 5533 units) and [Thr^1]MDP (25 or 50 µg) co-entrapped
with the toxoid in the same liposomes, generally reduced primary IgG_1 response to levels below those obtained with control liposomes containing the
antigen alone (Figs. 5 and 6). However, IL-2 (320 units) in separate liposomes mixed with the liposomal toxoid significantly improved immune response. On the other hand, the liposoluble St.[Thr^1]MDP (co-entrapped with
the toxoid or incorporated in separate liposomes and then mixed with the
liposomal toxoid) gave significantly higher immune response than those
observed with the water soluble [Thr^1]MDP in similar forms (D. Davis and G.
Gregoriadis, in preparation).

Whilst the significance of these findings is now under investigation,
recent work with pure liposomal mouse rIL-2 used in higher doses (10^3 and
10^4 Cetus units) revealed (in contrast to data above) an improvement of

Table 3. Immune Responses to Liposome-entrapped Poliovirus Peptide 3-VP2

Peptide formulation	Secondary immune response ($\log_{10}$ titres; Mean SD.)							
	IgG_1	p*	IgG_{2a}	p	IgG_{2b}	p	IgG_3	p
A. Free peptide	1.28 0.12		1.72 0.38		1.47 0.42		1.71 0.16	
B. DMPC, chol.	2.60 0.23	<0.001 B vs A	1.74 0.25	NS B vs A	1.94 0.23	<0.05 B vs A	1.95 0.18	<0.05 B vs A
C. DSPC, chol.	3.81 0.28	<0.001 C vs A,B	3.07 0.23	<0.001 C vs A,B	3.24 0.21	<0.001 C vs A,B	2.79 0.39	<0.001 C vs A,B

Balb/c mice were injected intramuscularly with 20 µg free or liposome-entrapped poliovirus 3-VP2 peptide on day 0 and day 28. Blood plasma was assayed by ELISA for anti-peptide IgG subclasses 10 days after the second injection. (From Xiao et al, 1989.) *Probability of significance was estimated by the t-test following analysis of variance.

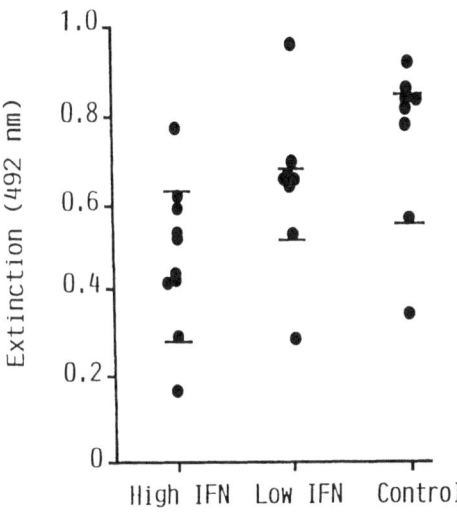

Fig. 5. The effect of IL-2 and [Thr^1]MDP co-entrapped with tetanus toxoid
in liposomes on the primary immune response. Balb/c mice (groups
of 5-10) were injected intramuscularly once with 1.0 ug tetanus
toxoid entrapped alone (control) or together with IL-2 (350 units),
50.3 µg [Thr^1]MDP (High MDP) or 25.2 µg [Thr^1]MDP (Low MDP) in DRV
liposomes with phospholipid to toxoid mass ratios of 318:1, 794:1,
221:1 and 226:1 respectively. A significantly diminishing response
(IgG$_1$) to toxoid co-entrapped with Low MDP, High MDP and IL-2 in
that order was observed according to the Kruskall Wallis test
(H=21.29, p<0.01). For other details see legend to Fig. 1.

primary response (IgG$_1$ and IgG$_{2b}$) at the higher dose of co-entrapped rIL-2
and no effect for the same dose of separately entrapped mediator (Lloyd and
Gregoriadis, 1989). Results in Table 4 show secondary immune responses
(IgG$_1$, IgG$_{2a}$ and IgG$_{2b}$) from the same experiment (Lloyd and Gregoriadis,
1989). It appears that the higher (10^4 units; all three subclasses) and the
lower (10^3 units; IgG$_1$) dose of rIL-2 co-entrapped with tetanus toxoid significantly reduces antibody response to the toxoid at the low dose (IgG$_{2a}$
and IgG$_{2b}$) and, for one of the subclasses (IgG$_{2a}$), at its high dose.

COMPARISON OF LIPOSOMES WITH OTHER IMMUNOLOGICAL ADJUVANTS

    In recent experiments (Gregoriadis and Panagiotidi, 1989), the immuno-
adjuvant action of PC DRV was compared with that of a variety of other adjuvants (Table 5). Under our experimental conditions, antibody readings in
both the primary and secondary responses were similar for DRV (with or without co-entrapped [Thr^1]MDP) and alum. Furthermore, significantly higher
responses were obtained with DRV without [Thr^1]MDP (primary) and DRV with
[Thr^1]MDP (primary and secondary) compared to those seen with CFA and
[Thr^1]MDP alone.

CONCLUSIONS

    Evidence presented here and elsewhere (Gregoriadis, 1988) strongly
supports further evaluation of liposomes as carriers of vaccines. From the
practical point of view, DRV liposomes are easy to prepare and entrap antigens quantitatively in the absence of conditions potentially damaging to the
antigen. Coupling procedures can also be employed for the attachment of
antigens (or ligands) to the surface of DRV liposomes. (It should be

Table 4. Effect of Liposomal Interleukin-2 on the Immune Response to Entrapped Tetanus Toxoid

Liposomes	IgG_1		IgG_{2a}		IgG_{2b}	
A. Entrapped toxoid only	0.184		0.205		0.153	
	0.299		0.251		0.184	
	0.359		0.661		0.352	
	0.505		0.879		1.543	
	0.546		0.910		1.553	
B. Co-entrapped toxioid and rIL-2 (10^3 units)	0.898		0.170		0.293	
	1.520		0.221		0.540	
	1.591	p 0.01	0.242	N.S.	0.776	N.S.
	1.740	(T=15)	0.368	(T=21)	0.790	(T=29)
	2.000		0.389		1.073	
C. Co-entrapped toxoid and rIL-2 (10^4 units)	1.135		0.722		1.260	
	1.648		1.063		1.560	
	1.755	p 0.01	1.484	p 0.05	2.000	p 0.05
	1.997	(T=15)	2.000	(T=17)	2.000	(T=17)
	2.000		2.000		2.000	
D. Separately entrapped toxoid and rIL-2 (10^3 units)	0.187		0.088		0.047	
	0.201		0.104		0.064	
	0.237	N.S.	0.130	p<0.05	0.073	p<0.05
	0.291	(T=24)	0.187	(T=17)	0.259	(T=17)
	0.418		0.319		0.301	
E. Separately entrapped toxoid and rIL-2 (10^4 units)	0.316		0.905		0.242	
	0.723		1.215		0.283	
	1.234	N.S.	1.495	p<0.05	0.986	N.S.
	1.416	(T=18)	2.000	(T=16)	1.440	(T=25)
	1.490		2.000		2.000	

Balb/c mice in groups of five were injected intramuscularly on day 0 and day 42 with 1 µg tetanus toxoid entrapped in DRV liposomes without (A) or with rIL-2 (co-entrapped) (B and C), or entrapped alone in DRV which were mixed with similar DRV containing rIL-2 (separately entrapped) (D and E). ELISA readings shown are from blood samples obtained 14 days after the booster injection (secondary response). The Mann-Whitney test was used to compare groups B, C, D and E with group A (control). Median readings are underlined. N.S., not significant. T, lower sum of the ranks in pairs of groups compared. (From Lloyd and Gregoriadis, 1989.)

pointed out that with antigens or ligands which are sensitive to the coupling procedures, these can be firstly linked to SUV which can then be used to generate DRV bearing much of the antigen or ligand on their surface (Garcon et al, 1986; Weissig et al, 1989; Senior and Gregoriadis, 1989)). DRV preparations with entrapped or surface-linked antigen can be freeze-dried with most of the antigen being recovered within or on intact vesicles on reconstitution in saline.

Results from experiments in which Balb/c mice were injected with DRV-incorporated tetanus toxoid, suggest that (a) liposomal adjuvanticity is reflected in most IgG subclasses, (b) there is no shift in subclasses when

Table 5. Comparison of DRV Liposomes with other Adjuvant Formulations

Adjuvant formulation	Time after first injection (days)			
	9	17	27	70[b]
(A) BSA in PC DRV[a]	0.097	0.278	0.572	0.795
(B) BSA in PC DRV with [Thr^1]MDP co-entrapped[a]	0.060	0.214	0.627	0.911
(C) BSA mixed with [Thr^1]MDP	0.037[c]	0.082[c,d]	0.290[c,d]	0.537[d]
(D) BSA in alum	0.100	0.151	0.564	0.806[e]
(E) BSA in CFA	0.019[c]	0.062[c,d]	0.365[c,d]	0.551[d]

Balb/c mice (in groups of five) were injected intramuscularly on day 0 and day 28 with 1 µg BSA in various adjuvant formulations. Animals were bled at time intervals and anti-BSA IgG_1 assayed in sera by ELISA (median readings shown). Comparison of ELISA values was made by the method of Kruskall and Wallis followed by the Mann-Whitney test for comparisons between groups. [a]Phospholipid to BSA mass ratio in A and B was 1000:1. [b]42 days after booster injection. [c]$p < 0.01-0.05$ (C and E vs. A). [d]$p < 0.01-0.05$ (C and E vs. B). [e]$p < 0.05$ (D vs. B). (From Gregoriadis and Panagiotidi, 1989.)

compared to the response obtained with the free antigen and (c) adjuvanicity is the outcome of events following primary immunization. Our studies also indicate that structural characteristics of liposomes such as membrane fluidity, amount of liposomal lipid relative to the antigen and ligands (specific for antigen-presenting cells) grafted on the liposomal surface influence adjuvanticity (albeit in different ways for different antigens). In addition, a number of physiological and non-physiological mediators appear to modulate (primary and secondary) immune responses depending on whether the mediator is given in the same or separate liposomes with the antigen. With rIL-2 in particular, the mediator appears to either augment or reduce antibody responses to the liposomal antigen depending on the mode of presentation of mediator and antigen in liposomes. Such actions of rIL-2 are different for the various IgG subclasses tested.

Future vaccines based on synthetic peptides and recombinant subunit antigens are expected to require a new generation of immunological adjuvants. In contrast to some of the adjuvants presently under investigation (for a review see Allison and Byars, 1986), liposomes exhibit unique structural versatility allowing freedom in vaccine design. When liposomes are made of appropriate lipids, they produce no side effects and are known to elicit both humoural and cell-mediated immunity. Moreover, many of the earlier technological difficulties have been resolved and production on industrial scale is now feasible (Gregoriadis, 1988). More recently, several phase one and phase two clinical trials on the use of liposomes in antimicrobial and cancer therapy have been initiated (Lopez-Berestein and Fidler, 1989) and this should facilitate similar trials with liposomal vaccines. It remains to be seen whether liposomes will prove sufficiently superior to other adjuvants (both technologically and in terms of immuno-adjuvant action) so as to warrant their development in vaccine formulation.

Acknowledgements

Work reported here was supported by a Medical Research Council project grant and grants from The British Council, Wellcome Biotechnology Ltd.,

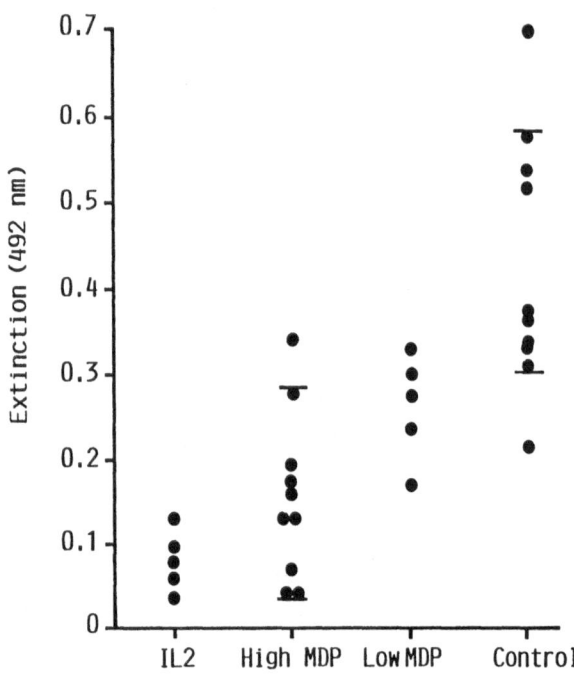

Fig. 6. The effect of IFN-γ co-entrapped with tetanus toxoid in liposomes
on the primary immune response. Balb/c mice (groups of 8-10) were
injected intramuscularly once with 0.75 μg tetanus toxoid entrapped
alone (control) or together with 5533 (High IFN) or 585 units (Low
IFN) of IFN-γ in DRV liposomes with phospholipid to toxoid mass
ratios 342:1, 201:1 and 325:1 respectively. A significantly diminishing response (IgG_1) to toxoid co-entrapped with the low and
high doses of IFN-γ in that order was observed according to the
Kruskall Wallis test (H=11.65, p<0.01). For other details see
legend to Fig. 1.

Government of Peoples Republic of Germany and the University of Singapore.
The author is indebted to the following additional past and present members
of the MRC Group for their participation in the work: A. Davies, D. Davis,
N. Garcon, C. Kirby, S. Selzer, J. Senior, H. da Silva, Z. Wang and V.
Weissig.

REFERENCES

Allison, A.C. and Byars, N.E., 1986, An adjuvant formulation that select-
    ively elicits the formation of antibodies of protective isotypes and
    of cell-mediated immunity, J.Immunol.Meth., 95:157.
Allison, A.C. and Gregoriadis, G., 1974, Liposomes as immunological adjuv-
    ants, Nature (Lond.), 252:252.
Alving, C.R., Banerji, B., Shiba, T., Kotani, S., Clements, J.D. and
    Richards, R.L., 1980, Liposomes as vehicle for vaccines, in: "New
    Developments with Human and Veterinary Vaccines", Alan R. Liss, Inc.,
    New York.
Bakouche, O. and Gerlier, D., 1986, Enhancement of immunogenicity of tumour
    virus antigen by liposomes. The effect of lipid composition,
    Immunology, 57:219.
Dancey, G.F., Yasuda, T. and Kinsky, S.C., 1978, Effect of liposomal model
    membrane composition on immunogenicity, J.Immunol., 120:1109.
Davis, D. and Gregoriadis, G., 1987, Liposomes as adjuvants with immuno-

purified tetanus toxoid: Influence of liposomal characteristics, Immunology, 61:229.

Davis, D., Davies, A. and Gregoriadis, G., 1986, Liposomes as immunological adjuvants in vaccines: Studies with entrapped and surface-linked antigen, Biochem.Soc.Trans., 14:1036.

Davis, D., Davies, A. and Gregoriadis, G., 1987, Liposomes as adjuvants with immunopurified tetanus toxoid: The immune response, Immunol.Lett. 14:341.

Desiderio, J.V. and Campbell, S.G., 1985, Immunization against experimental murine salmonellosis with liposome-associated O-antigen, Inf.Immun., 48:658.

Epstein, M.A., Morgan, A.J., Finerty, S., Randle, B.J. and Kirkwood, J.K., 1985, Protection of cottontop tamarins against Epstein-Barr virus-induced malignant lymphoma by a prototype subunit vaccine, Nature (Lond.), 318:287.

Francis, M.J., Fry, C.M., Rowlands, D.J., Brown, F., Bittle, J.L., Houghten, R.A. and Lernerr, R.A., 1985, Immunological priming with synthetic peptides of foot-and-mouth disease virus, J.Gen.Virol., 66:2347.

Garçon, N., Senior, J. and Gregoriadis, G., 1986, Coupling of ligands to liposomes before entrapment of agents sensitive to coupling procedures, Biochem.Soc.Trans., 14:1038.

Garçon, N., Gregoriadis, G., Taylor, M. and Summerfield, J., 1988, Mannose-mediated targeted immunoadjuvant action of liposomes, Immunology, 64:743.

Gerlier, D., Sakai, F. and Dore, J.F., 1978, Inclusion d'un antigene de surface cellulaire associe au virus de Gross dans des liposomes, C.R.Acad.Sc.Paris, 286, Serie D, 439.

Gregoriadis, G., ed., 1984, "Liposome Technology", Vol. 1, CRC Press, Inc., Boca Raton.

Gregoriadis, G., 1985, Liposomes as carriers for drugs and vaccines, Trends Biotechnol., 3:235.

Gregoriadis, G., 1986, Liposomal subunit vaccine against Epstein-Barr virus-induced malignant lymphoma, Nature (Lond.), 320:87.

Gregoriadis, G., ed., 1988, "Liposomes as Carriers of Drugs: Recent Trends and Progress", John Wiley and Sons, Chichester.

Gregoriadis, G. and Allison, A.C., 1974, Entrapment of proteins in liposomes prevents allergic reactions in preimmunized mice, FEBS Lett., 45:71.

Gregoriadis, G. and Panagiotidi, C., 1989, Immunoadjuvant action of liposomes: Comparison with other adjuvants, Immunol.Lett., 20:237.

Gregoriadis, G. Davis, D. and Davies, A., 1987, Liposomes as immunological adjuvants in vaccines: Antigen incorporation studies, Vaccine, 5:145.

Gregoriadis, G., Garçon, N., Senior, J. and Davis, D., 1988, The immuno-adjuvant action of liposomes: Nature of immune response and influence of liposomal characteristics, in: "Liposomes as Drug Carriers: Trends and Progress", G. Gregoriadis, ed., John Wiley and Sons, Chichester.

Hedlund, G., Jansson, B. and Sjogren, H.O., 1984, Comparison of immune responses induced by rat RT-1 antigens presented as inserts into liposomes, as protein micelles and as intact cells, Immunology, 53:69.

Kramp, W.J., Six, H.R. and Kasel, J.A., 1982, Postimmunization clearance of liposome-entrapped adenovirus type 5 hexon, Proc.Soc.Exp.Bioc.Med., 169:135.

Kahl, K.L., Scott, C.A., Lelchuk, R., Gregoriadis, G. and Liew, F.Y., 1989, Vaccination against murine cutaneous leishmaniasis using L. Major antigen/liposomes: Optimization and assessment of the requirmenet for intravenous injection, J.Immunol., in press.

Kinsky, S.C., 1978, Immunogenicity of liposomal model membranes, Ann.N.Y. Acad.Sci., 308:111.

Kirby, C. and Gregoriadis, G., 1984, Dehydration-rehydration vesicles (DRV): A new method for high yield drug entrapment in liposomes, Biotechnology, 2:979.

Manesis, E.K., Cameron, C.H. and Gregoriadis, G., 1979, Hepatitis B surface antigen-containing liposomes enhance humoral and cell-mediated immunity to the antigen, FEBS Lett., 102:107.

Mettler, L., Czuppon, A.B., Buckheim, W., Baukloh, V., Ghyczy, M., Etschenberg, J. and Holstein, A.F., 1983, Induction of high titre mouse-antihuman spermatozoal antibodies by liposome incorporation of spermatozoal membrane antigens, Am.J.Reprod.Immunol., 4:127.

Naylor, P.T., Larsen, H.S., Huang, L. and Rouse, B.T., 1982, In vivo induction of anti-herpes simplex virus immune response by type 1 antigens and lipid A incorporated into liposomes, Inf.Immun., 36:1209.

Perry, A. and Ofek, I., 1984, Inhibition of blood clearance and hepatic tissue binding of Escherichia coli by liver lectin-specific sugars and glycoproteins, Inf.Imm., 43:257.

Pierce, N.F. and Sacci, Jr., J.B., 1984, Enhancement by lipid A of mucosal immunogenicity of liposome-associated cholera toxin, Rev.Inf.Dis., 6:563.

van Rooijen, N. and van Nieuwmegen, R., 1980, Liposomes in immunology: Evidence that their adjuvant effect results from surface exposition of the antigens, Cell.Immunol., 49:402.

Seltzer, S., Gregoriadis, G. and Dick, R., 1988, DRV liposomes in contrast imaging, Invest.Radiol., 23:131.

Senior, J., and Gregoriadis, G., 1989, Dehydration-rehydration vesicle methodology facilitates a novel approach to antibody binding to liposomes, Biochim.Biophys.Acta, 1003:58.

Shek, P.N. and Sabiston, B.H., 1982, Immune response mediated by liposome-associated protein antigens. II. Comparison of the effectiveness of vesicle-entrapped and surface-associated antigen in immunopotentiation, Immunology, 47:627.

Snippe, H.,, van Dam., J.E.G., van Houte, A.J., Williers, J.M.N., Kamerling, J.P. and Vliegenthart, J.F.G., 1983, Preparation of a semisynthetic vaccine to Streptococcus pneumoniae Type 3, Inf.Immun., 42:842.

Snyder, S.L. and Vannier, W.E., 1984, Immunologic response to protein immobilised on the surface of liposomes via covalent azo-bonding, Biochim. Biophys.Acta, 772:288.

Tan, L. and Gregoriadis, G., 1989, The effect of interleukin-2 on the immunoadjuvant action of liposomes, Biochem.Soc.Trans., 17:693.

Tan, L., Loyter, A. and Gregoriadis, G., 1989, Incorporation of reconstituted influenza virus envelopes into liposomes: Studies of immune response in mice, Biochem.Soc.Trans., 17:129.

Weissig, V., Lasch, J. and Gregoriadis, G., 1989, Covalent coupling of sugars to liposomes, Biochim.Biophys.Acta, 1003:54.

Xiao, Q., Gregoriadis, G. and Ferguson, M., 1989, Immunoadjuvant action of liposomes for entrapped poliovirus peptides, Biochem.Soc.Trans., 17:695.

IMMUNOADJUVANT ACTION OF LIPOSOMES: MECHANISMS

Nico van Rooijen and Donghui Su[1]

Department of Histology, Medical Faculty, Free University
P.O. Box 7161, 1007 MC Amsterdam, The Netherlands
[1]Visiting Scientist from the Department of Microbiology
Fujian Medical College, Fuzhou, P.R. China

INTRODUCTION

Humoral Immune Reactions in vivo

A large variety of antigens may initiate a series of events, finally leading to the production of specific antibodies by defined cells belonging to the B cell lineage of the immune system. The immunogenicity of these antigens may depend on different cellular requirements. Many antigens (e.g. proteins and erythrocytes) need the help of T-lymphocytes to start an immune response. These are referred to as thymus-dependent (TD) antigens which evoke a TD immune response. A number of other antigens have the capability to start an immune response without the help of T-lymphocytes. These are the thymus-independent (TI) antigens which evoke a TI immune response. Apart from T-lymphocytes and B-lymphocytes, most antigens require the participation of cells of the mononuclear phagocyte system (dendritic cells or macrophages). Although many immunologists nowadays seem to consider the immune response as a purely in vitro phenomenon, the immune system belongs in the body (Marx, 1985), where antigens, cells and their products have the opportunity to meet each other under optimum conditions. Evidence is growing that it is the microenvironment in a lymphoid organ, for which there is no in vitro equivalent, that creates the optimum conditions for the cells of the immune system to cooperate either directly or through their products. The possible cellular interactions during an immune response in the spleen and the localization and migration of the participating cells in defined splenic compartments have been recently reviewed (Van Rooijen et al, 1986). Herein we also postulated that there is a single differentiation pathway for all antibody-forming cells in the spleen, which is independent of the antigen and the type of immune response. We believe that this pathway is running along all of the different cells which may be required for any particular type of response.

Immune Reactions against Liposome Associated Antigens or Haptens

The first report on the immunoadjuvant activity of liposomes in the immune response against diphtheria toxoid (Allison and Gregoriadis, 1974) has been followed by many studies confirming their capability to enhance immune responses to various antigens (Van Rooijen and Van Nieuwmegen, 1980, 1982, 1983a, b; Van Rooijen et al, 1981). The main goals of these studies were to investigate the possible applicability of liposomes as an adjuvant for the preparation of vaccines and for the presentation of tumor antigens to the immune system. Antigen presentation by liposomes may concern:

1. Presentation of haptens. In this case the liposome functions as a carrier for the hapten. The hapten is not able to elicit an immune response without a carrier molecule. As usual, the carrier determines the type of immune response, being a thymus-independent IgM response without generation of immunological memory for liposomal carriers (Yasuda et al, 1977; Van Houte et al, 1979; Claassen et al, 1987; see also Fig. 1.II).

2. Presentation of antigen (e.g. protein) molecules with various antigenic determinants. These antigens also elicit an immune response when given free in solution, but the response can be strongly enhanced when the antigen is given in a form associated with liposomes (Van Rooijen and Van Nieuwmegen, 1982; see also Fig. 1.III). Antigens may be associated with liposomes either by encapsulation (masked antigenic determinants) or by coupling to the surfaces of the liposomes (exposed antigenic determinants; see also Fig. 2). Contrary to liposome-associated haptens, liposome associated antigens elicit a thymus-dependent immune response including the production of both IgM and IgG antibodies and the generation of immunological memory (Van Rooijen, 1988).

It is the purpose of the present contribution to discuss the mechanisms by which liposomes modulate the immune response against associated haptens or antigens. Particularly, attention is given to the possible role of macrophages in immunopotentiation by liposomes.

CHARACTERISTICS OF THE LIPOSOME-MEDIATED IMMUNE RESPONSE

Thymus-Independent Response to Haptenated Liposomes

Haptenated liposomes evoke a hapten-specific humoral IgM immune response. Since it appeared not possible to evoke an IgG response after either primary or secondary immunization with haptenated liposomes, and in addition mice depleted of, or deficient in T-lymphocytes appeared to respond to haptenated liposomes, it was concluded that haptenated liposomes are TI antigens (Van Houte et al, 1979). TI antigens are further divided into two types, i.e. TI-1 and TI-2. The TI-2 antigens activate B lymphocytes from all the normal mouse strains except those of CBA/N mice, which carry an X-linked immune defect, and the B cells from neonatal mice of all strains. The TI-1 antigens, which generally have intrinsic mitogenic properties, stimulate the B cells of all mouse strains, including those with the X-linked immune defect. Antibody responses to TI-2 antigens like trinitrophenyl (TNP)-Ficoll obviously require the presence of mature B cells. Accessory cell-derived factors appeared not essential for responses to TI-1 antigens like TNP-lipopolysaccharide (LPS) and TNP-Brucella abortus although these factors could enhance the responses, whereas TI-2 responses require such factors (see e.g. Goud et al, 1988). Studies by Snippe et al (1982) showed that CBA/N mice respond to haptenated liposomes and for this reason, these antigens have to be considered TI-1 antigens. However, contrary to other known TI-1 antigens they appear to be non-mitogenic in vivo. Their immunogenicity can be completely blocked by prior injection of Ficoll bearing the same hapten. The TI-2 character of haptenated liposomes is further supported by the resemblance of the in vivo behaviour of the response against such antigens and that against TNP-Ficoll (Claassen et al, 1986a; 1987). For the above reasons it is not easy to escape the conclusion drawn by Snippe et al (1982) that haptenated liposomes may represent a distinct subclass of TI (-3?) antigens, that partly resemble TI-1 antigens and partly resemble TI-2 antigens.

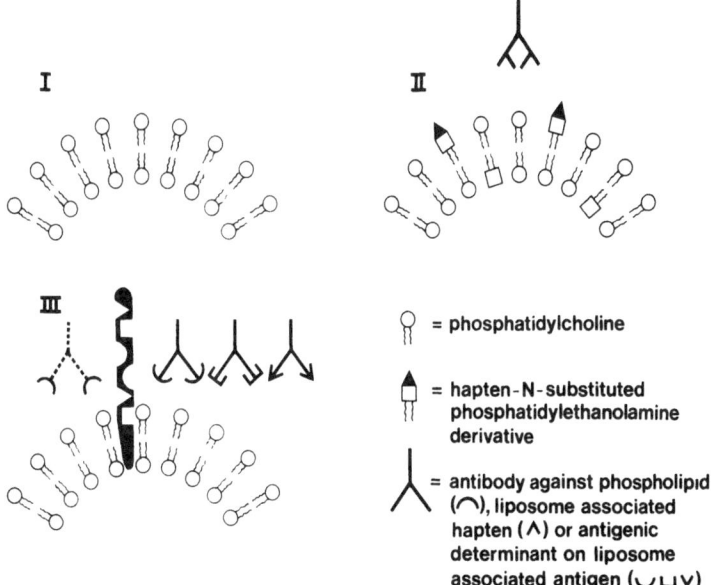

Fig. 1. Different immune reactions against liposomes and associated haptens or antigens.

I.   Phosphatidylcholine liposomes without antigens do not elicit antibody production.

II. Liposomes with haptens incorporated in their bilayers may elicit anti-hapten antibody production.

III. Liposomes with an antigen exposed on their surfaces may elicit the production of antibodies directed against various antigenic determinants. When lipid A is associated with liposomes, apart from antibodies against lipid A, antibodies may be elicited against the liposomal phospholipids. (From Van Rooijen and Van Nieuwmegen, 1982.)

Thymus Dependent Response to Liposome Associated Antigens

Contrary to haptenated liposomes, liposome associated (protein) antigens elicit an IgG response in addition to the preceding IgM response (Latif and Bachawat, 1987; Van Rooijen, 1988), start the generation of immunological memory (Van Rooijen and Van Nieuwmegen, 1981) and appeared to behave as real T-cell-dependent (TD) antigens (Shek and Sabiston, 1982; Beatty et al, 1984). Recent studies have focused on the antigen-specific interleukin-2 (IL-2) production which appeared to be enhanced when the rabies antigens used for priming were given either specifically incorporated in liposomal bilayers (Oth et al, 1987) or were non-specifically associated with liposomes (Mansour et al, 1988). Production of IL-2, mainly by T-helper lymphocytes is one of the early events, indicating that a thymus-dependent (TD) immune response has been initiated (Robb, 1984; Spitz et al, 1985). Since the immune response to haptenated liposomes appeared to consist exclusively of IgM antibodies (Yasuda et al, 1977; Van Houte et al, 1979), the question was raised if liposomes would also change the relative participation of IgM and IgG antibodies in the immune response to associated TD (protein) antigens. We demonstrated that the adjuvant effect of liposomes with respect to surface exposed albumin antigen concerned mainly the production of IgM antibodies, whereas only a moderate effect was found on the IgG response (Van Rooijen and Van Nieuwmegen, 1983b; Van Rooijen, 1988). In contrast with these results, Latif and Bachhawat (1987) recently reported a strong adjuvant effect of liposomes with respect to the IgG response against surface exposed lysozyme. Davis et al (1987) found that adjuvanticity of liposomes is reflected in most antibody subclasses and that there is no shift in subclasses compared to the response obtained with the free antigen.

Association of (Protein) Antigens with Liposomes: Entrapped and Surface-Exposed Antigens

To enhance the antibody response to antigens, these can be associated with liposomes in various ways (Fig. 2). Surface exposition can be achieved by a number of methods for covalent attachment of the antigen to one of the phospholipids to be used for preparation of the liposomes, by non-specific adsorption of (partly) hydrophobic antigens with the phospholipid bilayers or by 'hydrophobization' of hydrophilic antigens followed by non-specific adsorption (see for review Van Rooijen and Van Nieuwmegen, 1983a). Antigens can also be entrapped in the liposomes when dissolved in the aqueous solution used for preparation of the liposomes. For many antigens the latter procedure will result in liposomes, with antigen both entrapped in their aqueous compartments and exposed on their outer surfaces. The ratio between entrapped and exposed antigen may vary to a large extent and will depend on the structure of the antigen e.g. the presence of hydrophobic groups. This may have influenced the results of studies comparing the effectiveness of liposome-entrapped and surface-exposed antigens in immunopotentiation (Van Rooijen and Van Nieuwmegen, 1980; Shek and Sabiston, 1982; Snyder and Vannier, 1984). There is a lot of evidence now, that liposomes act as an adjuvant for both entrapped and surface-exposed antigens (Davis and Gregoriadis, 1987; Gregoriadis et al, 1987). An important difference between liposome-entrapped antigens and surface-exposed antigens is that the

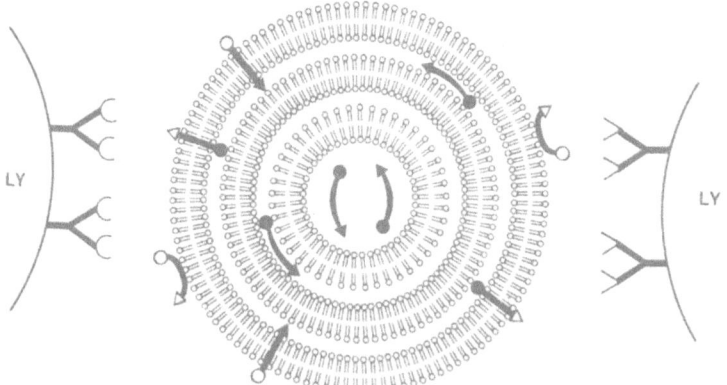

Fig. 2. Schematic representation of a liposome with entrapped and exposed antigen molecules. Open triangles and circles represent exposed antigenic determinants. Closed triangles and circles represent masked antigenic determinants. Receptors on immunocompetent cells (LY) that meet the liposome may recognize the exposed determinants only. Macrophages seem to be required for processing of the entrapped antigen molecules, by ingestion of the liposome, digestion of the liposomal membranes, and exposing of the released antigen molecules (or parts thereof) on their surfaces. (Reproduced by permission of Alan R. Liss, Inc; from Van Rooijen et al, 1982.)

antigenic determinants (epitopes) of the latter antigen are exposed and are directly accessible to cells of the immune system, whereas those of the entrapped antigen are masked (Fig. 2). For that reason it seems to be a logical conclusion that macrophages are required for the processing of liposome-entrapped antigens, since only these cells are able to release the antigens after digestion of the liposomal bilayers by phospholipases in their lysosomal compartments.

The Fate of Liposome Associated Antigen, the Development of Antibody Producing Cells and the Formation of Immune Complexes

Apart from an enhancement of the antibody response, which was more impressive for IgM antibodies than for IgG antibodies in our studies (Van Rooijen, 1988), we found no important differences between the characteristics of the humoral immune response against liposome associated antigens and similar antigens when given free in solution. The localization pattern of specific antibody forming cells developing in the spleen after intravenous administration of liposome associated or free antigens did not differ, neither for the primary (Van Rooijen et al, 1983) nor for the secondary response (Van Rooijen et al, 1984), in spite of the fact that free and liposome associated antigens will have a different distribution pattern in the spleen. Anti-human serum albumin (HSA) antibody forming cells could be detected in the spleen of rabbits from about 4 days after an intravenous injection with either free HSA (10 mg) or liposome associated HSA (0.05 mg). The bulk of these cells developed in the periphery of the periarteriolar lymphocyte sheaths and in the small sheaths of lymphoid tissue around the terminal arterioles (see Fig. 3 and Van Rooijen et al, 1983a, b). Compared to the fate of free (protein) antigens during the primary immune response, a large part of liposome associated antigens will be ingested by macrophages in the liver and in the marginal zone and red pulp of the spleen. This is also the normal fate of antigens given as a booster injection, when antigens meet an excess of antibodies in the circulation, followed by the consecutive ingestion of the bulk of the formed immune complexes by macrophages. Moreover, small (protein) antigens will penetrate much easier e.g. in the lymphoid cell compartments of the white pulp of the spleen than their larger liposome associated counterparts, and they also have to be given in considerably larger amounts in order to elicit a comparable strong immune response. There is a lot of evidence that immune complexes formed from the antigen administered, and specific antibody evoked in response to it, play an important role in the generation of immunological memory (Klaus et al, 1980) and/or the maintenance of a long term antibody production (Tew et al, 1980). This is believed to be mediated by a small part of these immune complexes, which escape degradation by macrophages and are retained for long periods on the cell processes of follicular dendritic cells in lymphoid follicles of spleen and lymph nodes (Mandel et al, 1980; Klaus et al, 1980; Van Rooijen, 1980). From a viewpoint of application of liposomes for vaccination (i.e. generation of immunological memory) it may thus be important that immune complexes are formed during the response. There is no reason to believe that association of antigens with liposomes completely prevents the formation of immune complexes. Apart from leakage of the antigen from the liposomes, antigen released from unstable liposomes or liposomal bilayers affected by e.g. albumins (Zborowski et al, 1977) part of the antigen may be released from the liposomes by complement-dependent immune lysis of the liposomes (see, for example, Legros et al, 1988), and macrophages may release (partly degraded) antigens after disruption of liposomal bilayers (Weissmann and Dukor, 1970; Unanue, 1984). Indeed, trapping of immune complexes in follicles of the spleen was found during the primary immune response against liposome associated human serum albumin, intravenously given to rabbits (see Fig. 3 and Van Rooijen et al, 1983).

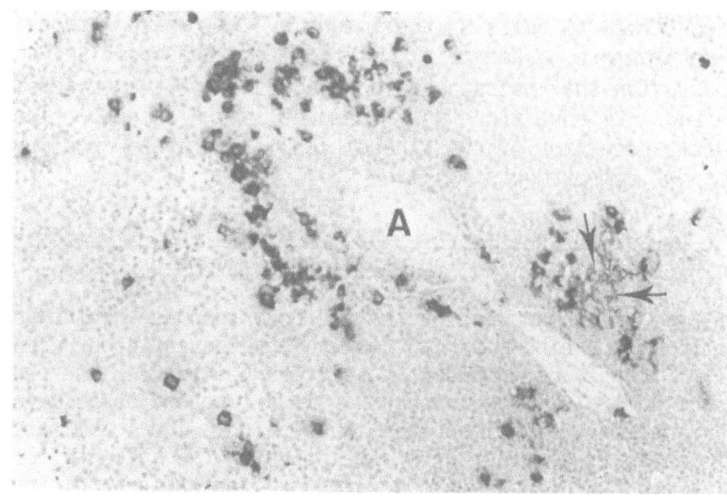

Fig. 3. Anti-human serum albumin (HSA) antibody producing cells are demonstrated in spleen tissue by incubation of cryostat sections with an
HSA-Horseradish peroxidase (HRP) conjugate followed by enzymehisto-
chemical demonstration of the HRP. Anti-HSA producing cells are
present in the periphery of a periarteriolar lymphocyte sheath in
the spleen of a rabbit at seven days after intravenous injection of
50 µg liposome-associated HSA. Note the presence of extracellular
anti-HSA surrounded by some anti-HSA producing cells in a follicle
center (arrows). Extracellular anti-HSA antibody is trapped in the
follicles as an HSA-anti-HSA immune complex and can be demonstrated
due to its free antigen binding sites. A = arteriole. (Reproduced
by permission of Alan R. Liss, Inc; from Van Rooijen et al, 1983).

THE ROLE OF MACROPHAGES IN THE IMMUNE RESPONSE AGAINST LIPOSOME ASSOCIATED
ANTIGENS

Macrophages in Humoral Immune Reactions

Various types of non-lymphoid cells have been suggested to be involved
in different immune phenomena. So, dendritic cells are considered antigen
presenting cells in TD immune responses (Steinman and Nussenzweig, 1980).
Marginal zone macrophages in the spleen, which were shown to ingest and
retain TI-2 antigens (polysaccharides) selectively (Humphrey and Grennan,
1981) are considered antigen presenting cells in TI-2 immune responses
(Humphrey, 1985). Follicular dendritic cells which have the capability to
retain immune complexes in an undegraded form on their cell processes for
long periods (Mandel et al, 1980; Tew et al, 1980) are considered to be involved in the generation of immunological memory (Klaus et al, 1980). The
early lines of evidence for a role of macrophages in the inductive stage of
immune reactions have been reviewed by Weissmann and Dukor (1970) and are
summarized as follows:

1. In a variety of experimental models, the presence of macrophages appeared
to be a prerequisite for the induction of antibody formation.

2. The immunogenicity of several protein antigens appeared to be directly
related to their state of aggregation and, therefore, to their relative
palatability to macrophages.

3. Several proteins have been found to be more (up to 1,000 times) potent
immunogens when administered to recipients within macrophages than when

given in free form. This difference especially concerned antigens which are rather slowly taken up by phagocytic cells in vivo, but was absent for aggregated antigens. Free and liposome associated albumin antigens seem to be a representative example of this group of antigens.

More recent lines of evidence for a role of macrophages in the induction of humoral immune responses have been reviewed by Unanue (1984). Protein antigens taken up by macrophages undergo a series of changes, aside from their complete degradation to amino acids. The fate of these proteins is complex, and the folowing states of the antigens could be identified:

1. Molecules that remain bound to the cell surface but were not presented directly to T cells.

2. Molecules that are released in soluble form but were not operative in antigen presentation.

3. Molecules that have undergone an intracellular processing and appeared to be directly involved in the major histocompatibility complex (MHC) restricted antigen presentation. Intracellular processing required that the antigen was internalized into an acid vesicle, handled there for about one hour and consecutively recycled to the surface for presentation to the T-lymphocytes. Fragmentation of the antigen appeared to be an essential step in the intracellular processing.

It should be kept in mind that most of the above results were obtained during in vitro studies, in the absence of the normal cell populations and their microenvironment. In vivo verification is required to study an obligatory role for macrophages in immune reactions.

An Approach to Study the Role of Macrophages in the Immune Response to Liposome Associated Antigens

Antigens that are exposed on the outer surfaces of liposomes may be recognized by receptors on immunocompetent cells. On the other hand it is difficult to see how antigens that are entrapped in the aqueous compartments of liposomes (see Fig. 2) can elicit an immune response without the help of macrophages. The macrophages could ingest the liposomes, digest the liposomal phospholipid bilayers with the help of lysosomal phospholipases (Barrett, 1984) and process the released antigens as reviewed by Unanue (1984) and summarized above. However, there is no simple approach to study whether macrophages are obligatory for immune reactions against liposome associated antigens in vivo. The main problem is that macrophages are found in nearly all organs of the body and show large differences in their capability to present antigens to T-lymphocytes (Unanue, 1984). Although a number of different approaches for macrophage depletion have been described (see, for example, Pinto et al, 1988; Table 1), none of these will result in complete depletion of all macrophages in the body. The implication is that macrophages elsewhere in the body may take over (part of) the tasks of macrophages that are depleted. So, after complete depletion of macrophages in spleen and liver, we found no effect on both the 'in situ' and the total antibody response against a consecutively, intravenously administered TD antigen (Claassen et al, 1986b). The conclusion may be drawn that local macrophages were not obligatory. However, the antigen may have reached, for instance, a number of lymph nodes, stimulated the local T cells there, followed by migration of these stimulated T cells to the spleen, where they in turn activated the B cells to become antibody forming cells.

Another problem may be that different approaches for in vivo macrophage depletion effect the macrophages to a different extent. So, intravenously given carrageenan did not result in the physical elimination of macrophages

Table 1. The Primary Antibody Response in Mice against HSA after Intravenous Injection with HSA-associated Liposomes

Intravenous injection at day -2	Intravenous HSA-liposomes at day 0 (per mouse)	Location of liposome-associated HSA	IgM anti-HSA antibodies at		IgG anti-HSA antibodies at	
			day 5	day 7	day 5	day 7
Empty liposomes	50 µg HSA, 1.2 mg phospholipid	entrapped (masked)	5.33+0.52 (6)	7.75+0.42 (6)	9.58+0.38 (6)	11.08+0.74 (6)
DMDP liposomes	"	"	4.30+0.82 (10)	7.00+0.75 (10)	7.40+1.88 (10)	8.72+1.56 (10)
			p<0.05	p<0.05	p<0.05	p<0.01
Empty liposomes	16 µg HSA, 2.4 mg phospholipid	surface (exposed)	8.60+0.62 (5)	8.38+0.70 (5)	4.40+1.18 (5)	9.02+1.00 (5)
DMDP liposomes	"	"	6.43+0.82 (7)	7.11+0.94 (7)	ND (7)	4.63+1.92 (7)
			p<0.01	p<0.05	p<0.01	p<0.01

Liposomes with entrapped HSA were prepared by the freeze-drying method starting with multilamellar vesicles (MLV). The resulting dehydration-rehydration vesicles (DRV; Kirby and Gregoriadis, 1984) containing both entrapped HSA and HSA on their surface were treated with trypsin to remove the surface-HSA. Liposomes with surface exposed HSA were prepared by incubating empty MLV with HSA. Both of these HSA-associated liposomes were composed of phosphatidylcholine (PC) and cholesterol (CHOL) in a molar ratio of 7:2. In each experiment, two groups of B6D2 mice were injected intravenously with empty or DMDP containing liposomes composed of PC and CHOL (molar ratio 6:1). Two days later, some of the mice in both groups were injected intravenously with HSA entrapped in liposomes and the remaining mice injected with HSA exposed on liposomal surfaces. IgM and IgG responses at day 5 and day 7 after priming were analysed by ELISA and expressed as $\log_2$ antibody titer (mean ± SD). Numbers in parentheses denote mice used. p value (probability of significance) was derived from Student's t-test. ND = not detectable.

in the spleen in our studies (unpublished results). Although phagocytosis of carbon particles was abolished by this treatment, macrophages remained present in normal numbers in their splenic compartments and showed an enhanced (lysosomal) acid phosphatase activity. Since the exact mechanism by which macrophages handle the antigens during the stage of macrophage processing is by no means clear (Unanue, 1984) the best approach to study the role of macrophages in in vivo immune reactions to liposome associated antigens seems to be based on their physical elimination, without affecting other cells at the same time. To prevent the complications described above as much as possible, we have chosen an approach of physical elimination of macrophages, based on drug containing phosphatidylcholine-cholesterol liposomes for our in vivo studies on the role of macrophages. Similar liposomes were used consecutively as antigen carriers. This approach in which liposomes with entrapped dichloromethylene diphosphonate (Cl2MDP) were used for physical elimination of macrophages in spleen and liver has been described in detail elsewhere (Van Rooijen and Claassen, 1988). Since liposome-entrapped drugs and liposome-entrapped antigens will be targeted to macrophages in the same compartments of the same organs, problems related to a different distribution pattern of macrophage-eliminated drugs and antigens are also avoided. Because liposomes made of phosphatidylcholine and cholesterol may be considered as immunologically inert vesicles (Schuster et al, 1979) we do not believe that the drug containing liposomes create some kind of immunological memory with respect to the liposomes, used for administration of the antigen.

The Effect of Elimination of Macrophages on the Immune Response against Liposome Associated Antigens

Using the above approach we studied the IgM and IgG immune responses against human serum albumin (HSA) in mice which were intravenously primed with liposome associated HSA, two days after intravenous injection with either empty (PBS containing) liposomes or dichloromethylene diphosphonate (DMDP) containing liposomes. The elimination of splenic macrophages by treatment with liposome-encapsulated DMDP which is generally completed within one day was checked at the end of the experiment (Van Rooijen and Claassen, 1988; Van Rooijen et al, 1988). In the sera of control mice injected with comparable amounts of non-liposome associated HSA, no anti-HSA antibodies were detected within two weeks after antigen administration. It may thus be concluded that the positive responses to both the liposome-entrapped HSA and the HSA exposed on the surfaces of empty liposomes (see Table 1) are due to the adjuvant activity of liposomes. Table 1 summarizes the results of two preliminary experiments in mice on the effect of pretreatment with liposome-encapsulated DMDP on (a) the immune response against liposome-entrapped HSA and (b) the immune response against HSA exposed on the surfaces of liposomes. Results clearly confirm our recent suggestion that antigens coated on liposomes primarily enhance the IgM response whereas liposome-entrapped antigens primarily enhance the IgG response (Van Rooijen, 1988). Considering the IgM response, HSA exposed on the outer surfaces of liposomes elicited an eight-fold higher serum antibody concentration at day 5 than HSA entrapped in liposomes. On the contrary, at the same time interval after antigen administration, liposome-entrapped HSA produced a 32-fold higher IgG anti-HSA antibody concentration than did surface exposed HSA. Considering the normal kinetics of IgM and IgG production during the primary immune response against TD antigens, it is not surprising that these differences were less pronounced at day 7 after antigen injection. The conclusion may be drawn that apart from their application for immunopotentiation, liposomes also seem to be promising when modulation of the immune response is wanted with respect to the ratio between IgM and IgG production. Elimination of macrophages in the spleen by treatment with liposome encapsulated DMDP two days before antigen injection resulted in markedly reduced IgM and IgG anti-HSA antibody titers both at day 5 and day 7. A surprising finding was that this reduction was more impressive for antigen exposed on the outer surfaces of liposomes than for the entrapped antigen. Obviously the immunogenicity of the entrapped antigen is not completely dependent on the unmasking capability of the macrophage (disruption of the phospholipid bilayers) as we expected. Since free antigen was not able to elicit an antibody response within two weeks, antigen molecules released from liposomes by leakage, influence of serum components etc. may not be responsible. If we assume that antigen, once entrapped in the liposomes, has not the capability to move to their outer surface, the remaining options are (a) macrophages somewhere in the body, reached by liposome-encapsulated DMDP in too small amounts for their elimination were consecutively involved in the processing of liposome-entrapped antigens, also in small amounts, but sufficient to evoke a detectable immune response; (b) other cells e.g. particular subsets of B lymphocytes (Unanue, 1984) are able to unmask and process part of the entrapped antigens (probably in the aqueous compartments between the peripheral phospholipid bilayers). So, on the one hand it is clear that our results confirm those of earlier investigators (Van Rooijen and Van

Nieuwmegen, 1979; Shek and Lukovich, 1982), that macrophages are involved in the _in vivo_ processing of liposome associated antigens. On the other hand, it is clear that local macrophages (i.e. macrophages in the compartments of e.g. the spleen where the development of antibody-forming cells takes place after intravenous injection of the antigen) are not obligatory for the induction of an immune response against such antigens. However, compared to free protein TD antigens, for which we found no reduction in the antibody responses in similarly macrophage-depleted animals (Claassen et al, 1986b; and unpublished results), local macrophages may be considered the most important candidates for unmasking and processing of liposome associated antigens. It should be realized, however, that processing of antigens may be performed by various cells in different compartments of the body and only requires the presence of T cells close to the processing cell (Unanue, 1984). The T cells may then migrate to other lymphoid organs or compartments in order to stimulate particular subsets of B cells to become antibody-forming cells.

REFERENCES

Allison, A.C. and Gregoriadis, G., 1974, Liposomes as immunological adjuvants, _Nature_, 252:252.

Barrett, A.J., 1984, Proteolytic and other metabolic pathways in lysosomes, _Bioch.Soc.Trans._, 12:899.

Beatty, J.D., Beatty, G.B., Paraskevas, F. and Froese, E., 1984, Liposomes as immune adjuvants: T cell dependence, _Surgery_, 96:345.

Claassen, E., Kors, N. and Van Rooijen, N., 1986a, Influence of carriers on the development and localization of anti-trinitrophenyl-antibody-forming cells in the murine spleen, _Eur.J.Immunol._, 16:271.

Claassen, E., Kors, N. and Van Rooijen, N., 1986b, Influence of carriers on the development and localization of anti-2,4,6-trinitrophenyl (TNP) antibody-forming cells in the murine spleen. II. Suppressed antibody response to TNP-Ficoll after elimination of marginal zone cells, _Eur.J.Immunol._, 16:492.

Claassen, E., Kors, N. and Van Rooijen, N., 1987, Immunomodulation with liposomes: The immune response elicited by liposomes with entrapped dichloromethylene-diphosphonate and surface-associated antigen or hapten, _Immunology_, 60:509.

Davis, D., Davies, A. and Gregoriadis, G., 1987, Liposomes as adjuvants with immunopurified tetanus toxoid: The immune response, _Immunol. Lett._, 14:341.

Davis, D. and Gregoriadis, G., 1987, Liposomes as adjuvants with immunopurified tetanus toxoid: Influence of liposomal characteristics, _Immunology_, 61:229.

Goud, S.N., Muthusamy, N. and Subbarao, B., 1988, Differential responses of B cells from the spleen and lymph node to TNP-Ficoll, _J.Immunol._, 140:2925.

Gregoriadis, G., Davis, D. and Davies, A., 1987, Liposomes as immunological adjuvants: Antigen incorporation studies, _Vaccine_, 5:145.

Humphrey, J.H. and Grennan, D., 1981, Different macrophage populations distinguished by means of fluorescent polysaccharides. Recognition and properties of marginal zone macrophages, _Eur.J.Immunol._, 11:221.

Humphrey, J.H., 1985, Splenic macrophages: Antigen presenting cells for TI-2 antigens, _Immunol.Lett._, 11:149.

Kirby, C. and Gregoriadis, G., 1984, Dehydration-rehydration vesicles: A simple method for high yield drug entrapment in liposomes, _Biotechnology_, 2:979.

Klaus, G.G.B., Humphrey, J.H., Kunkl, A. and Dongworth, D.W., 1980, The follicular dendritic cell: Its role in antigen presentation in the generation of immunological memory, _Immunol.Rev._, 53:3.

Latif, N.A. and Bachhawat, B.K., 1987, The effect of surface-coupled antigen of liposomes in immunopotentiation, Immunol.Lett., 15:45.

Legros, F., Cashillo, M., Praet, M., Vandenbranden, M., Lemoine, O., Cabiaux, V., Van Vooren, J.P., Nyabenda, J., Dierckx, P., Turneer, M. and Ruysschaert, J.M., 1988, Detection of lytic antimycobacterial antibodies by immune lysis of liposomes sensitized to tuberculin, J.Immunol.Meth., 108:223.

Mandel, T.E., Phipps, R.P., Abbot, A. and Tew, J.G., 1980, The follicular dendritic cell: Long-term antigen retention during immunity, Immunol.Rev., 53:29.

Mansour, S., Thibodeau, L., Perrin, P., Sureau, P., Mercier, G., Joffret, M.L. and Oth, D., 1988, Enhancement of antigen-specific interleukin 2, production by adding liposomes to rabies antigens for priming, Immunol.Lett., 18:33.

Marx, J.L., 1985, The immune system belongs in the body, Science, 227:1190.

Oth, D., Mercier, G., Perrin, P., Joffret, M.L., Sureau, P. and Thibodeau, L., 1987, The association of the rabies glycoprotein with liposome induces an in vitro specific release of interleukin 2, Cell.Immunol., 108:220.

Pinto, A.J., Stewart, D., Volkman, A., Jendrasiak, G., Van Rooijen, N. and Morahan, P.S., 1988, Selective depletion of macrophages using toxins encapsulated in liposomes: Effect on antimicrobial resistance, in: "Liposomes in the Therapy of Infectious Diseases and Cancer", G. Lopez-Berestein and I. Fidler, eds., Alan R. Liss, New York, USA, in press.

Robb, R.J., 1984, Interleukin 2: The molecule and its function, Immunol. Today, 5:203.

Schuster, B.G., Neidig, M., Alving, B.M. and Alving, C.R., 1979, Production of antibodies against phosphocholine, phosphatidylcholine, sphingo-myelin and lipid A by injection of liposomes containing lipid A, J.Immunol., 122:900.

Shek, P.N. and Lukovich, S., 1982, The role of macrophages in promoting the antibody response mediated by liposome-associated protein antigens, Immunol.Lett., 5:305.

Shek, P.N. and Sabiston, B.H., 1982, Immune response mediated by liposome-associated protein antigens. I. Potentiation of the plaque-forming cell response, Immunology, 45:349.

Shek, P.N. and Sabiston, B.H., 1982, Immune response mediated by liposome-associated protein antigens. II. Comparison of the effectiveness of vesicle-entrapped and surface-associated antigen in immunopotentiat-ion, Immunology, 47:627.

Snippe, H., Van Houte, A.J., Lizzio, E.F., Willers, J.M.N. and Merchant, B., 1982, Characterization of immunogenic properties of haptenated liposomal model membranes in mice. VI. Response in B cell-defective CBA/N mice, Immunology, 45:545.

Snyder, S.L. and Vannier, W.E., 1984, Immunologic response to protein immobilized on the surface of liposomes via covalent azo-binding, Biochim.Biophys.Acta, 772:288.

Spitz, M., Gearing, A., Callus, M., Spitz, L. and Thorpe, R., 1985, Inter-leukin-2 in vivo: Production of and response to interleukin-2 in lymphoid organs undergoing a primary immune response to erythrocytes, Immunology, 54:527.

Steinman, R.M. and Nussenzweig, M.C., 1980, Dendritic cells: Features and functions, Immunol.Rev., 53:127.

Tew, J.G., Phipps, R.P. and Mandel, T.E., 1980, The maintenance and regul-ation of the humoral immune response: Persisting antigen and the role of follicular antigen-binding dendritic cells as accessory cells, Immunol.Rev., 53:175.

Unanue, E.R., 1984, Antigen-presenting function of the macrophage, Ann.Rev. Immunol., 2:395.

Van Houte, A.J., Snippe, H. and Willers, J.M.N., 1979, Characterization of immunogenic properties of haptenated liposomal model membranes in mice. I. Thymus independence of the antigen, Immunology, 37:505.

Van Rooijen, N. and Van Nieuwmegen, R., 1979, Liposomes in immunology: Impairment of the adjuvant effect of liposomes by incorporation of the adjuvant lysolecithin and the role of macrophages, Immunol. Commun., 8:381.

Van Rooijen, N. and Van Nieuwmegen, R., 1980, Liposomes in immunology: Evidence that their adjuvant effect results from surface exposition of the antigens, Cell Immunol., 49:402.

Van Rooijen, N., 1980, Immune complex trapping in lymphoid follicles: A discussion on possible functional implications, in: "Phylogeny of Immunological Memory", J. Manning, ed., Elsevier, Amsterdam, The Netherlands.

Van Rooijen, N., Van Nieuwmegen, R. and Kors, N., 1981, The secondary immune response against liposome associated antigens, Immunol.Commun., 10:59.

Van Rooijen, N., Van Nieuwmegen, R., 1982, Immunoadjuvant properties of liposomes. in: "Targeting of Drugs", G. Gregoriadis, J. Senior and A. Trouet, eds., Plenum Press, New York.

Van Rooijen, N., Van Nieuwmegen, R. and Kors, N., 1982, Liposomes as immunological adjuvants, in: "Cell Function and Differentiation", G. Evangelopoulos, ed., Alan R. Liss Inc., New York, part A.

Van Rooijen, N. and Van Nieuwmegen, R., 1983a, Use of liposomes as biodegradable and harmless adjuvants, Meth.Enzymol., 93:83.

Van Rooijen, N. and Van Nieuwmegen, R., 1983b, Association of an albumin antigen with phosphatidylcholine liposomes alters the nature of immunoglobulins produced during the immune response against the antigen, Biochim.Biophys.Acta, 755:434.

Van Rooijen, N., Kors, N., Van Nieuwmegen, R. and Eikelenboom, P., 1983, The development of specific antibody-containing cells and the localization of extracellular antibody in the follicles of the spleen of rabbits after administration of free or liposome associated albumin antigen, Anat.Rec., 206:189.

Van Rooijen, N., Van Nieuwmegen, R., Kors, N. and Eikelenboom, P., 1984, The development of specific antibody-containing cells in the spleen of rabbits during the secondary immune response against free or liposome associated albumin antigen, Anat.Rec., 208:579.

Van Rooijen, N., Claassen, E. and Eikelenboom, P., 1986, Is there a single differentiation pathway for all antibody-forming cells in the spleen? Immunol.Today, 7:193.

Van Rooijen, N., 1988, Liposomes as immunological adjuvants: Recent developments, in: "Liposomes as Drug Carriers: Recent Trends and Progress", G. Gregoriadis, ed., John Wiley and Sons, Chichester.

Van Rooijen, N. and Claassen, E., 1988, In vivo elimination of macrophages in spleen and liver, using liposome-encapsulated drugs: Methods and applications. in: "Liposomes as Drug Carriers: Recent Trends and Progress", G. Gregoriadis, ed., John Wiley and Sons, Chichester.

Van Rooijen, N., Kors, N. and Kraal, G., 1988, Macrophage subset repopulation in the spleen: Differential kinetics after liposome-mediated elimination. Submitted for publication.

Weissman, G. and Dukor, P., 1970, The role of lysosomes in immune responses, Adv.Immunol., 12:283.

Yasuda, T., Dancey, G.F. and Kinsky, S.C., 1977, Immunologenic properties of liposomal model membranes in mice, J.Immunol., 119:1863.

Zborowski, J., Roerdink, F. and Scherphof, G., 1977, Leakage of sucrose from phosphatidylcholine liposomes induced by interaction with serum albumin, Biochim.Biophys.Acta, 497:183.

LIPOSOMAL VACCINE TO <u>STREPTOCOCCUS PNEUMONIAE</u> TYPE 3 AND 14

H. Snippe, A.F.M. Verheul and J.E.G. van Dam

Laboratory of Microbiology, Faculty of Medicine, Utrecht
University, Utrecht, The Netherlands

INTRODUCTION

<u>Streptococcus pneumoniae</u> is responsible for lower-respiratory-tract in-
fections in humans and also the most common cause of otitis media (bacterial
middle ear infection) in children. Pneumococcal pneumonia is rarely a primary infection of the lung but such factors as damage to the respiratory
tract, fatigue, chilling of the body and general debilitation predispose to
infection. <u>S. pneumoniae</u> can be isolated from the pharynx of 30-70% of apparently normal humans.

The first isolations of the pneumococcus were done independently by
Pasteur and Sternberg in 1881. Pneumococci are ovoid, non-motile, non-spor-
ing organisms which usually occur in pairs (old nomenclature: <u>Diplococcus
pneumoniae</u>) enveloped by a capsule. The capsule is the most striking
morphologic feature of the organism and the most voluminous when the organism is at its most virulent stage. <u>Streptococcus pneumoniae</u> is divided into
types that are characterized by these carbohydrate capsules which differ in
chemical composition and structure (Larm and Lindberg, 1976; Kenne and
Lindberg, 1983; Jennings, 1983). There are 85 different capsular types of
the pneumococcus and two systems of nomenclature: the Danish, now generally
used, and Eddy's, which was in use in the United States (Kaufman et al,
1960).

The capsular polysaccharides play an important role in infection. They
render the bacteria resistant to non-specific host defence. Under unfavourable growth conditions, the organisms become avirulent and lose their capsules and immunological specificity. Structural properties must account for
the role of polysaccharide capsules in bacterial pathogenesis. Heavily encapsulated type 3 pneumococci are extremely virulent in mice, while type 37
pneumococci, having the same degree of encapsulation, are not. On the other
hand, type 12 pneumococci, which have very small capsules, are extremely
virulent in humans (Austrian, 1981).

The organism possesses no endotoxin but produces disease and ultimate-
ly death through its capacity to multiply in the tissues. The high rate of
mortality from <u>S. pneumoniae</u> infection despite the availability of appropriate antibiotic therapy (with about 5% case fatality rate, which is even
higher among patients with bacteremia and meningitis; 20-40%, Health and
Public Policy Committee, 1986), prompted for preventive approaches.

CAPSULAR POLYSACCHARIDE VACCINES

One approach to obtain a carbohydrate vaccine is the isolation of capsular polysaccharides from a culture of living bacteria. Early successful experiments with the type specific capsular polysaccharides of Streptococcus pneumoniae as a vaccine were overshadowed by the discovery of sulfonamides and penicillin. Renewed interest for preventive vaccination was due to increasing incidence of antibiotic resistant bacterial strains (Finland, 1978). The prophylaxis of human infections with isolated polysaccharides from encapsulated bacteria has been the subject of profound and expanding research. Besides immunological experiments on animals and clinical trials with human volunteers, the chemical structure of capsular antigens of many pathogens were elucidated.

A polysaccharide vaccine that is built up of the capsular polysaccharide from 14 serotypes (Robbins, 1978; Austrian, 1981) has been licensed (Pneumovax^R). Out of the 85 known pneumococcal serotypes these are the causative agents for 70-80% of the pneumoccoccal infections in the United States. A 23 valent vaccine that will cover nearly 90% of bacteremic infections in the U.S. was introduced in 1983 (Robbins, 1983). Epidemiological studies in different geographical areas showed some variation in prevalence of serotypes (e.g. serotypes 45 and 46 in South Africa and Asia) which necessitates an adaptation of the composition of a vaccine per region.

Although many positive reports have been published and pneumococcal vaccination is recommended for patients with a high risk for pneumonia (Finland, 1978) the effect of the vaccine on these groups and small children is often still unsatisfactory, due to the inherent immunological character of polysaccharide antigens. Moreover, it has not been possible to demonstrate any efficacy of the pneumococcal vaccine in preventing pneumonia in high-risk patients.

IMMUNE RESPONSE

The (human) immune response to infections caused by encapsulated bacteria consists of a highly complex interplay of different cells and molecular compounds. Three factors play a role in the process of eliminating invading encapsulated bacteria, leading to ingestion and killing of the invading organisms: 1) phagocytosis by macrophages and granulocytes, which adhere either directly to microbes or via a mechanism mediated by the complement system or by specific antibodies, leading to ingestion and killing of the invading organisms; 2) activation of the complement system; 3) production of antibodies. In the early acute stages of infections the humoral immune response is the most important factor. Antibodies or immunoglobulins are produced by the interaction of two types of lymphocytes. For most antigens, thymus (T) derived lymphocytes stimulate the bone-marrow derived B-cells to produce specific antibodies of immunoglobulin type G (IgG), but polysaccharides are T-cell independent (TI) antigens, which activate B-cells directly to produce mainly IgM antibodies upon immunization. This TI-character of carbohydrates is a disadvantage of the existing polysaccharide vaccines. Moreover the immune response against bacterial polysaccharides is only developing in later stages of the ontogenesis, with consequently limited use for capsular polysaccharide vaccines in children. TI-antigens do not lead to "memory" induction which is necessary for a long lasting protection. When helper T-cells can be induced, which in turn stimulate B-cells to produce IgG antibodies (thymus dependent, TD-response) durable protection should be obtained. Conjugation of carbohydrates to protein as a macromolecular carrier converts the antigen from a TI- into a TD-immunogen and, as a consequence, results in a significant increase in immunogenicity.

Immunological studies on carbohydrate haptenated proteins and liposomes (Svenson and Lindberg, 1978; Wood and Kabat, 1981; Snippe et al, 1983a,b; Zigterman et al, 1985; Snippe et al, 1988) showed the induction of different classes of antibodies directed against the carbohydrate part. Protein-carbohydrate conjugates give rise to IgM and IgG antibodies, while liposomes conjugated with carbohydrate haptens or polysaccharides induce mostly IgM antibodies (Snippe et al, 1983a,b; Snippe et al, 1988).

CONJUGATES

Carriers

Proteins (Svenson and Lindberg, 1978; Snippe et al, 1983a; Zigterman et al, 1985), liposomes (Wood and Kabat, 1981; Snippe et al, 1983b) and synthetic polymers (Chernyak et al, 1985) have been used as carriers for carbohydrate haptens and found to give rise to antibody production with specificity for the carbohydrate moiety. With these antibodies, protection is obtained against bacterial infection. The choice of the carrier is influenced by the purpose for which the conjugate is studied. Frequently used as carriers for immunological studies are albumins (BSA and HSA, bovine and human serum albumin, respectively), toxoids (diphtheria and tetanus toxoids) and (insoluble) hemocyanin (KLH, keyhole limpet hemocyanin). Besides the production of carbohydrate-specific antibodies, an immunological response is found also for the carrier-protein, which limits the application of these conjugates as human vaccine.

In contrast to proteins, liposomes are non-immunogenic carriers, i.e. no immune response is observed against the artificial lipid membrane. Anti-liposome antibodies which react with phospholipid constituents of the membrane have been reported only after lipid A incorporation (Fogler et al, 1987). Liposomes are artificial lipid bilayer vesicles usually composed of phospholipids, cholesterol, and a charged amphiphile (e.g. phosphatidic acid or octadecylamine). They are prepared by suspending the lipids in aqueous solution by which multilamellar concentric bilayer vesicles (MLV) are formed spontaneously (diameter 400-3500 nm). Ultrasonic irradiation (sonication) of such a suspension results in small unilamellar vesicles (SUV, with a diameter of 25-40 nm). In order to obtain liposomal membranes with an incorporated carbohydrate derivative (Fig. 1), conjugation of the carbohydrate part to a lipophilic molecule is necessary. The lipid portion in these glycoconjugates acts as an anchor to fix the glycon chains on the membrane surfaces.

Oligosaccharides

Anti-polysaccharide antibodies are protective for the host and therefore the production of vaccines with defined oligosaccharide determinants which induce these antibodies is a matter of considerable interest. In an approach to prepare a vaccine against Streptococcus pneumoniae, studies have been made of oligosaccharides derived from the capsular polysaccharide (Hotchkiss, 1937; Campbell and Pappenheimer, 1966). Low molecular weight compounds that are representative parts of the polysaccharide structure have been used to inhibit the antibody-antigen precipitine reaction. Immuno-dominant groups will inhibit the precipitine reaction more strongly than less important structural components. This was studied extensively on dextran-antidextran models (Kabat, 1960) to establish the location and size of determinant groups. Maximum inhibition was found for oligosaccharides with dimensions of a hexasaccharide. Information about the specificities, sizes and shapes of the antibody combining sites will be useful for the design of (semi-)synthetic vaccines. When covalently attached to appropriate macromolecular carriers, oligosaccharides are known to elicit carbo-

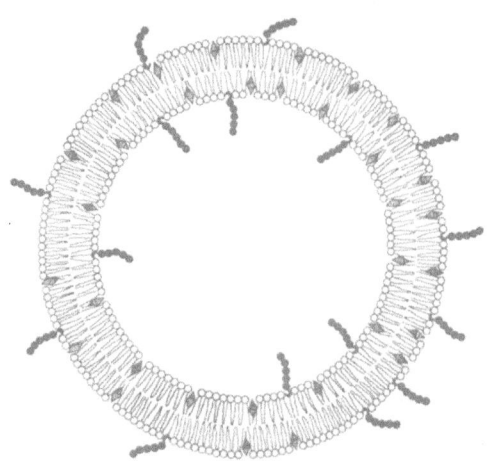

Fig. 1. Schematic drawing of an oligosaccharide haptenated liposome
(SUV), prepared from a phospholipid, cholesterol and a neoglycolipid.

hydrate specific antibody responses (Goebel, 1940; Svenson and Lindberg,
1978; Wood and Kabat, 1981).

Oligosaccharides can be obtained by partial degradation of the native
polysaccharide. For this depolymerization, chemical as well as enzymic
approaches are available. The chemical methods that can be used for specific depolymerization of a polysaccharide are very much dependent on the
chemical prop[erties of the building units. The enormous variation in repeating structures makes it impossible to apply the same procedure to all
polymers. Skilful exploitation of differences in reactivity towards chemical reagents allows for a number of possibilities in obtaining oligosaccharides from natural sources.

Enzymic hydrolysis of bacterial polysaccharides is usually very specific for only one glycosidic bond, splitting off one or more repeating units.
Most hydrolytic enzymes with specificity for baterial polysaccharides are
bacteriophage-derived glycanases. For S. pneumoniae no such bacteriophages
have been described.

SYNTHETIC OLIGOSACCHARIDES

Another approach is based on the preparation of synthetic oligosaccharide units. Advantages of articifical synthetic antigens over those obtained by isolation procedures are independence of biological material and preparation without contaminating bacterial products (for example synthesis of
lipid A-free O-antigen (Proctor, 1984). Besides the technical limitations
of synthetic carbohydrate chemistry, working with hazardous chemicals and
the production of chemical waste material can be considered as disadvantages
of this method.

Much progress has been made in the synthesis of oligosaccharide structures in the last decade. The repeating oligosaccharide structures have
been synthesized from O-antigens or capsular polysaccharides of Shigella,
Salmonella, Meningococci, Streptococci, Pneumococci, E. coli, Pseudomonas,
and even through polymerization of the oligosaccharide units.

ADJUVANTS

Preliminary studies on the immunogenicity of synthetic vaccines and subunit vaccines (viruses) indicate that these types of antigens are poorly immunogenic. This phenomenon is probably due to the fact that these antigens consist of polymers of a restricted number of (mono)saccharides or aminoacids. Therefore, one of the main hopes for useful applications of synthetic vaccines lies in the field of adjuvants. Adjuvants are substances that give no immune response by themselves but enhance or change the type of response when administered together with an antigen. By using appropriate adjuvants, vaccines can be designed with (semi-)synthetic haptens coupled to a carrier or incorporated in the liposomal membrane. In searching for an adjuvant that changes the immune response from a TI (IgM production) into a TD response (IgG production and "memory" induction), immunostimulating or immunomodulating compounds such as bacterial surface components (eg. lipid-A and lipopolysaccharides from different bacterial species e.g. <u>Klebsiella</u> O-type 3) and muramyl dipeptide (MDP), the active compound of Freund's adjuvants have been investigated. Because of mitogenic activity of lipid A and pyrogenic and somnogenic side effects of MDP, many synthetic analogs have been prepared and tested for their immunopotentiating activity. Other adjuvants studies are insoluble aluminum salts, (synthetic) surface active compounds, polyanions (Hilgers, 1986) and, recently, polyols (Zigterman et al, 1987).

GLYCOLIPID SYNTHESIS

The realization that polysaccharides or oligosaccharide determinants when covalently attached to macromolecular carrier-proteins are able to elicit carbohydrate specific IgG antibodies, has led to the development of many coupling procedures (Aplin and Wriston, 1981; Stowell and Lee, 1980). The conjugation methods employ random activation of the many functional groups of the polysaccharides. More specific coupling can be obtained when oligosaccharide-protein conjugates are produced, which are chemically more defined. In most cases, the carbohydrate-protein-lipid coupling results in the introduction of undesirable chemical groups in the conjugate. These chemical groups might behave as new antigenic determinants and therefore evoke specific immune reactions or, on the other hand, be toxic eg. carcinogenic. In human vaccines, conjugates such as aromatic "spacer"-arm groups are highly undesirable.

Synthetic methods to prepare covalently linked glycolipid conjugates face the problem of coupling water soluble carbohydrates to hydrophobic lipids. Different methods of tackling this problem have been described (Gigg, 1980). Most preparative methods make use of chemical derivatization of the sugar hydroxyl groups which results in increased solubility of the compounds in organic solvents. Coupling of a carbohydrate to a lipid, e.g. fatty acids, sphingosine bases, glycero(phospho-)lipids or alkyl- and steroid alcohols, has to be performed in homogeneous medium. It involves activation of the sugar derivative (via halogens or imidate) and the use of an appropriate catalyst, as developed in synthetic carbohydrate chemistry. Deprotection gives the desired conjugates. Only mono- or small oligosaccharides have been coupled in this way. Other methods, for which no chemical manipulation of the carbohydrate part is needed, exploit the different reactivity of the terminal monosaccharide unit. This reducing sugar is either oxidized to an aldonic acid and subsequently converted into a N-alkyl aldonamide (Williams et al, 1979), or reductively aminated in one step to form a N-alkyl-1 deoxy-alditol (Wiegandt and Ziegler, 1974; Read et al, 1977; Hoagland et al, 1979; Gosh et al, 1981; Wood and Kabat, 1981; Snippe et al, 1983). The ring structure of the reducing sugar residue is lost but a spacer molecule is obtained instead. The choice of the reaction medium and catalysts play an important role for successful coupling by these methods, especially when the alkyl-chain is longer and the oligosaccharide is built up of more monosaccharide residues.

PREPARATION OF OLIGOSACCHARIDES

The capsular polysaccharide of Streptococcus pneumoniae serotype 3 (S3) is built up of cellobiuronic acid repeating units, a disaccharide consisting of D-glucuronic acid (D-GlcpA) which is β (1→4) linked to D-glucose (D-Glcp). The disaccharide units are β(1→3) linked to each other.

→3)-βD-GlcpA(1→4)-β-D-Glcp(1→

For the depolymerisation of the S3 polysaccharide chemical and enzymic approaches are available (Scheme 1). By partial acid hydrolysis the β(1→3) bond is cleaved preferentially, resulting in the formation of a mixture of cellobiuronic oligomers H [→3)-β-D-GlcpA(1→4)-β-D-Glcp(1→]$_n$ OH (route A). Specific cleavage of the β(1→4) bond between GlcA and Glc by enzymic depolymerisation with Bacillus palustris-derived glycanase leads to oligosaccharides with pseudo-laminaribiuronic-acid units H[→4)-β-DGlcp(1→3)-β-D-GlcpA-(1→]$_n$ OH (route B, Campbell et al, 1966). We isolated a tetra (TS$_3$) and a hexasaccharide (HS$_3$) by column chromatography after a partial acid hydrolysis of polysaccharide S$_3$.

The capsular polysaccharide of Streptococcus pneumoniae serotype 14 (S14) is built up of branched tetrasaccharide repeating units (consisting of D-glucose, D-galactose and N-acetyl-D-glucosamine in the molar ratio of 1:2:1) which are all present in a β-linkage (Lindberg et al, 1977);

[→3)-β-D-Galp-(1→4)-β-D-Glcp-(1→6)-β-D-GlcNAcp-(1→]$_n$

$\overset{\frown}{}$
4
↑
1
$\underset{\smile}{}$
β

D-Galp

Partial depolymerisation of S14 by the deamination of 2-amino-2-deoxy-D-glucopyranosyl residues with nitrous acid results in the formation of 2,5-anhydro-D-mannose and splits the glycosidic linkage (Scheme 2, Defaye, 1970;

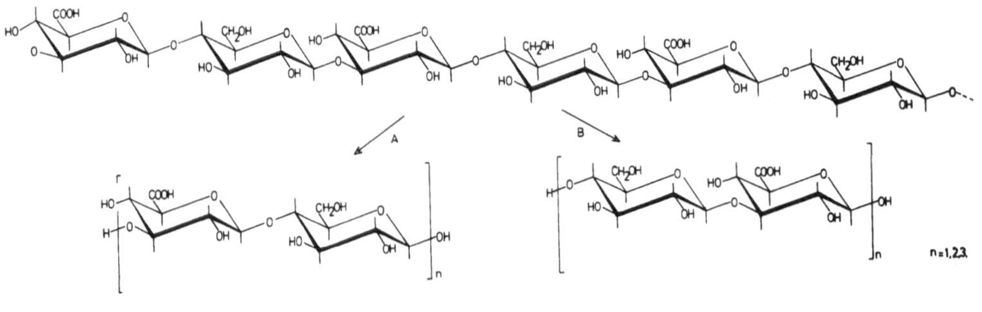

Scheme 1. Depolymerization of Streptococcus pneumoniae serotype 3 capsular polysaccharide S3 by partial acid hydrolysis (route A) (HS n=3) or by incubation with Bacillus palustris derived glycanase (route B) (HS' n=3).

Erbing et al, 1979; Horton and Philips, 1979). The obtained oligosaccharides, having a reactive aldehyde function, are unstable unless 2,5-anhydro-D-Man is reduced with NaBH₄. The reactivity of the aldehyde group (not engaged in intramolecular hemiacetal information) can also be used for covalent coupling of these oligosaccharides to primary aminogroups of carriers (Bjork et al, 1982; Hoffman et al, 1983) by reductive amination with sodium cyanoborohydride.

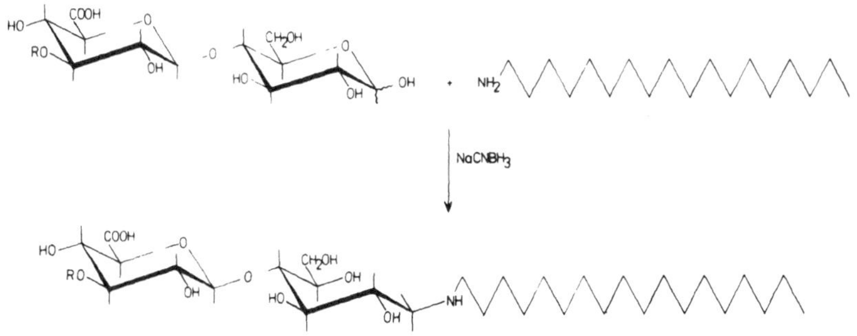

Scheme 2. Depolymerization of <u>Streptococcus pneumoniae</u> serotype 14 capsular polysaccharide and coupling of a tetrasaccharide to stearylamine (TS$_{14}$-S, see text).

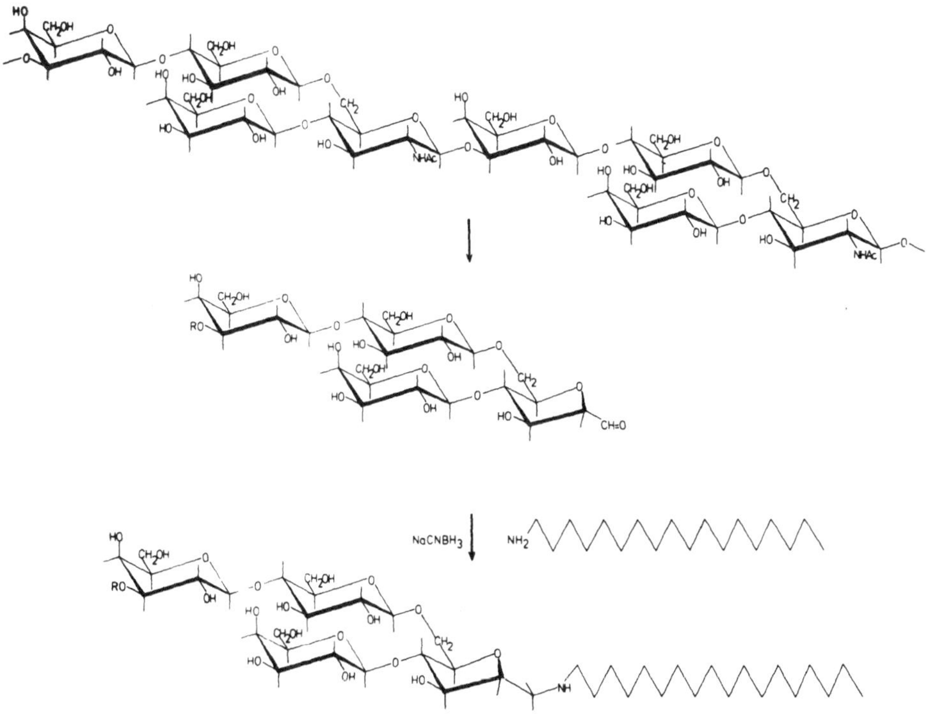

Scheme 3. Coupling of S3-derived hexasaccharides to stearylamine by means of reductive amination with sodium cyanoborohydride, HS-S: R=H[→3)β-GlcpA-(1→4)β-D-Glcp-(1→]$_2$.

Before deamination with nitrous acid can be performed, the N-acetyl groups in S14 polysaccharide must be converted into an amino function by de-N-acetylation with anhydrous hydrozine at $100^{\circ}C$. By varying the time of hydrazinolysis, the size of the oligosaccharides can be manipulated. We isolated a tetrasaccharide (TS_{14}) by column chromatography.

PREPARATION OF TS_3-S, HS_3-S AND TS_{14}-S

The isolated hexa- and tetrasaccharides are coupled to octadecylamine ($NH_2(CH_2)_{17}$) by the reductive amination method with sodium cyanoborohydride ($NaCNBH_3$) in tetrahydrofuran/water (5:2 v/v) for 2-5 days at $45^{\circ}C$ (Schemes 2 and 3).

The oligosaccharide-lipid (HS_3-S and TS_{14}-S) conjugates thus obtained are purified by LH-20 column chromatography and identified by sugar analysis followed by combined gas chromatography/mass spectrometry for the identification of reductively aminated sugar residue (Tsai, 1970; Kamerling et al, 1975).

INCORPORATION OF OLIGOSACCHARIDE-LIPID CONJUGATES INTO LIPOSOMAL MEMBRANES

Liposome preparations are haptenated with oligosaccharides by incorporation of the neoglycolipids into a lipid bilayer membrane consisting of dipalmitoyl-L-α-phosphatidylcholine (DPPC) and cholesterol (Uemera et al, 1974; Van Houte et al, 1979; Snippe et al, 1983b). The ratio of neoglyco-lipid, DPPC, and cholesterol was fixed at X:(90-X):10 wherein X represents the epitope density of neoglycolipid in mol %. Liposomes were prepared by dispersion of the lipid film, formed after evaporation of the organic solvent, into phosphate buffered saline by vortexing. Subsequently, sonication of the lipid mixture leads to small unilamellar vesicles haptenated with S3 or S14 derived oligosaccharides.

IMMUNOLOGICAL RESPONSE TO VACCINATION WITH LIPOSOMES HAPTENATED WITH OLIGOSACCHARIDES

The Immunogenicity of HS_3-S-liposomes

Female Balb/c mice immunized intravenously with various amounts of liposomes haptenated with neoglycolipids consisting of conjugates of hexa-saccharide and stearylamine (HS_3-S, 0.1-30 nmol HS_3-S per mouse; epitope density was kept constant at 5 mol %) gave rise to varying levels of circulating antibodies (Fig. 2), as determined by a haemagglutination assay with indicator erythrocytes optimally derivatized with S3 (Baker et al, 1969). Liposomes with a dose of 0.1 nmol HS_3-S were not immunogenic, but higher doses of HS_3-S in liposomes induced circulating immunoglobulin M (2-mercaptoethanol (2-ME)-sensitive) antibodies starting 5 days after immunization (Fig. 3). Mice immunized with the neoglycolipid HS_3-S only, also developed anti-S3 immunoglobulin M antibodies, although a much higher dose (30-fold increase) of HS-S was required. The hexasaccharide (HS_3) was not immunogenic when tested over a whole dose range (Fig. 2).

The effect of varying epitope densities, using a fixed dose of HS_3-S (1 nmol) was also studied (Fig. 4). In these experiments the epitope density was varied by changing the DPPC/HS-S molar ratio of the liposomes. Maximal haemagglutination titers were obtained with liposomes containing 20 mol % HS_3-S (Fig. 4). Immunization of mice with HS-S liposomes at day 0 and boosted nine weeks later with the same antigen indicated that (1) a maximum antibody titer is attained one week after immunization, (2) no enhanced

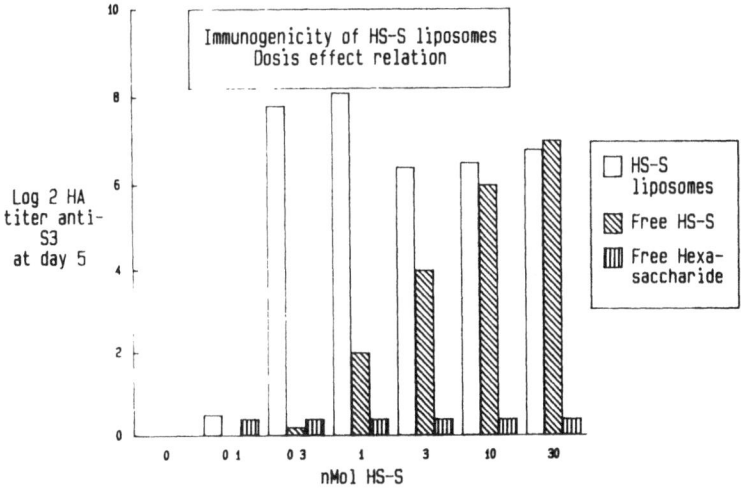

Fig. 2. Serum antibody titers at day 5 after immunization with graded amounts of HS_3-S liposomes (▢), free HS-S (▨), or HS (▥).

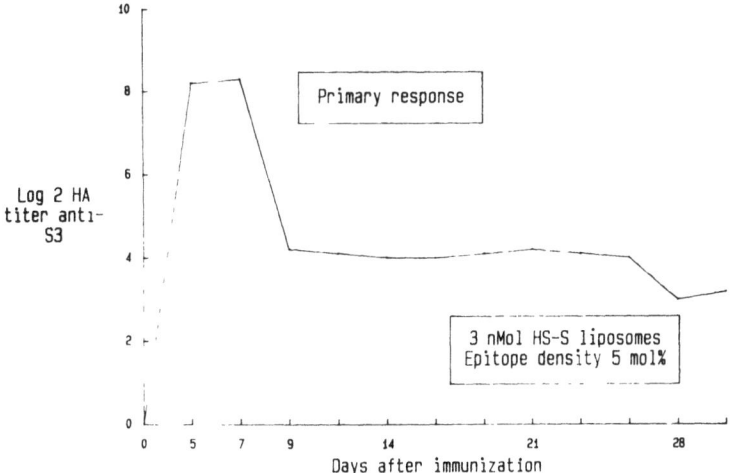

Fig. 3. Serum antibody titers after immunization with HS_3-S-liposomes.

secondary response occurred and (3) only very limited (no detectable) amounts of 2-ME resistant (IgG) antibody were produced (data not shown). Nonionic block polymers (NBPs) have proved to be potent adjuvants for the humoral immune response against liposomes haptenated with tripeptide enlarged dinitrophenyl groups (Zigterman et al, 1987). In the following experiment the capacity of several NBPs (see Table 1) to enhance the immune response to TS_3-S-liposomes was tested. The NBPs were administered one day prior to immunization with TS-liposomes (Fig. 5). L121, 31R1, 130R1 and 130R2 enhanced HA-titers while 1501 did not affect HA-titers significantly under these conditions.

The Immunogenicity of TS_{14}-S Liposomes

Female Balb/c mice were immunized i.p. with TS_{14}-S-liposomes (5 nmol TS_{14}-S per mouse; epitope density 5 mol %). The mice were bled at regular intervals and the amount of S14 specific IgM and IgG antibodies were determined by ELISA. The amount of antibody of a certain isotype present in the sera was compared with the amount present in a reference serum (Effective

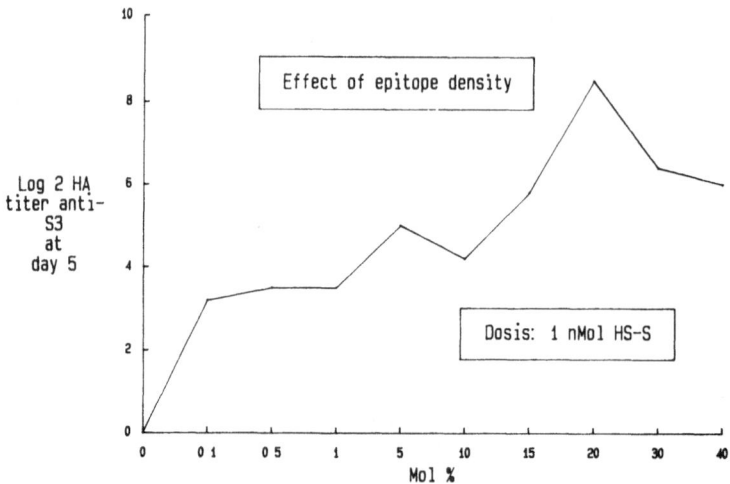

Fig. 4. Effect of HS_3-S epitope density on the response in Balb/c mice.

Dose value, ED; Zigterman et al, 1988). After immunization with TS_{14}-S liposomes a slow raise of anti-S_{14} specific antibodies of the IgM class occurs, starting 14 days after immunization (Fig. 6). At day 5 the response is still below values present at day 0 (naturally occurring antibodies).

THYMUS INDEPENDENCE

The characteristics of the primary and secondary response to HS_3-S-liposomes correspond with those of thymus-independent antigens. This finding was further confirmed in mice deficient in thymus-derived (T) lymphocytes. Immunization of nude Balb/c mice (nu/nu) and their heterozygous littermates (nu/+) with HS-S-liposomes gave rise to similar anti-S3 circulating antibody levels over a longer period (Fig. 7).

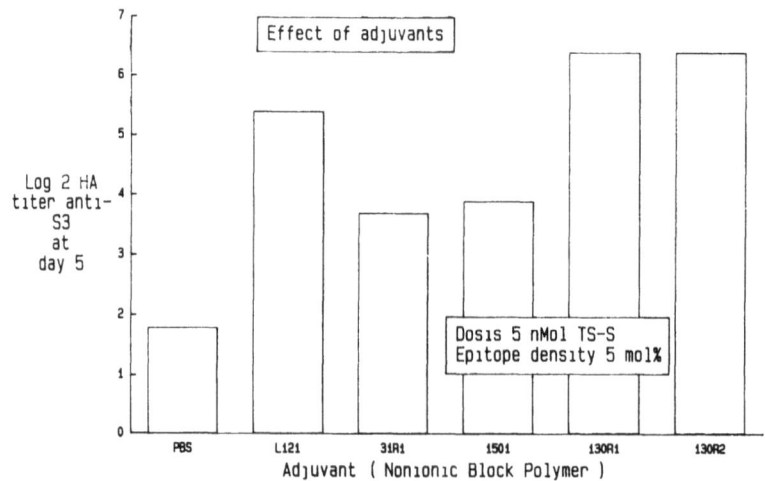

Fig. 5. Modulation of TS_3-S liposome induced antibody levels by NBPs.

Table 1. Chemical Properties of NBP Surfactants

Compound	Average[a] Molecular Weight	% POE[a]	Hydrophile-lipophile balance[b]	Structure
Triblock				
L121	4400	10	0.5	POE-POP-POE
L122	5000	20	4.0	
Reversed Triblock				
31R1	3200	10	1.7	POP-POE-POP
Octablock				
1501	7900	10	1.0	POE-POP⟍ ⟋POP-POE POE-POP⟋ ⟍POP-POE
Reversed Octablock				
130R1	6800	10	1.4	POP-POE⟍ ⟋POE-POP POP-POE⟋ ⟍POE-POP
130R2	7740	20	2.9	

a Average molecular weight and % POE according to the manufacturer.

b Hydrophile-lipophile balance-values were obtained from Hunter et al (1984).

RESPONSE OF VARIOUS INBRED MOUSE STRAINS TO HS_3-S-LIPOSOMES

HS-S-liposomes are immunogenic in numerous inbred strains of mice. In most mouse strains, the anti-S3 response evoked by HS-S-liposomes is comparable with that obtained after immunization with 0.5 μg S3 polysaccharide. However, the anti-S3 antibody levels reached are somewhat lower and the response fades away within 35 days. In addition, a remarkable observation was made: although CBA/N mice do not mount an immune response to S3 polysaccharide (which is genetically determined (Amsbaugh et al, 1972; Sher, 1982), they do respond to HS-S-liposomes, 5 days after immunization.

INDUCTION OF PROTECTION

Protective immunity to a lethal dose of S. pneumoniae type 3 (Reed and Munch, 1938; Snippe et al, 1983a) was determined after intraperitoneal injection of 25 x 50% lethal doses (LD_{50} for type 3, 4 x 10^3 CFU). Immunization with 0.5 μg S3 served as a positive control, and injections with either phosphate-buffered saline or non-haptenated liposomes served as a negative control. Mice immunized with various amounts of HS_3-S-liposomes (1.0 nmol of HS_3-S; epitope density, 5 nmol %) were protected against 25

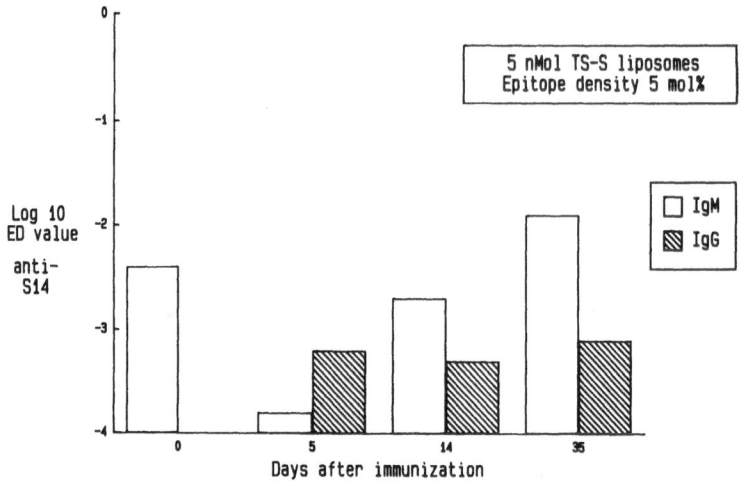

Fig. 6.   Immunogenicity of TS_{14}-S-liposomes.

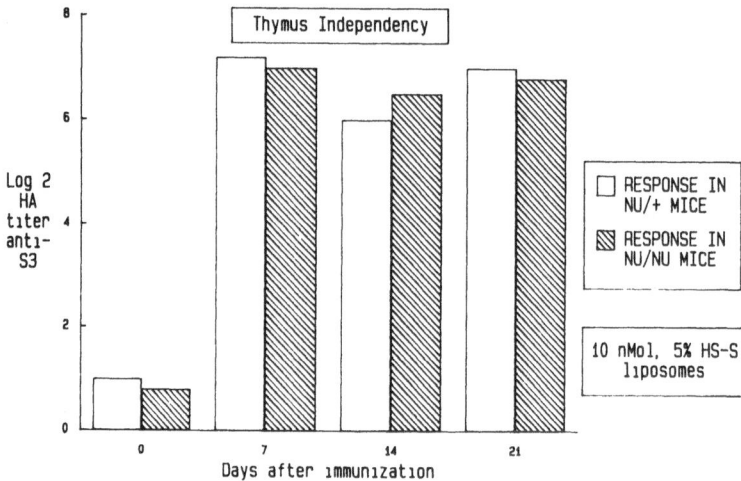

Fig. 7.   Serum antibody response of nude mice to immunization with HS_3-S-liposomes.

LD_{50} of <u>S. pneumoniae</u> given seven days after intravenous injection (Table 2). The specificity of the vaccine (HS$_3$-S-liposomes) is also presented in this Table. No cross-protection was observed against <u>S. pneumoniae</u> type 11 (LD_{50} for type 11, 4×10^5 CFU) indicating the specificity of the HS-S-liposome induced protection.

CONCLUSIONS

Oligosaccharides isolated from <u>S. pneumoniae</u> type 3 or type 14 poly-saccharide can be linked to both carrier proteins as well as liposomal mem-branes in order to induce specific antibodies and protection. Liposomal membranes are non-immunogenic and therefore have the advantage of avoiding undesired immune reactions against the carrier. On the other hand liposomal membranes lack determinants which could render the oligosaccharide-liposome thymus dependent. As all results presented indicate that oligosaccharide-liposomes behave as thymus independent antigens, adjuvants might be valuable in changing the immunogenic characteristics of liposomes. Preliminary res-

Table 2. Induction of Protection to <u>S. pneumoniae</u> Type 3 after Immunization with HS_3-S-liposomes, Showing Specificity of the Reaction[a]

Group	Immunizing agent	Route[b]	Challenge organism (at 25 LD_{50})	No. of survivors/ no injected
1	HS_3-S-liposomes (1 nmol HS_3-S)	i.v.	Type 3	6/6
2	HS_3-S-liposomes (1 nmol HS_3-S)	i.v.	Type 11	0/6
3	Phosphate-buffered saline	i.v.	Type 3	0/6
4	Phosphate-buffered saline	i.v.	Type 11	0/6
5	S3 (0.5 µg)	i.p.	Type 3	6/6
6	S3 (0.5 µg)	i.p.	Type 11	0/6

a Groups of six BALB/c mice were immunized as indicated. At day 7, the mice were challenged intraperitoneally with 25 LD_{50} of *S. pneumoniae* type 3 or 11, and 2 weeks later the number of survivors was recorded.

b i.v.: intravenous; i.p.: intraperitoneal.

ults on the adjuvant activity of the group of Nonionic Block Polymers (NBPs) suggest that these adjuvants are serious candidates to include in a synthetic pneumococcal vaccine. Although a synthetic vaccine based on the liposomal membrane model seems not feasible in the near future, results presented in this chapter will help to develop a new pneumococcal vaccine based on oligosaccharide-protein conjugates. The immunochemical approach to using synthetic oligosaccharides allows us to define more precisely the important epitopes on the polysaccharide antigens.

REFERENCES

Amsbaugh, D.F., Hansen, C.T., Prescott, B., Stashak, P.W., Barthold, D.R. and Baker, P.J., 1972, Genetic control of the antibody response to type III pneumococcal polysaccharide in mice. I. Evidence that an X-linked gene plays a decisive role in determining responsiveness, J.Exp.Med., 136:931.
Aplin, J.D. and Wriston, J.C., 1981, Preparation, properties, and applic-

ations of carbohydrate conjugates of proteins and lipids, <u>C.R.C. Crit.Rev.Biochem.</u>, 5:295.

Austrian, R., 1981, Some observations on the pneumococcus and on the current status of pneumococcal disease and its prevention, <u>Rev.Infect.Dis.</u>, 3:S1.

Baker, P.J., Stashak, P.W. and Prescott, B., 1969, Use of erythrocytes sensitized with purified pneumococcal polysaccharide for the assay of antibody and antibody-producing cells, <u>Appl.Microbiol.</u>, 17:422.

Björk, I., Larm, O., Lindberg, U., Norling, K. and Riquelme, M-E., 1982, Permanent activation of antithrombin by covalent attachment of heparin oligosaccharides, <u>FEBS Lett.</u>, 143:96.

Campbell, J.H. and Pappenheimer, A.M., 1966, Quantitative studies of the specificity of anti-pneumococcal polysaccharide antibodies types III and VIII. I. Isolation of oligosaccharides from acid and from enzymic hydrolysates, <u>Immunochemistry</u>, 3:195.

Chernyak, A.Y., Antonov, K.V., Kochetkov, N.K., Padyukov, L.N. and Tsvetkova N.V., 1985, Two synthetic antigens related to <u>Streptococcus pneumoniae</u> type 3 capsular polysaccharide, <u>Carbohydr.Res.</u>, 141:199.

Defaye, J., 1970, 2,5-Anhydrides of sugars and related compounds, <u>Adv. Carbohydr.Chem.Biochem.</u>, 25:181.

Erbing, C., Lindberg, B. and Svensson, S., 1979, Deamination of methyl 2-amino-2-deoxy-a- and -b-D-glucopyranosides, <u>Acta Chem.Scand.</u>, 27: 3699.

Finland, M., 1978, And the walls come tumbling down. More antibiotic resistance, and now the pneumococcus, <u>N.Engl.J.Med.</u>, 299:770.

Fogler, W.E., Swartz, G.M. and Alving, C.R., 1987, Antibodies to phospholipids and liposomes: binding of antibodies to cells, <u>Biochim. Biophys.Acta</u>, 903:265.

Gigg, R., 1980, Synthesis of glycolipids, <u>Chem.Phys.Lipids</u>, 26:287.

Goebel, W.F., 1940, Studies on antibacterial immunity induced by artificial antigen II immunity to experimental pneumococcal infection with an antigen-containing cellobiuronic acid, <u>J.Exp.Med.</u>, 69:353.

Gosh, P., Bachkawat, B.K. and Surolia, A., 1981, Synthetic glycolipids: interaction with galactose-binding lectin and hepatic cells, <u>Arch. Biochem.Biophys.</u>, 206:454.

Health and Public Policy Committee (Am. Coll. Phys.) 1986, Pneumococcal vaccine, <u>Ann.Int.Med.</u>, 104:118.

Hilgers, L.A.T., 1986, Immunomodulating properties of synthetic adjuvants, Thesis, R.U. Utrecht.

Hoagland, P.D., Pfeffer, P.E. and Valentine, K.M., 1979, Reductive amination of lactose: unusual 13C-n.m.r. spectroscopic properties of N-alkyl-(1-deoxy-lactitol-lyl)amines, <u>Carbohydr.Res.</u>, 74:135.

Hoffman, J., Larm, O. and Schollander, E., 1983, A new method for covalent coupling of heparin and other glycosaminoglycans to substances containing primary aminogroups, <u>Carbohydr.Res.</u>, 117:328.

Horton, D. and Phillips, K.D., 1979, The nitrous acid deamination of glycosides and acetates of 2-amino-2-deoxy-D-glucose, <u>Carbohydr.Res.</u>, 30:367.

Hotchkiss, R.D. and Goebel, W.F., 1937, Chemo-immunological studies on the soluble specific substance of Pneumococcus. III. The structure of aldobionic acid from the type III polysaccharide, <u>J.Biol.Chem.</u>, 121:195.

Hunter, R.L. and Bennett, B., 1984, The adjuvant activity of nonionic block polymer surfactants. II. Antibody formation and inflammation related to the structure of triblock and octablock copolymers, <u>J.Immunol.</u>, 133:3167.

Jennings, H.J., 1983, Capsular polysaccharides as human vaccines, <u>Adv. Carbohydr.Chem.Biochem.</u>, 41:155.

Kabat, E.A., 1960, The upper limit for the size of the human antidextran combining site, <u>J.Immunol.</u>, 84:82.

Kamerling, J.P., Gerwig, G.J. and Vliegenthart, J.F.G., 1975, Characterization by gas-liquid chromatography-mass spectrometry and proton magnetic resonance spectroscopy of pertrimethylsilyl methyl glycosides obtained in the methanolysis of glycoproteins and glycopeptides, Biochem.J., 151:491.

Kaufmann, F., Lund, E. and Eddy, B.E., 1960, Proposal for a change in the nomenclature of Diplococcus pneumoniae and a comparison of the Danish and American type designations, Int.Bull.Bacteriol.Nomencl.Taxon., 10:31.

Kenne, L. and Lindberg, B., 1983, Bacterial polysaccharides, in: "The Polysaccharides", Vol. 2, G.O. Aspinall, ed., Academic Press.

Larm, O. and Lindberg, B., 1976, The pneumococcal polysaccharides, a reexamination, Adv.Carbohydr.Chem.Biochem., 33:295.

Lindberg, B., Lnngren, J. and Powell, D.A., 1977, Structural studies on the specific type-14 pneumococcal polysaccharide, Carbohydr.Res., 58:177.

Proctor, R.A., 1984, "Handbook of Endotoxin", Vol. 1, Chemistry of Endotoxin, E.Th. Rietschel, ed., Elsevier, Amsterdam.

Read, B.D., Demel, R.A., Wiegandt, H. and Van Deenen, L.L.M., 1977, Specific interaction of concanavalin A with glycolipid monolayers, Biochim. Biophys.Acta, 470:325.

Reed, L.J. and Muench, H., 1938, A simple method for estimating fifty percent endpoints, Am.J.Hyg., 27:493.

Robbins, J.B., 1978, Vaccines for the prevention of encapsulated bacterial diseases: current status, problems and prospects for the future, Immunochemistry, 15:839.

Robbins, J.B., Austrian, R., Lee, C-J., Rastogi, S.C., Schiffman, G., Henrichsen, J., Mkel, P.H., Broome, C.V., Facklam, R.R., Tiesjema, R.H. and Parke, J-C., 1983, Considerations for formulating the second-generation pneumococcal capsular-polysaccharide vaccine with emphasis on the cross-reactive types within groups, J.Infect.Dis., 148:1136.

Scher, I., 1982, The CBA/N mouse strain: an experimental model illustrating the influence of the X-chromosome on immunity, Adv.Immunol., 33:1.

Snippe, H., Van Houte, A.J., Van Dam, J.E.G., de Reuver, M.J., Jansze, M. and Willers, J.M.N., 1983a, Immunogenic properties in mice of hexasaccharide from the capsular polysaccharide of Streptococcus pneumoniae type 3, Infect.Immun., 40:856.

Snippe, H., Van Dam, J.E.G., Van Houte, A.J., Willers, J.M.N., Kamerling, J.P. and Vliegenthart, J.F.G., 1983b, Preparation of a semisynthetic vaccine to Streptococcus pneumoniae type 3, Infect.Immun., 42:842.

Snippe, H., Zigterman, J.W.J., Van Dam, J.E.G. and Kemerling, J.P., 1988, Oligosaccharide-haptenated liposomes used as a vaccine to Streptococcus pneumoniae, in: "Liposomes as Drug Carriers", G. Gregoriadis, ed., J. Wiley & Sons Ltd., Chichester.

Stowell, C.P. and Lee, Y.C., 1980, Neoglycoproteins, the preparation and application of synthetic glycoproteins, Adv.Carbohydr.Chem.Biochem., 37:225.

Svenson, S.B. and Lindberg, A.A., 1978, Immunochemistry of Salmonella O-antigens. Preparation of an octasaccharide-bovine serum albumin immunogen representative of Salmonella serogroup B O-antigen and characterization of the antibody response, J.Immunol., 120:1750.

Tsai, C.S., 1970, Determination of the degree of polymerization of N-acetyl chitooligoses by chromatographic methods, Anal.Biochem., 36:114.

Uemera, K., Nicolotti, R.A., Six, H.R. and Kinskey, S.C., 1974, Antibody formation in response to liposomal membranes sensitized with N-substituted phosphatidyl ethanolamine derivatives, Biochemistry, 13:1572.

Van Houte, A.J., Snippe, H. and Willers, J.M.N., 1979, Characterization of immunogenic properties of haptenated liposomal model membranes in mice. I. Thymus independence of the antigen, Immunology, 37:505.

Wiegandt, H. and Ziegler, W., 1974, The use of reductaminated sugars for the preparation of oligosaccharide conjugates. I. Synthetic glycolipids containing glycosphingolipid-derived oligosaccharides, Hoppe-Seyler's Z.Physiol.Chem., 355:11.

Williams, T.J., Plessas, N.R., Goldstein, I.J. and Lnngren, J., 1979, A new class of model glycolipids: synthesis, characterization, and interaction with lectins, <u>Arch.Biochem.Biophys.</u>, 195:145.

Wood, C. and Kabat, E.A., 1981a, Immunochemical studies of conjugates of isomaltosyl oligosaccharide to lipid. I. Antigenicity of the glycolipids and the production of specific antibodies in rabbits, <u>J.Exp.Med.</u>, 154:432.

Wood, C. and Kabat, E.A., 1981b, Immunochemical studies of conjugates of isomaltosyl oligosaccharide to lipid. II. Specificities and reactivities of the antibodies formed in rabbits to stearyl-isomaltosyl oligosaccharides, <u>Arch.Biochem.Biophys.</u>, 212:262.

Wood, C. and Kabat, E.A., 1981c, Immunochemical studies of conjugates of isomaltosyl oligosaccharide to lipid. III. Fractionation of rabbit antibodies to stearyl-isomaltosyl oligosaccharides and a study of their combining sites by a competitive binding assay, <u>Arch.Biochem. Biophys.</u>, 212:277.

Zigterman, J.W.J., Van Dam, J.E.G., Snippe, H., Rotteveel, F.T.M., Jansze, M., Willers, J.M.N., Kamerling, J.P. and Vliegenthart, J.F.G., 1985, Immunogenic properties of octasaccharide-protein conjugates derived from Klebsiella serotype II capsular polysaccharide, <u>Infect.Immun.</u>, 47:421.

Zigterman, J.W.J., Snippe, H., Jansze, M. and Willers, J.M.N., 1987, Adjuvant effects of nonionic block polymer surfactants on liposome induced humoral immune response, <u>J.Immunol.</u>, 138:220.

Zigterman, J.W.J., Verheul, A.F.M., Ernste, E.B.H.W., Rombouts, R.F.M., de Reuver, M.J., Jansze, M., Snippe, H. and Willers, J.M.N., 1988, Measurement of the humoral immune response against <u>Streptococcus pneumoniae</u> type 3 capsular polysaccharide and oligosaccharide containing antigens by ELISA and ELISPOT techniques, <u>J.Immunol.Meth.</u>, 106:101.

LIPOSOMES AS CARRIERS OF VACCINES: DEVELOPMENT OF A LIPOSOMAL MALARIA VACCINE

Carl R. Alving[1], Robert L. Richards[1], Michael D. Hayre[1]
Wayne T. Hockmeyer[2] and Robert A. Wirtz[3]

Departments of [1]Membrane Biochemistry, [2]Immunology, and
[3]Entomology, Walter Reed Army Institute of Research
Washington, DC 20307-5100, USA

FATE OF LIPOSOMES IN VIVO

Liposomes that have been injected parenterally into animals have a well-known natural tendency to be ingested rapidly and in large amounts by macrophages. Uptake of liposomes by macrophages has often been cited as a potential hurdle that could theoretically block applications of liposomes as drug carriers for certain purposes. However, the macrophage itself has served as a target for delivery of liposome-encapsulated drugs and immuno-modulators, particularly for treatment of infectious diseases and cancer (Alving, 1983, 1989; Fidler, 1985; Swenson et al, 1988). It is certainly true that overcoming of the macrophage as an "obstacle" can be difficult, but several reports have indicated that increased blood circulation time and distribution of liposomes to certain tissues can be achieved by the use of special biophysically or biochemically tailored liposomes (Hwang et al, 1980; Gregoriadis et al, 1982; Allen and Chonn, 1987; Papahadjopoulos and Gabizon, 1987; Gabizon and Papahadjopoulos, 1988).

DELIVERY OF LIPOSOMES TO ANTIGEN PRESENTING CELLS

Although delivery of liposomes to cells other than macrophages may pose a considerable challenge, in the field of immunology and particularly in the area of vaccine development capture of liposomes by macrophages is not a detrimental phenomenon. In fact, uptake of liposomes by macrophages is highly advantageous to the immune response and greatly facilitates the use of liposomes as carriers of antigens and vaccines. The macrophage is one of a limited number of cell types that are known as "antigen presenting cells" (APC) or "accessory cells", and accumulation of liposomal antigens and ad-juvants within macrophages stimulates the humoral immune response to many antigens (Alving, 1987).

Many of the details about the characteristics and functions of APCs are still controversial, but compelling arguments have led to widespread agree-ment that the humoral immune response to most antigens requires the inter-vention of APCs (reviewed by Unanue, 1984, and Unanue and Allen, 1987). The APC is characterized by the presence of class II major histocompatibility antigen molecules (Ia antigen) on its surface, and only three major cell types express Ia antigen: macrophages, B cells, and dendritic cells of lymphoid tissue and skin. The immunogenic material is internalized by the APC where it may be partially degraded, although the absolute requirement for such degradation has not been fully resolved. The immunogenic epitopes form a complex with Ia antigen molecules, and the complex is transported to the cell surface. In the next stage of the immune response the antigen-Ia complex (or individual constituents thereof) on the APC surface interact specifically with CD-4-positive T-helper cell lymphocytes. The APC and the T-helper cell both are secretory cells, and among the important resulting immunologic products that influence the humoral immune response are included interleukin-1 (from the APC) and interferon-γ (from T-helper cells).

As mentioned earlier, three major cell types express Ia antigen and can serve as APCs. There has been debate on the relative importance or prominence of each type of APC; for example, different relative emphasis has been placed by different investigators on the role of macrophages (Unanue, 1984; Unanue and Allen, 1987) or dendritic cells (Steinman and Nussenzweig, 1980; Tew et al, 1980). Regardless of the relative importance of macrophages compared to dendritic cells in the immunologic scheme of the body, it seems likely that macrophages serve predominently, or even exclusively, as APCs for liposomal antigens. Several lines of evidence support this conclusion. First, as noted earlier liposomes are ingested avidly by macrophages, but are taken up only slightly, if at all, by most other cells under normal in vivo circumstances. In contrast, dendritic cells do not serve as endocytic cells (Steinman and Nussenzweig, 1980) and it is unlikely therefore that dendritic cells would capture liposomes to any significant extent. Second, depletion of macrophages by treatment of animals with carageenan suppressed the ability of the animals to produce an immune response to a liposomal antigen (Shek and Lukovich, 1982). Third, in an interesting experiment liposomes containing a poorly immunogenic antigen (bovine serum albumin) were first ingested by macrophages in vitro. The macrophages containing the liposome-encapsulated albumin were then injected into mice and a murine humoral immune response to the liposomal albumin was observed (Beatty et al, 1984).

IMMUNE RESPONSE TO LIPOSOMAL ANTIGENS

Liposomes have been used to induce humoral immunity to numerous liposomal protein antigens (reviewed by Alving, 1987). There has been considerable investigation of the biophysical parameters that influence the immune response. For example, it is likely that liposome-encapsulated and surface-bound antigens are both effective for inducing antibodies. It is also possible that very small liposomes are somewhat more effective for inducing antibodies than large ones, but obtaining very small liposomes can sometimes pose inconvenient technical problems in formulating vaccines because care must be taken to avoid altering the chemical structure of the antigen. With individual antigens differences in the potency of the immune response have also been noted with liposomes containing different net surface charges.

In most but not all cases the antibody response induced by liposomes is enhanced if the liposomes contain an adjuvant (Alving, 1987). Immunity is also increased by mixing or emulsifying the liposomes with certain nonliposomal adjuvants. In our own experience either Freund's complete adjuvant or aluminum hydroxide (alum) can enhance immunity. In the case of alum, additive or synergistic immunostimulating effects were observed when nonliposomal alum was used in combination with liposomal lipid A to stimulate humoral immunity to a liposomal malaria antigen (Richards et al, 1988).

In summary, a considerable amount of information has now been developed on APCs, and it is well-known that liposomes are avidly delivered to at least one APC, the macrophage. The physical, immunologic, and metabolic interactions of liposomes with macrophages have also been carefully studied. It is therefore logical that liposomes should be proposed as carriers of antigens for vaccines. In the past, limited supplies of useful antigens posed barriers to development of practical vaccines. However, we are currently examining synthetic antigens derived from malaria parasites as potential candidates for a liposomal malaria vaccine.

MALARIA SPOROZOITE VACCINE

In the life cycle of the malaria parasite, the malaria organism is injected into the bloodstream of the mammalian host by the bite of the female anopheline mosquito. The form of the organism that is injected by the mosquito is known as the sporozoite. After entry into the blood the sporozoite rapidly disappears from the blood due to uptake by the liver. Over a period of days it then develops into a different form before reappearance in the blood.

During the past few years, several laboratories have undertaken programs to manufacture and test potential synthetic vaccines to induce humoral immunity against the sporozoite stage of the human malaria parasite Plasmodium falciparum. The feasibility of developing an effective vaccine against the sporozoite form of P. falciparum has been demonstrated in several ways. A successful clinical trial utilizing irradiated sporozoites as antigens demonstrated that protective immunity could be achieved in humans (Clyde et al, 1975). Furthermore, protection against sporozoite-induced infection in animals or humans was correlated with the presence of a high titer of polyclonal antibodies to sporozoites (Clyde et al, 1975; Nussenzweig et al, 1969; Rieckmann et al, 1979) and protection was even obtained by passive infusion of monoclonal antibodies in animals (Potocnjak et al, 1980). The major antigen responsible for inducing protective immunity is a protein that covers the surface of the sporozoite, the circumsporozoite (CS) protein. A positive correlation was observed between anti-CS protein antibodies among individuals living in endemic areas and protection against sporozoite challenge (Hoffman et al, 1986).

The gene encoding the CS protein of P. falciparum has been cloned (Dame et al, 1984). A region in the middle of the protein which contains repeating tetrapeptides is thought to be capable of inducing protective immunity (Dame et al, 1984; Young et al, 1985; Ballou et al, 1985; Zavala et al, 1985). A positive correlation was observed between antibody titers and protection against sporozoite challenge after immunizing humans with a vaccine containing a recombinant protein (R32tet$_{32}$) derived from the structure of the repetitive region of the CS protein (Ballou et al, 1987).

The available evidence therefore suggests that protective immunity could theoretically be achieved by inducing high titers of antibodies to the CS protein. However, considerable complexity has been observed in the immune response to the CS protein. Titers of naturally-occurring antibodies to CS protein in endemic areas do not correlate well with protection against new malaria infections that occur after treatment for malaria (Hoffman et al, 1987). Lack of protection may be related to immunosuppression that apparently is induced against the repeating tetrapeptide region of the CS protein by intracellular parasites (Webster et al, 1987). When mice were immunized with either irradiated sporozoites or with recombinant or peptide vaccines against the repeat region of the CS protein from P. berghei, protection against experimental infection with malaria was greater, irrespective of antibody titer, after immunization with the irradiated sporozoites (Egan et al, 1987). The latter study suggests that cell-mediated immunity could contribute to an immune response against sporozoites.

Although the antibody titer against the sporozoite or CS protein is not necessarily predictive of protection in the course of natural infections, the feasibility of obtaining a useful vaccine by artifically inducing a strong humoral immune response has not yet been tested successfully in humans. Vaccine trials with a recombinant antigen ("falciparum sporozoite vaccine number 1", or FSV-1) or with a peptide antigen containing epitopes derived from the repeat region of the CS protein did not induce high titers of antibodies in human volunteers (Ballou et al, 1987; Herrington et al, 1987). The numbers of sporozoites experimentally used to test the efficacies of the vaccines in the immunized volunteers were quite high, but despite high sporozoite doses partial or complete protection was still observed among individuals having the highest antibody titers (Ballou et al, 1987; Herrington et al, 1987). The current research challenge, therefore, is to devise methods that can induce consistently high titers of anti-CS protein antibodies by using synthetic vaccines containing repetitive peptides from the CS protein. Recent evidence has suggested that potential vaccines containing liposomes with recombinant or peptide sporozoite antigens can induce titers of antibodies in rabbits and monkeys to CS protein that are substantially higher than those obtained with current vaccines (Alving et al, 1986; Richards et al, 1988).

DEVELOPMENT OF A LIPOSOMAL SPOROZOITE VACCINE

The initial studies in our program were performed with liposomes containing an antigen consisting of bovine serum albumin (BSA) conjugated to 16 mer peptides derived from the structure of the repetitive region of the CS protein. The 16 mer peptides contained 4 asparagine-alanine-asparagine-proline tetrapeptides. Murine antibodies against this antigen were obtained by utilizing Freund's complete adjuvant during the immunization procedure. The antibodies that were produced by this method blocked invasion of human hepatoma cells <u>in vitro</u> (Ballou et al, 1985).

Upon injection of the peptide-BSA conjugate by itself into rabbits (without adjuvant or liposomes), little or no immunogenic activity was detected as measured by ELISA (Fig. 1). In contrast, liposoomes containing the peptide-BSA conjugate induced a potent humoral immune response, and the activity was increased further when the liposomes contained lipid A (Fig. 1).

When the sporozoite is injected by the bite of a mosquito into the blood of a host animal it enters the liver within a very short period, usually within a few minutes. It is currently presumed that it is only during this brief period when the sporozoite resides in the blood that the antibodies can gain access to the sporozoite and prevent continuation of the malaria infection. Therefore a major goal of a sporozoite immunization scheme is to maintain a high titer of antibodies over a long period of time, preferably for at least 6 to 12 months. As shown in Fig. 2, after immunization with liposomes containing peptide-BSA and lipid A at 0, 4 and 28 weeks, activity was still present at maximal levels more than 55 weeks after the primary immunization.

Further work in this program has utilized the same recombinant protein antigen ($R32tet_{32}$) that was employed in the original FSV-1 clinical trial (Ballou et al, 1987). The $R32tet_{32}$ antigen contains 32 tetrapeptide repeats from the CS protein, and also contains a tail consisting of 32 amino acids derived from the tetracycline resistance gene of the cloning vector (Young et al, 1985; Wirtz et al, 1987). In a series of experiments with $R32tet_{32}$ involving both rabbits and rhesus monkeys we have now reached the conclusion that among the combinations that we have tested, the most potent immunogenic formulation consists of an alum-liposome mixture in which the liposomes

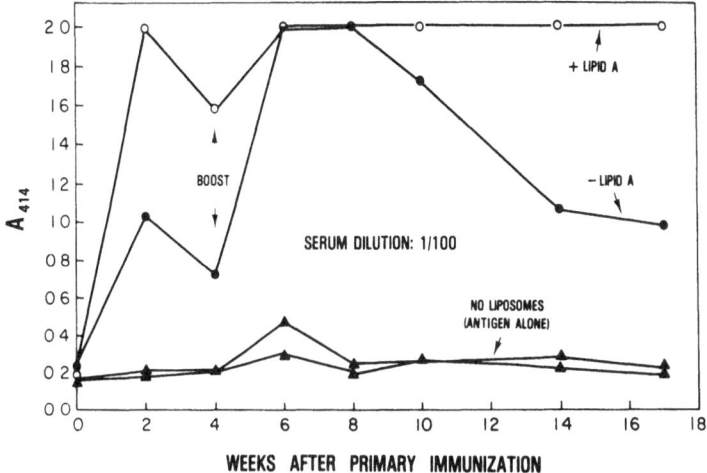

Fig. 1. Immune response qgainst malaria sporozoite peptide-BSA conjugate in rabbits. Antibody activity was detected by enzyme-linked immunosorbent assay (ELISA) with absorbance at 414 nm using a 1/100 dilution of serum. The maximum measurable response (A414) was 2.0. (See Alving et al, 1986 for details of immunization and assay.)

contain both the R32tet$_{32}$ and lipid A [abbreviated as L(Ag + LA) + alum] (Richards et al, 1988). An example of results that were typical both for rabbits and monkeys is shown in Fig. 3. The highest response occurred with L(Ag + LA) + alum and the lowest response occurred with Ag (R32tet$_{32}$) alone. The vaccine containing only alum (Ag + alum) gave immunogenic effects at an intermediate level. In monkey experiments, direct comparison of the liposomal formulation with the actual vaccine that was used in the FSV-1 human trial confirmed the same relative ranking of activities. Depending on the dose of vaccine used, and on the date of observation, in monkeys the optimal liposomal vaccine was 15 times to 500 times more potent as an immunogenic formulation than the vaccine used in the FSV-1 human trial.

In summary, it appears from the above results that the liposomes have an ability to serve as an effective carrier for synthetic antigens that are derived from the structure of the CS protein on the surface of the P. falciparum sporozoite. The liposomes also have an inherent adjuvant activity; that is, they induce a higher immune response to synthetic peptides or proteins from sporozoites than would otherwise be observed by other methods of immunization. It is reasonable to presume that the adjuvant activity of liposomes may be due to a combination of factors. These factors may include a focussed and enhanced delivery of the antigen to an APC (macrophage) and protection of the antigen from metabolic destruction at other sites in the body that do not participate in the immune response.

There is another possible mechanism by which the liposome might serve as an adjuvant, namely by overcoming immunosuppression that might otherwise be induced by injection of the antigen. Careful inspection of Figs. 2 and 3 reveals that in each case the animals injected with antigen alone (without liposomes) only expressed antibody activity starting between 35 and 40 weeks after initial immunization. It is unlikely that this antibody response was due to a recent immunization dose since the previous immunization dose was either at 28 weeks (Fig. 2) or 4 weeks (Fig. 3) after initial immunization. A similar pattern has also been observed in a preliminary experiment with monkeys. It is possible that initial immunization with either the peptide (Fig. 2) or the recombinant protein (Fig. 3) induces a normal immune response but also simultaneously induces proliferation of suppressor lympho-

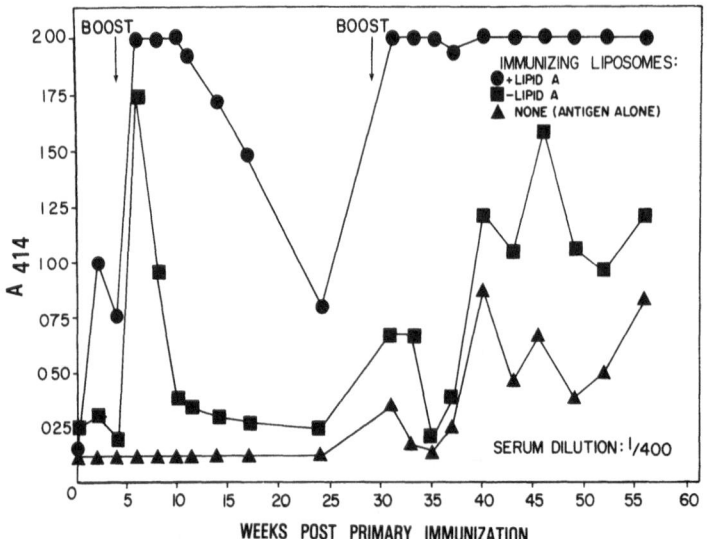

Fig. 2. Long term antibody activity after immunizing rabbits with peptide-BSA conjugate. The rabbits were the same as those in Fig. 1, but a 1/400 dilution of serum was used instead of 1/1000.

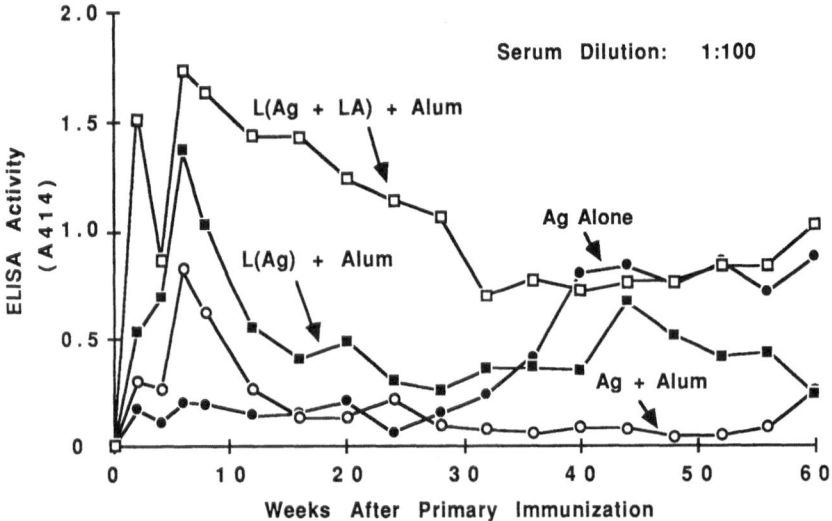

Fig. 3. Effects of alum, lipid A, and liposomes on mean antibody activities of rabbits injected at 0 and 4 weeks. Each point represents the mean ELISA antibody response after subtraction of the preimmune value of a serum dilution of 1:100 for six (Ag + Alum) or four (all other groups) rabbits. Ag alone, free R32tet$_{32}$; Ag + Alum, 32tet$_{32}$ absorbed with alum; L(Ag) + alum, alum absorbate of R32tet$_{32}$ encapsulated in liposomes lacking lipid A; L(Ag + LA) + alum, alum absorbate of R32tet$_{32}$ encapsulated in liposomes containing lipid A. (From Richards et al, 1988.)

cytes. The spontaneous disappearance of the suppressor effect 35 to 40 weeks later may explain the unexpected expression of antibody activity that was observed at that time.

The unique way in which liposomal antigens are handled, by rapid entry into a phagocytic APC, may cause a bypass of the normal mechanism that the body uses for induction of suppressor lymphocytes. Although this suppressor cell concept is still speculative, it is compatible with immunosuppression against $R32tet_{32}$ that is seen in the course of natural infections with malaria in humans (Webster et al, 1987), and it could help to explain the great difficulty that has been observed in producing high titers of antibodies to synthetic sporozoite antigens in humans.

REFERENCES

Allen, T.M. and Chonn, A., 1987, Large unilamellar liposomes with low uptake into the reticuloendothelial system, FEBS.Lett., 223:42.

Alving, C.R., 1989, Macrophages as targets for delivery of liposome-encapsulated antimicrobial drugs, Advanced Drug Deliv.Rev. (in press).

Alving, C.R., 1987, Liposomes as carriers of vaccines, in: "Liposomes from Biophysics to Therapeutics", M.J. Ostro, ed., Marcel Dekker, Inc., New York.

Alving, C.R., Richards, R.L., Moss, J., Alving, L.I., Clements, J.D., Shiba, T., Kotani, S., Wirtz, R.A. and Hockmeyer, W.T., 1986, Effectiveness of liposomes as potential carriers of vaccines: applications to cholera toxin and human malaria sporozoite vaccine, Vaccine, 4:166.

Alving, C.R., 1983, Delivery of liposome-encapsulated drugs to macrophages, Pharmacol.Therap., 22:407.

Ballou, W.R., Rothbard, J., Wirtz, R.A., Gordon, D.M., Williams, J.S., Gore, R.W., Schneider, I., Hollingdale, M.R., Beaudoin, R.L., Maloy, W.L., Miller, L.H. and Hockmeyer, W.T., 1985, Immunogenicity of synthetic peptides from circumsporozoite protein of Plasmodium falciparum, Science, 228:996.

Ballou, W.R., Hoffman, S.L., Sherwood, J.A., Hollingdale, M.R., Neva, F.A., Hockmeyer, W.T., Gordon, D.M., Schneider, I., Wirtz, R.A., Young, J.F., Wasserman, G.F., Reeve, P., Diggs, C.L. and Chulay, J.D., 1987, Safety and efficacy of a recombinant DNA Plasmodium falciparum sporozoite vaccine, Lancet i (June 6):1277.

Beatty, J.D., Beatty, B.G., Paraskevas, F. and Froese, E., 1984, Liposomes as immune adjuvants: T cell dependence, Surgery, 96:345.

Clyde, D.F., McCarthy, V.C., Miller, R.M. and Woodward, W.E., 1975, Immunization of man against falciparum and vivax malaria by use of attenuated sporozoites, Am.J.Trop.Med.Hyg., 24:397.

Dame, J.B., Williams, J.L., McCutchan, T.F., Weber, J.L., Wirtz, R.A., Hockmeyer, W.T. Maloy, W.L., Haynes, J.D., Schneider, I., Roberts, D., Sanders, G.S., Reddy, E.P., Diggs, C.L. and Miller, L.H., 1984, Structure of the gene encoding the immunodominant surface antigen on the sporozoite of the human malaria parasite Plasmodium falciparum, Science, 225:593.

Egan, J.E., Weber, J.L., Ballou, W.R., Hollingdale, M.R., Majarian, W.R., Gordon, D.M., Maloy, W.L., Hoffman, S.L., Wirtz, R.A., Schneider, I., Woollett, G.R., Young, J.F. and Hockmeyer, W.T., 1987, Efficacy of murine malaria sporozoite vaccines: Implications for human vaccine development, Science, 236:453.

Fidler, I.J., 1985, Macrophages and metastasis-A biological approach to cancer therapy: Presidential address, Canc.Res., 45:4714.

Gabizon, A. and Papahadjopoulos, D., 1988, Liposome formulations with prolonged circulation time in blood and enhanced uptake by tumors, Proc.Natl.Acad.Sci.U.S.A., 85:6949.

Gregoriadis, G., Kirby, C., Large, P., Meehan, A. and Senior, J., 1982, Targeting of liposomes: Study of influencing factors, in: "Targeting of Drugs", G. Gregoriadis, J. Senior and A. Trouet, eds., Plenum, New York.

Herrington, D.A., Clyde, D.F., Losonsky, G., Cortesia, M., Murphy, J.R., Davis, J., Baqar, S., Felix, A.M., Heimer, E.P., Gillessen, D., Nardin, E., Nussenzweig, R.S., Nussenzweig, V., Hollingdale, M.R. and Levine, M.R., 1987, Safety and immunogenicity in man of a synthetic peptide malaria vaccine against Plasmodium falciparum sporozoites, Nature, 328:257.

Hoffman, S.L., Wistar, Jr., R., Ballou, W.R., Hollingdale, M.R., Wirtz, R.A., Schneider, I., Marwoto, H.A. and Hockmeyer, W.T., 1986, Immunity to malaria and naturally acquired antibodies to the circumsporozoite protein of Plasmodium falciparum, New Eng.J.Med., 315:601.

Hoffman, S.L., Oster, C.N., Plowe, C.V., Woollett, G.R., Beier, J.C., Chulay, J.D., Wirtz, R.A., Hollingdale, M.R. and Mugambi, M., 1987, Naturally acquired antibodies to sporozoites do not prevent malaria: Vaccine development implications, Science, 237:639

Hwang, K.J., Luk, K.S. and Beaumier, P.L., 1980, Hepatic uptake and degradation of unilamellar sphingomyelin/cholesterol liposomes: A kinetic study, Proc.Natl.Acad.Sci., 77:4030.

Nussenzweig, R., Vanderberg, J. and Most, H., 1969, Protective immunity produced by the injection of X-irradiated sporozoites of Plasmodium berghei. IV. Dose response, specificity and humoral immunity, Milit. Med., 134:1176.

Papahadjopoulos, D. and Gabizon, A., 1987, Targeting of liposomes to tumor cells in vitro, Ann.N.Y.Acad.Sci., 507:64.

Potocnjak, P., Yoshida, N., Nussenzweig, R.S. and Nussenzweig, V., 1980, Monovalent fragments (Fab) of monoclonal antibodies to a sporozoite surface antigen (Pb44) protect mice against malarial infection, J.Exp.Med., 151:1504.

Richards, R.L., Hayre, M.D., Hockmeyer, W.T. and Alving, C.R., 1988, Liposomes, lipid A, and aluminum hydroxide enhance the immune response to a synthetic malaria sporozoite antigen, Infect.Immun., 56:682.

Rieckmann, K.H., Beaudoin, R.L., Cassells, J.S. and Sell, K.W., 1979, Use of attenuated sporozoites in the immunization of human volunteers against falciparum malaria, Bull.W.H.O., 57(Suppl. 1):261.

Shek, P.N. and Lukovich, S., 1982, The role of macrophages in promoting the antibody response mediated by liposome-associated antigens, Immunol. Lett., 5:305.

Steinman, R.M. and Nussenzweig, M.C., 1980, Dendritic cells: Features and functions, Immunol.Rev., 53:129.

Swenson, C.E., Popescu, M.C. and Ginsberg, R.S., 1988, Preparation and use of liposomes in the treatment of microbial infections, C.R.C.Crit. Rev.Microbiol., 15(Suppl. 1):S1.

Tew, J.G., Phipps, R.P. and Mandel, T.E., 1980, The maintenance and regulation of the humoral immune response: persisting antigen and the role of follicular antigen-binding dendritic cells as accessory cells, Immunol.Rev., 53:175.

Unanue, E.R., 1984, Antigen-presenting function of the macrophage, Ann.Rev. Immunol., 2:395.

Unanue, E.R. and Allen, P.M., 1987, The basis for the immunoregulatory role of macrophages and other accessory cells, Science, 236:551.

Webster, H.K., Boudreau, E.F., Pang, L.W., Permpanich, B., Sookto, P. and Wirtz, R.A., 1987, Development of immunity in natural Plasmodium falciparum malaria: Antibodies to the falciparum sporozoite vaccine 1 antigen (R32tet$_{32}$), J.Clin.Microbiol., 25:1002.

Wirtz, R.A., Ballou, W.R., Schneider, I., Chedid, L., Gross, M.J., Young, J.F., Hollingdale, M., Diggs, C.L. and Hockmeyer, W.T., 1987, Plasmodium falciparum: Immunogenicity of circumsporozoite protein constructs produced in Escherichia coli, Exp.Parasitol., 63:166.

Young, J.F., Hockmeyer, W.T., Gross, M., Ballou, W.R., Wirtz, R.A., Trosper, J.H., Beaudoin, R.L., Hollingdale, M.R., Miller, L.H., Diggs, C.L. and Rosenberg, M., 1985, Expression of Plasmodium falciparum circumsporozoite proteins in Escherichia coli for potential use in a human malaria vaccine, Science, 228:958.

Zavala, F., Tam, J.P., Hollingdale, M.R., Cochrane, A.H., Quakyi, I., Nussenzweig, R.S. and Nussenzweig, V., 1985, Rationale for development of a synthetic vaccine against Plasmodium falciparum malaria, Science, 228:1436.

NONIONIC BLOCK COPOLYMER SURFACTANTS AS IMMUNOLOGICAL ADJUVANTS: MECHANISMS OF ACTION AND NOVEL FORMULATIONS

Robert L. Hunter*, Beth Bennett*, Devery Howerton*
Steve Buynitzky** and Irene J. Check*

*Department of Pathology, Emory University, Atlanta, Georgia
30322 and **Cytrx Corporation, Norcross, Georgia 30368, USA

INTRODUCTION

Modern molecular biology has provided us with the ability to produce custom tailored antigens with an ease undreamed of a few years ago. However, experience has also shown that most nonreplicating purified subunit antigens are weak immunogens that will require immunopotentiation to become the hoped for effective vaccines of the future. The dominant paradigm of modern immunology is that immunogenicity and immune responses are controlled by specific interactions between various ligands and their receptors. From our perspective, producing effective adjuvants for subunit vaccines will require looking beyond this paradigm to a consideration of nonspecific physicochemical factors which subtly but powerfully influence immune responses.

Our work, like that of many investigators in the field of vaccine adjuvants, began by reviewing the literature on Mycobacteria. Injections of live Mycobacteria into animals induces immune responses against tubercule proteins which differ in class and intensity from those which are induced by injection of purified tubercule proteins themselves. Obviously, investigations of the components of the organisms in addition to the purified protein were necessary to understand the immunogenicity of Mycobacteria. Two observations from the literature provided the opening for our investigation of this problem. First, Ribi et al (1986) demonstrated that BCG cell walls prepared in an oil emulsion reproduced many of the immunologic effects of live mycobacteria. Secondly, we found that protein antigens covalently conjugated with fatty accids selectively induced delayed type hypersensitivity (Coon and Hunter, 1973). Several studies suggested that the critical property of the lipid-conjugated proteins which determined immunogenicity was their surface activity.

Reviewers had frequently noted that many adjuvants are surface active agents while most common surface active agents are not adjuvants (Waksman, 1979; Hunter, 1980; Warren et al, 1986). We hypothesized that surface activity was a key factor in adjuvant activity and began a search for potentially informative experimental models. The dominant contrahypothesis was that each adjuvant has unique reactive sites which interact with specific cell receptors. The synthetic block copolymers turned out to be uniquely suited to evaluate relationships between surface activity and adjuvant activity. The block copolymers are composed of hydrophilic blocks or chains of

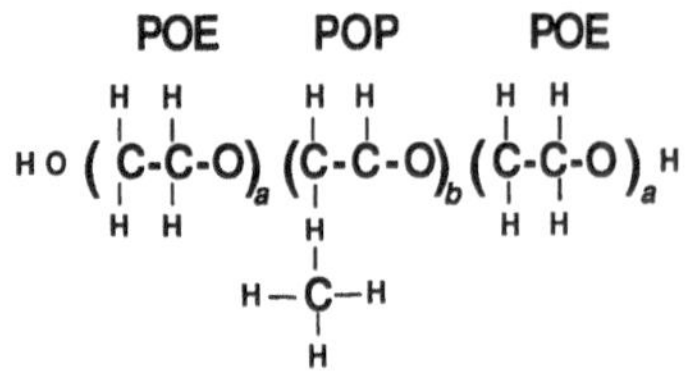

POE   POP   POE

$$HO\left(\underset{\underset{H}{|}\,\underset{H}{|}}{\overset{\overset{H}{|}\,\overset{H}{|}}{C\text{-}C\text{-}O}}\right)_a \left(\underset{\underset{H}{|}}{\overset{\overset{H}{|}\,\overset{H}{|}}{C\text{-}C\text{-}O}}\right)_b \left(\underset{\underset{H}{|}\,\underset{H}{|}}{\overset{\overset{H}{|}\,\overset{H}{|}}{C\text{-}C\text{-}O}}\right)_a H$$

$$\underset{\underset{H}{|}}{\overset{\overset{H}{|}}{H\text{-}C\text{-}H}}$$

Fig. 1.   Structure of a triblock copolymer. a and b are integers which are
         varied to produce a series of surface active copolymers composed of
         polyoxethylene (POE) and polyoxypropylene (POP). The structure of
         the copolymers discussed in this paper are described in more detail
         (Hunter and Bennett, 1984 and 1986; Atkinson et al, 1988).

polyoxyethylene (POE) and hydrophobic blocks of polyoxypropylene (POP)
(Hunter et al, 1981) (Fig. 1). They are not degraded in vivo but are eventually excreted intact. By varying the lengths and arrangement of the two
types of chains, a series of surface active agents which span nearly the
entire range of physicochemical properties of known nonionic surfactants was
produced. Since all of these preparations are composed of identical units,
it is unlikely that any have specific reactive sites which are not present
on all others. Consequently, differences in biologic activity among them
are more likely, due to variation in physical chemistry.

What at first seemed like a relatively straightforward problem has now
become rather complex. The first group of adjuvant active copolymers discovered consisted of a large central hydrophobic chain of POP flanked by two
short chains of POE. In time, it was determined that the ability to mediate
adhesion among proteins and surfaces was critical to the adjuvant activity
of these copolymers. Next, we learned that certain members of a different
series of copolymers which are ionophores for monovalent cations were also
powerful adjuvants when used in appropriate protocols. The copolymers had
physicochemical and biologic activities quite different from those of the
adhesive adjuvant copolymers. We proposed that their adjuvant and ionophore
activities were related. Next, we learned that several of the copolymers
could induce the expression of membrane class II (Ia) antigen. This was
associated with a change in membrane fluidity. Finally, subunit vaccines
require carriers and a vehicle in addition to adjuvants. Salmonella flagella was evaluated as a particularly promising carrier for low molecular
weight antigens. We conducted studies which demonstrated the potential of
copolymer adjuvants to produce synergistic responses with each other and
with nontoxic lipid A derivatives, and with quartz in appropriate formulations. Representative observations from each of these areas of study are
reviewed in this paper.

ADHESION MECHANISM OF ADJUVANT ACTIVITY

In initial studies, one of the copolymers, L121, proved to be a powerful adjuvant for inducing antibody formation to BSA when injected in an oil-
in-water emulsion with 5% oil (Hunter et al, 1981). A closely related
copolymer, L101, was less effective in inducing antibody but more effective
in stimulating DTH. Most other copolymers were not adjuvants at all.
Subsequently, studies with L121 have been carried out with many antigens.
In most situations, the antibody response was at least equivalent to that of
Freund's complete adjuvant if the emulsions were properly prepared. It
peaks at approximately six weeks after a single injection and persists for
several months (Hunter et al, 1981; Hunter and Bennett, 1984 and 1986;
Snippe et al, 1981; unpublished).

Studies were carried out with many other block copolymers in an effort to identify the critical properties responsible for adjuvant activity. We found that each of the copolymers which were adjuvants in this model had a hydrophile-lipophile balance (HLB) of less than 2 (Hunter et al, 1981). The HLB is an empiric measure of the relative strengths of the hydrophilic and hydrophobic portions of the molecule. Agents with HLBs of less than 2 are poorly soluble and are known as spreading agents. Unlike detergents, they are completely unable to disperse hydrophobic deposits. They bind avidly to hydrophobic surfaces and promote the adhesion of other molecules to them. We proposed that this ability to promote adhesion of antigen molecules to the surface of oil droplets was an essential feature for the adjuvant activity of these compounds. Ongoing studies have suggested that an HLB of less than 2 is a necessary, but not sufficient, criterion for adjuvant activity of surface active agents. This correlation extends to several classes of agents in addition to the block copolymers.

Analysis of the structure-function relationship of existing copolymer adjuvants suggested a rationale for the synthesis of more effective preparations. The trend line for adjuvant activity points toward copolymers larger than any available preparations. We predicted that copolymers with larger POP hydrophobes would be even more effective adjuvants. Preparations L141 and L1801.5 with structures identical to that of L121 except with proportionately larger hydrophobic and hydrophilic chains were synthesized. Copolymer L141 contained a POP chain of an average molecular weight of 4800 and L181.5 a POP chain of 5100 daltons. They each contained approximately 10% by weight POE. Mice were injected with 1 mg of each copolymer with trinitrophenyl conjugated hen egg albumin (TNP_{10}-HEA) in an oil-in-water emulsion (Fig. 2). Copolymer L141 was the more effective adjuvant preparation. Its structure and physicochemical properties suggest that its mechanism of action is similar to that of the other adhesive adjuvants.

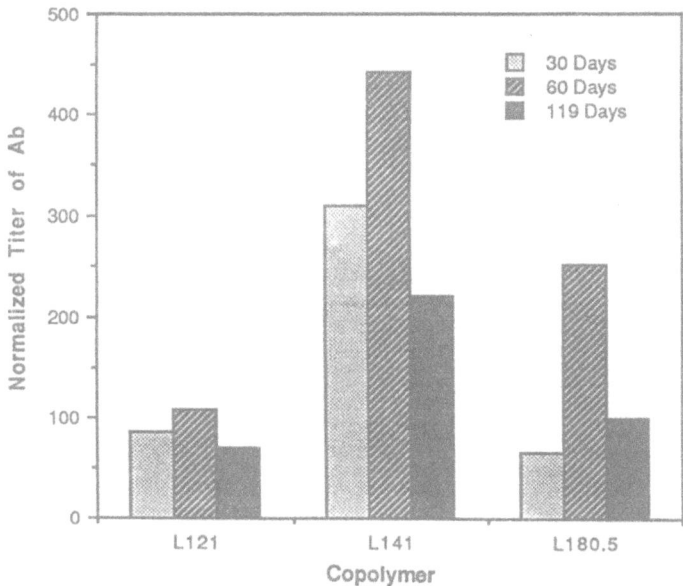

Fig. 2. Comparison of copolymer adjuvants administered with soluble antigen. Female CD-1 mice, 5 or 6 per treatment, were injected in the footpad with 50 μg aqueous TNP_{10}-HEA and 1.0 mg copolymer in a 2% oil-in-water emulsion. Samples of serum were taken at 1, 2 and 4 months and assayed for IgG antibody to the TNP hapten by solid phase ELISA using TNP-BSA as the target antigen.

Complement

Antigen

Copolymer

Fig. 3.   Proposed mechanism of adhesive coplymer adjuvants. The adhesive
copolymers form a high energy (low surface tension) interface with
aqueous media which binds antigen molecules to the surface of oil
drops in close proximity to activated mediators such as complement.
The copolymers need to be large and insoluble yet able to fold to
form hydrophilic adhesive surfaces.

Extensive physicochemical, electron microscopic, and biologic studies
were carried out in an effort to identify the critical features of the block
copolymers responsible for adjuvant activity (Hunter and Bennett, 1984 and
1986). The availability of a large series of chemically related polymers
facilitated these studies. Since they are poorly soluble, hydrophobic surfactants, they form macroscopic structures when placed in saline. We found
that each of the copolymers which were adjuvants formed fibers when suspended in saline while most of those which were not adjuvants formed spherical
masses resembling oil droplets (Hunter and Bennett, 1986). Subsequent studies demonstrated that two properties were responsible for the formation of
fibers. First, the agents which formed fibers were able to form hydrophilic
surfaces with a low surface tension. The hydrophobic portions of the adjuvants were able to fold and pack in such a way that the surface was completely covered by hydrophilic POE moieties (Fig. 3). If the hydrophobic POP
chains were either too short to make a complete fold or were sterically inhibited, then hydrophobic domains remained exposed at the surface resulting
in increased surface tension and forcing the material into a spherical
shape. Secondly, the formation of fibers implied an underlying tendency
towards linear organization. This was verified with electron microscopy. A
consistent property which distinguished the adjuvant copolymers was an ability to form organized hydrophilic adhesive surfaces.

The adhesive surfaces of the adjuvant copolymers bound protein antigens in vitro and promoted the retention of antigen in association with oil
droplets in vivo. The antigen bound to such surfaces in vitro was held in a
conformation that made it available for interaction with antibody to a much
greater extent than antigen bound to polystyrene plastic (Hunter and
Bennett, 1986). In addition, the fibrous surfaces of the adjuvant copolymers activated complement, induced chemotaxis and enhanced phagocytosis
without the necessity for complement (Hunter and Bennett, 1984). We proposed that the ability to concentrate antigen on a surface and to activate
host mediator systems such as complement and macrophages are essential components of adjuvant activity of the adhesive copolymers.

IONOPHORES AND ADJUVANT ACTIVITY

The reverse octablock copolymers, T150R1 and T130R2 differ in structure
in two important ways from the adhesive adjuvants L101, L121 and L141
(Atkinson et al, 1988). First, they have a tetrafunctional ethylene diamine
initiator and a four chain rather than a two chain structure. Secondly, the
POE chains are located in the center of the molecule flanked by hydrophobic
chains of POP. The reverse octablock copolymers were found to have distinctive properties. They are ionophore-selective for monovalent cations
(Atkinson et al, 1988). They induce release of histamine from mast cells

and basophils by an energy dependent process (Atkinson et al, a and b). They induce differentiation of HL60 cells along the monocytoid pathway (Al-Qawasmeh et al, 1988). In certain protocols, they suppress immune responses such as experimental allergic encephalomyelitis (Mezrow et al, 1987). In other protocols, they are adjuvants. We proposed that the distinctive activities of the reverse octablock copolymers are related to their activity as ionophores.

Cation fluxes are essential regulatory mechanisms in virtually all types of excitable cells. It seems highly likely that the ability of the reverse octablock copolymers to serve as ionophores could affect the excitability of lymphoid cells and that this might increase immune responses under appropriate circumstances.

The ionophore activity of the copolymer was thoroughly characterized in a red blood cell model (Atkinson et al, 1988). Certain copolymers were found to increase the influx of Na^+ and the efflux of K^+ from human erythrocytes. They were, however, ineffective at promoting the transport of Ca^{++}. The magnitude of the ion fluxes induced by the copolymers correlated with their efficacy in stimulating inflammation following injection into the footpads of animals. These compounds were also found to induce electrical conductance increases in planar lipid bilayers in a nonvoltage dependent and nonstepwise manner. The accumulated data suggest that aqueous monomers of these surface active agents partition into the membrane, where they facilitate the conductive movement of monovalent cations by means of a carrier type mechanism.

Several experiments to evaluate the adjuvant activity of the reverse octablock copolymers in oil-in-water emulsions with BSA or TNP_{10}-HEA in protocols similar to those used with adhesive adjuvants demonstrated that they had little or no ability to enhance immune responses (Hunter and Bennett, 1984 and 1986). However, one such copolymer, when injected the particulate antigen, TNP-flagella, effectively increased immune responses. This copolymer T150R1, is the largest and least soluble of the ionophore copolymers (Fig. 4).

Zigterman et al recently demonstrated that other block copolymers which are not adjuvants for soluble proteins in oil-in-water emulsions may be effective adjuvants for sheep red blood cells or other particulate antigens (Zigterman et al, 1987). It seems likely that the essential function of adhesive copolymers of holding soluble antigen molecules together in a particle is not necessary for particulate antigens. Consequently, other types of adjuvant activity may become effective. The most potent ionophore in our studies, copolymer T130R2, is not nearly as effective an adjuvant as copolymer T150R1. It is also considerably more toxic. It seems likely that differences in toxicity and the ability to partition to other compartments of the body underlie the differences in biologic activity among the ionophore copolymers.

INDUCTION OF IA AND MEMBRANE FLUIDITY

The modulation of expression of class II molecules of the histocompatibility complex (Ia) on the macrophage membrane plays a central role in immune regulation (Unanue, 1984). BCG organisms and immunoadjuvants such as beryllium sulfate, complete Freund's adjuvant and lipopolysaccharide induce Ia on macrophages. It has been proposed that the induction of Ia contributes to the adjuvant activity of these materials.

We tested twelve polymers with varying degress of adjuvant activity and a range of physicochemical properties for their ability to induce macrophage

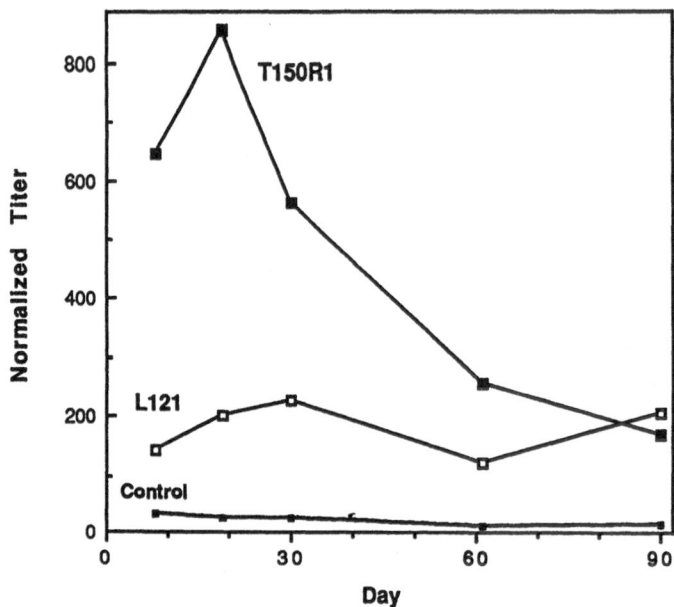

Fig. 4.   Copolymer adjuvants with flagella-bound antigen. Mice were in-
          jected in the footpad with 25 μg TNP-flagella (5 TNP per flagella
          in monomer (MW 40,000), 4 mice/group) with or without 1.0 mg co-
          polymer in emulsions of 2% squalane in 0.1% Tween-80 saline. Anti-
          hapten IgG was assayed periodically for 90 days by ELISA. Titers
          were normalized by comparison with a standard antiserum to facili-
          tate comparison among days.

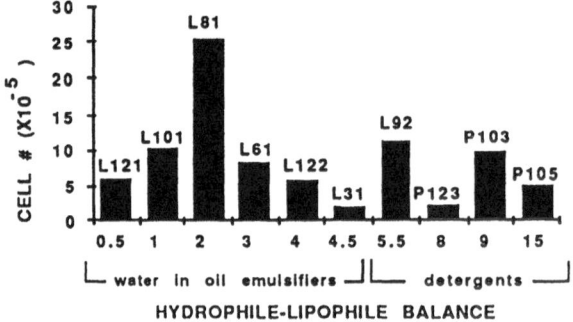

Fig. 5.   Induction of macrophage Ia vs. hydrophile-lipophile balance of co-
          polymers. C3H/HeN mice were injected i.p. with 2.5 mg copolymers,
          100 μg of Con A, or 1 ml of phosphate buffer saline. Peritoneal
          exudate cells were harvested 5 days later and the %Ia positive
          macrophages were determined by flow cytometry. The numbers of Ia
          positive macrophages were calculated on the total cell numbers and
          percentages of macrophages delineated by flow cytometry. The
          results shown are the mean values ± S.E.M. obtained from 3 to 6
          mice.

Ia expression in vivo (Howerton et al, 1987; Howerton, 1988). Mice were
injected intraperitoneally with copolymers and three to seven days later the
peritoneal exudate cells were assayed for Ia expression of immunofluores-
cence with flow cytometery. The copolymers had a range of Ia-inducing activity from no induction to a four-fold increase in the percentages of Ia
positive macrophages (Fig. 5). Comparisons of the structures of the co-

polymers with their ability to induce the expression of Ia revealed that activity centered around a POP molecular weight of 2250, 10% POE, and an HLB of 2, the properties of copolymer L81. Either increasing or decreasing these physicochemical parameters resulted in decreased Ia-inducing activity. Other copolymers with an HLB of 2 did not induce Ia to the same extent as L81. Those copolymers which have adjuvant activity in either of two published models induced Ia. The degree of Ia expression did not correlate with their potency as adjuvants. Therefore, Ia induction may contribute to, but is not sufficient for, the adjuvant activity of these compounds.

Macrophages induced by L81 in vivo had an 18-fold increase in Ia biosynthetic capacity. They were ten-fold more effective than normal macrophages in presenting antigen to a T cell hybridoma. L81 did not directly induce Ia expression by the WEHI-3 murine myelomonocytic cell line in vitro but did enhance Ia induction by a suboptimal concentration of gamma interferon up to ten-fold. In addition, L81 inhibited the loss of Ia from gamma interferon-induced Ia positive cells. Therefore, L81 may act by enhancing the effects of gamma interferon and delaying the turnover of Ia. The cells cultured with L81 also showed increased cell membrane fluidity as assessed by fluorescence polarization.

SALMONELLA FLAGELLA AS A VACCINE CARRIER

The adjuvant activity of various block polymers varies with the type of antigen used. With low molecular weight antigens, one potentially has the option of choosing a carrier in addition to the adjuvant and vehicle. Salmonella flagella is an unusually potent and nontoxic immunogen and it is readily purified. It does not produce a depot of antigen at the site of injection as do oil emulsions or alum preparations. It remains for long periods of time only on dendritic reticulum cells within germinal centers. There is little local inflammation at the site of injection. We propose that the prolonged retention of flagella within germinal centers is responsible for the prolonged antibody formation.

Flagella can be selected from strains of organisms which rarely induce natural antibody in humans. We carried out studies to determine if Salmonella flagella would be an effective carrier for the TNP hapten, and to investigate the effects of block copolymer adjuvants.

The TNP hapten was covalently bound to polymeric Salmonella flagella and administered without adjuvant to mice one time only via subcutaneous injection (Buynitzky and Hunter, 1988). Flagella proved to be a much more effective carrier than HEA and was significantloy better than keyhole limpet hemocyanin (Fig. 6). Titers of IgG developed in eight days and persisted over one year after a single injection of TNP-flagella in saline. The response was dose-dependent over the range of 4-50 μg of flagella. A second injection at 14 days produced a large increase in titer. The antibody response was augmented by block copolymers. L121 produced an increase in titer which was sustained for at least 90 days; T150R1 produced a greater but shorter lived response. Salmonella flagella carrier combined with block copolymer adjuvants appears to be a superior preparation for inducing antibody to low molecular weight antigens. It produces sustained high titer responses with low toxicity.

FORMULATIONS WITH ADHESIVE COPOLYMERS AND IONOPHORE COPOLYMERS OR LIPID A DERIVATIVES

Since the block copolymer adjuvants appear to act via distinct mechanisms, there is a possibility of incorporating them in more complex formul-

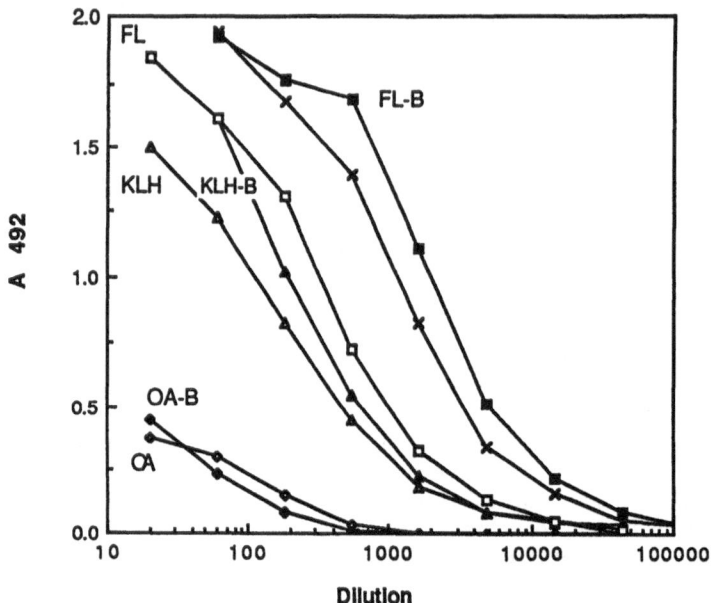

Fig. 6. Comparison of carriers for the TNP hapten. Thirty mice, 10 per TNP conjugated carrier, were given subcutaneous footpad injections of 25 µg TNP-flagella (FL) or the equivalent amount of hapten bound to keyhole limpet hemocyanin (KLH) or hen egg albumin (OA). (Doses were normalized such that each animal received 3.4 nmoles TNP in 40 µl PBS.) After 14 days, half the mice in each group were boosted with a repeat of the same immnization given on day 0 (designated as B). Antibody was assayed on days 14, 28, 90 and 196. The day 90 titers are shown as adsorbance in an ELISA assay.

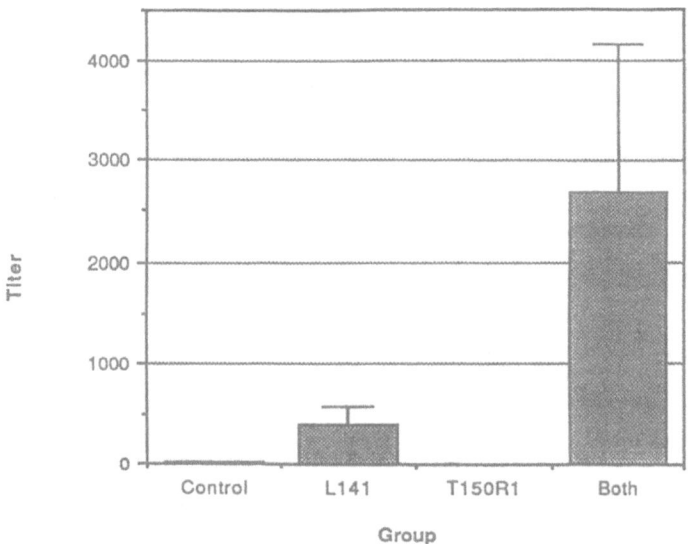

Fig. 7. Adjuvant effects of copolyers with lyophilized antigen in oil-in-water emulsions of 2% squalane. Oil and copolymer were mixed with dry TNP_{10}-HEA and subsequently homogenized in PBS, pH 7.4, with 0.2% Tween-80. Mice were given 50 µl divided between both rear footpads. The doses per animal were 50 µg antigen, 0.6 mg L141, and 0.1mg T150R1. Titers were measured at day 30 by ELISA as shown.

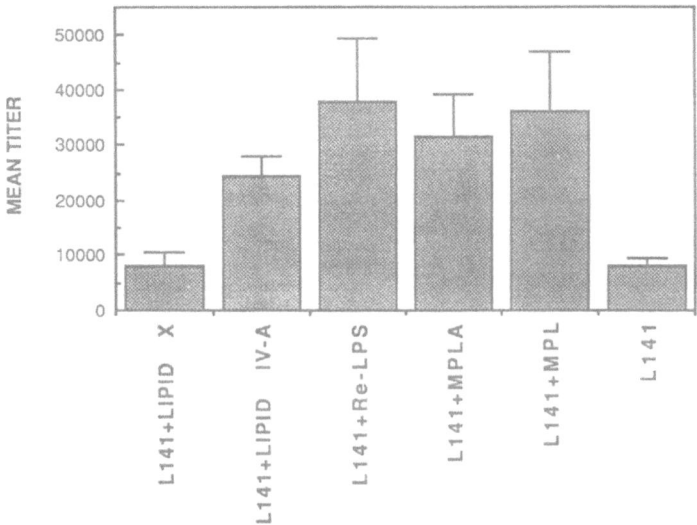

Fig. 8. Formulations of copolymers with lipid A derivatives. Groups of six ICR outbred female 7-week-old mice were injected with an oil-in water emulsion of 2% squalane, 50 µg TNP_{10}-HEA, 1 mg L141, and 100 µg of each of the test adjuvants lipid X, lipid IV-A, ReLPS and MPL. Lipid X is the smallest subunit of lipid A and lipid IV-A is an intermediate precursor of lipid A. Monophosphoryl lipid A (MPLA) is a nontoxic derivative of lipid A. MPL is monophosphoryl lipid A from Ribi Immunochem Research, Inc. ReLPS, lipid IV-A, lipid X and MPLA were prepared as described (Takayama et al, 1983).

ations to optimize activity for particular applications. TNP_{10}-HEA was prepared in oil-in-water emulsions with 1.0 mg of the adhesive triblock copolymer L141. To this was added the ionophore copolymer T150R1. The combination of adhesive and ionophore copolymers produced a marked increase in antibody response over that of either alone (Fig. 7). In preliminary experiments, copolymer L141 was prepared in oil-in-water emulsions with TNP_{10}-HEA and a variety of lipid A derivatives including the Re mutant lipopolysaccharide (LPS) and monophosphoryl lipid A from two sources. In addition, two precursors of lipid A, lipid IV-A and lipid X were evaluated. The LPS and both lipid A preparations produced a striking increase in antibody response over that of L141 alone (Fig. 8). The lipid IV-A was less effective and the lipid X had no detectable activity. In these experiments, as in our previous work, the most effective preparations involve the incorporation of the protein antigen with the adhesive adjuvant and lipid A or ionophore moiety into the oil droplet. Addition of antigen to the aqueous phase of the emulsion produced much lower antibody responses.

Finally, experiments were carried out with Freund's type water-in-oil emulsions containing adhesive and ionophore copolymers in place of Mycobacteria. Several of the preparations produced much higher titers than Freund's complete adjuvant with a less severe and less persistent inflammation at the site of injection (Fig. 9).

DISCUSSION

Most surface active agents disperse hydrophobic deposits. This accounts for their activity as detergents and emulsifiers. The surface active adjuvants are neither detergents nor emulsifiers. They interact with hydro-

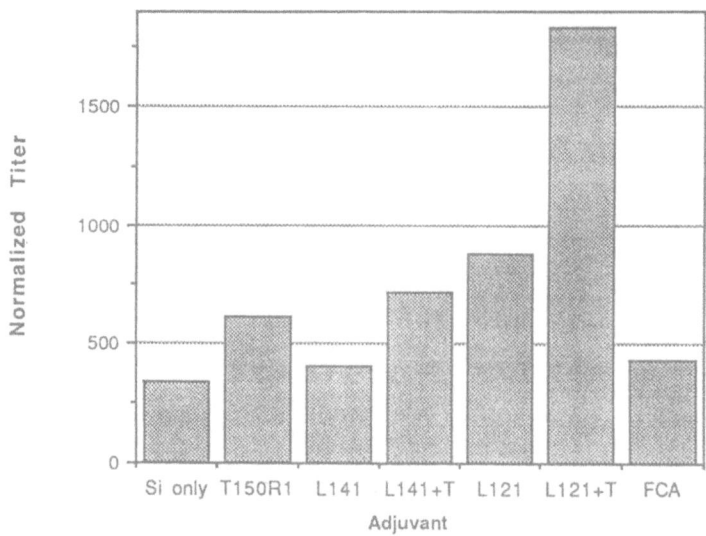

Fig. 9. Adjuvant effects of silica with selected copolymers. 50 µg lyophilized TNP_{10}-HEA (10.4 TNP per mole) was administered to mice in
50 µl doses split between both footpads. The dry antigen was mixed
with oil prior to emulsification with saline. The Freund's complete adjuvant (FCA) (Grand Island Biologicals) was made up as 60%
oil in saline with no additives. All other preparations were 60%
oil with 50 µl Span-80, 10µl Tween-80, and 15 mg silica (5 µm
Minusil) in 1.6 ml emulsion. Where used, copolymers were included
at a concentration of 0.6 mg (L121 and L141) or 0.1 mg (T150R1) per
mouse. The data is a composite from two experiments; 5-15 mice/
group. Group labels: Si only - Silica emulsion, T150R1 - Silica
emulsion + T150R1, L141 - Silica emulsion + 141, L141+T - Silica
emulsion + L141 + T150R1, L121 - Silica emulsion + L141, L121+T -
Silica emulsion + L121 + T150R1.

phobic deposits in diverse ways, but they do not disperse or solubilize
them. The block copolymers are chemically simple molecules composed of only
two subunits in a limited range of configurations. Nevertheless, they exhibit a very broad range of biologic activities. The large hydrophobic
preparations with the POE hydrophiles on the ends of the molecules are the
adhesive adjuvants which appear to act by binding antigen molecules and host
mediator proteins (complement) to the surface of oil drops and/or to the
surface of cells. Even within this group, there is variability in strength
of adjuvant activity and tendency to elicit humoral vs. cell-mediated immunity. The block copolymers act synergistically with other adjuvant moieties including other copolymers.

The ionophore copolymers are able to interact with cations because of
the cage-like four chain configuration of POE. They are able to stimulate
many types of excitable cells. This is not unexpected, since intracellular
concentrations of cations play regulatory roles in most cell processes. We
have not been able to detect differences among the copolymer ionophores in
ion selectivity even though they differ markedly in potency of ionophore
activity. We believe that the differences in ionophore activity and in the
various biologic responses derive from the ability of various copolymers to
partition among tissue and cell components in the body.

All of the copolymers studied to date which have adjuvant activity in
any of the systems used are able to induce expression of increased amounts

of Ia on macrophages. The mechanisms are not well understood, but there appears to be a correlation with an effect on fluidity of membrane lipids.

The ability to synthesize novel copolymers to optimize particular properties and the existence of at least three probable mechanisms of immunoadjuvant activity provides numerous opportunities for synthesizing preparations which will induce more effective immune responses.

Acknowledgements

We gratefully acknowledge Dr. Kuni Takayama for supplying the lipid A derivatives, Ms. Margaret Olsen for excellent technical assistance and Ms. Linda McGuire for editorial and secretarial service.

REFERENCES

Al-Qawasmeh, M., Check, I.J. and Hunter, R.L., 1988, Induction of monocytic differentiation of HL60 cells by ionophore copolymers, FASEB J., 2:A727

Atkinson, T.P., Bullock, J.O., Smith, T.F., Mullins, R.E. and Hunter, R.L., 1988, Ion transport mediated by copolymers composed of polyoxyethylene and polyoxypropylene, Am.J.Physiol., 254:C20

Atkinson, T.P., Smith, T.F. and Hunter, R.L., 1988a, In vitro release of histamine from murine mast cells by block copolymers composed of polyoxyethylene and polyoxypropylene, J.Immunol., in press.

Atkinson, T.P., Smith, T.F. and Hunter, R.L., 1988b, Histamine release from human basophils by synthetic block copolymers composed of polyoxyethylene and polyoxypropylene and synergy with immunologic and nonimmunologic stimuli, J.Immunol., in press.

Buynitzky, S. and Hunter, R.L., 1988, Improved copolymer adjuvants and polymeric carrier for enhancement of humoral immunity, FASEB J., 2:A1254.

Coon, J. and Hunter, R.L., 1973, Selective induction of delayed hypersensitivity of a lipid-conjugated protein antigen which is localized in thymus dependent lymphoid tissue, J.Immunol., 110:183.

Howerton, D., 1988, Effects of synthetic block copolymer adjuvants on macrophages: Induction of Ia expression in vivo and mechanisms of action in vitro, Ph.D. Thesis, Emory University, Atlanta, Georgia.

Howerton, D., Check, I.J. and Hunter, R.L., 1987, Induction of macrophage class II MHC antigen expression by synthetic copolymer surfactants: Relation to physicochemical structure and adjuvant activity, Fed. Proc., 46:939.

Hunter, R.L., 1980, An overview of the role of lipid in the induction of delayed hypersensitivity and recent studies on the effect of surface active agents on selecting immune responses, in: "Liposomes and Immunobiology", B.H. Tom and H.R. Six, eds., Elsevier, North Holland, New York.

Hunter, R.L. and Bennett, B., 1984, The adjuvant activity of nonionic block polymer surfactants. II. Antibody formation and inflammation related to the structure of triblock and octablock copolymers, J.Immunol., 1343:1367.

Hunter, R.L. and Bennett, B., 1986, The adjuvant activity of nonionic block polymer surfactants. III. Characterization of selected biologically active surfaces, Scand.J.Immunol., 23:287.

Hunter, R.L., Strickland, F. and Kedzy, F., 1981, Studies on the adjuvant activity of nonionic block polymer surfactants. I. The role of hydrophile-lipophile balance, J.Immunol., 127:1244.

Mezrow, C.K., Bennett, B., Check, I.J. and Hunter, R.L., 1987, Suppression of experimental allergic encephalomyelitis (EAE) by a synthetic block copolymer which induces thymic hyperplasia, Fed.Proc., 46:732.

Ribi, E. 1986, Structure-function relationship of bacteria adjuvants, in: "Advances in Carriers and Adjuvants for Veterinary Biologics", R.M. Nervig, P.M. Gough, M.L. Kaeberle and C.A. Whetstone, eds., Iowa State University Press, Ames, Iowa.

Snippe, H., DeReuver, M.J., Strickland, F., Willers, J.M.N. and Hunter, R.L., 1981, Adjuvant effect of nonionic block polymer surfactants in humoral and cellular immunity, Int.Arch.Allergy Appl.Immunol., 65: 390.

Takayama, K., Qureshi, N., Mascagni, P., Nashed, M.A., Anderson, L. and Raetz, C.R.H., 1983, Fatty acyl derivatives of glucosamine 1-phosphate in Escherichia coli and their relation to lipid A, J.Biol. Chem., 258:7379.

Unanue, E.R., 1984, Antigen-presenting function of the macrophage, Ann.Rev. Immunol., 2:395.

Waksman, G.H. 1979, Adjuvants and immune regulation by lymphoid cells, Springer Seminars in Immunopathology, 2:5.

Warren, H.S., Vogel, F.R. and Chedid, L.A., 1986, Current status of immunological adjuvants, Ann.Rev.Immunol., 4:369.

Zigterman, G.J.W.J., Snippe, H., Jansze, M. and Willers, J.M.N., 1987, Adjuvant effects of nonionic block copolymer surfactants on liposome-induced humoral response, J. Immunol., 138:220.

USE OF SYNTEX ADJUVANT FORMULATION TO AUGMENT HUMORAL RESPONSES TO HEPATITIS B VIRUS SURFACE ANTIGEN AND TO INFLUENZA VIRUS HEMAGGLUTININ

Noelene E. Byars, Gayle Nakano, Mary Welch and Anthony C. Allison

Department of Immunology, Institute of Biological Sciences
Syntex Research, Palo Alto, CA 94304, USA

The need for a safe, efficacious, widely applicable adjuvant formulation for use in vaccines has long been apparent. Living viruses can produce severe infections in persons with immunodeficiency, and possible undesirable long-term effects of viral nucleic acids are being recognized. For these reasons there is a trend towards use of subunit vaccines extracted from viruses or produced by recombinant DNA technology and synthetic antigens. Many of these antigens are relatively weak immunogens so the requirement for an effective formulation is even more critical. For some bacterial and viral diseases, humoral responses provide adequate protection. However, in the case of a number of viruses, parasites, fungi and tumors, cell-mediated immunity is the major protective mechanism of the host response. The only adjuvants currently approved in the United States for use in human vaccines are aluminium salts (alum). While these compounds often induce adequate antibody responses to antigens, they are ineffective with some antigens, e.g. influenza virus hemagglutinin: furthermore, alum does not consistently augment cell-mediated immune responses. Syntex adjuvant formulation (SAF) has been shown to be efficacious with several antigens in various animal species (Byars and Allison, 1987; Byars et al, 1989; Robey et al, 1986; Marx et al, 1986; Letvin et al, 1987). Both humoral and cell-mediated responses are increased when vaccines are formulated with SAF. To provide experimental evidence justifying the use of SAF in humans, our first objective was to show that the adjuvant formulation can increase the efficacy of established vaccines. A second goal is the use of SAF for novel vaccines.

The hepatitis B virus (HBV) vaccines presently in use consist of hepatitis B virus surface antigen (HBsAg) adsorbed to alum. The HBsAg is purified from plasma of hepatitis virus carriers or produced by recombinant DNA technology. Three doses of the alum-adjuvanted vaccines elicit protective antibody responses in 90-95% of normal adults in North America and Europe (Hilleman, 1985). The seroconversion rate is much lower in some groups of people especially susceptible to HBV infection, such as intravenous drug users and promiscuous homosexuals. In many developing countries (e.g. China and most of Africa) the rate of persistent HBV carriage is high. In Taiwan, this has been shown to be due to perinatal infection from infected mothers. Cirrhosis and hepatocellular carcinoma are strongly associated with persistent HBV infections (Beasely, 1981; Szmuness, 1978). Persistent HBV viremia can be prevented by immunization of neonates (Beasely et al, 1983), but improved efficacy of vaccines is again required. For all HBV immunizations, it would be convenient to reduce the amount of antigen required, and to obtain protective immunity with two rather than three doses of vaccine.

Influenza virus infections produce severe morbidity and excess mortality in the elderly and the very young (Lui and Kendal, 1987; Alling et al, 1981). The current influenza vaccines are saline suspensions of the hemagglutinin (HA) and neuraminidase, or of whole inactivated virions. Vaccination with these preparations protects only about 60–80% of the recipients. However, the proportion of responses in elderly persons living in nursing homes is even lower; only 27% were protected (Arden et al, 1986).

We have tested SAF with both HBsAg and influenza virus HA in mice and guinea pigs and found that use of SAF significantly increases immune responses to these antigens, compared to currently available vaccines. The response of aged mice to HA was consistently improved.

MATERIALS AND METHODS

(a) SAF preparation. A 2X concentrated SAF emulsion is prepared by emulsifying 10% (v/v) squalane and 5% (v/v) Pluronic L121^R in pH 7.2 phosphate buffered saline (PBS) containing 0.4% Tween 80. The vehicle components must be completely emulsified, so that no residual Pluronic L121 remains unincorporated. Small volumes of up to 10 ml can be readily emulsi-

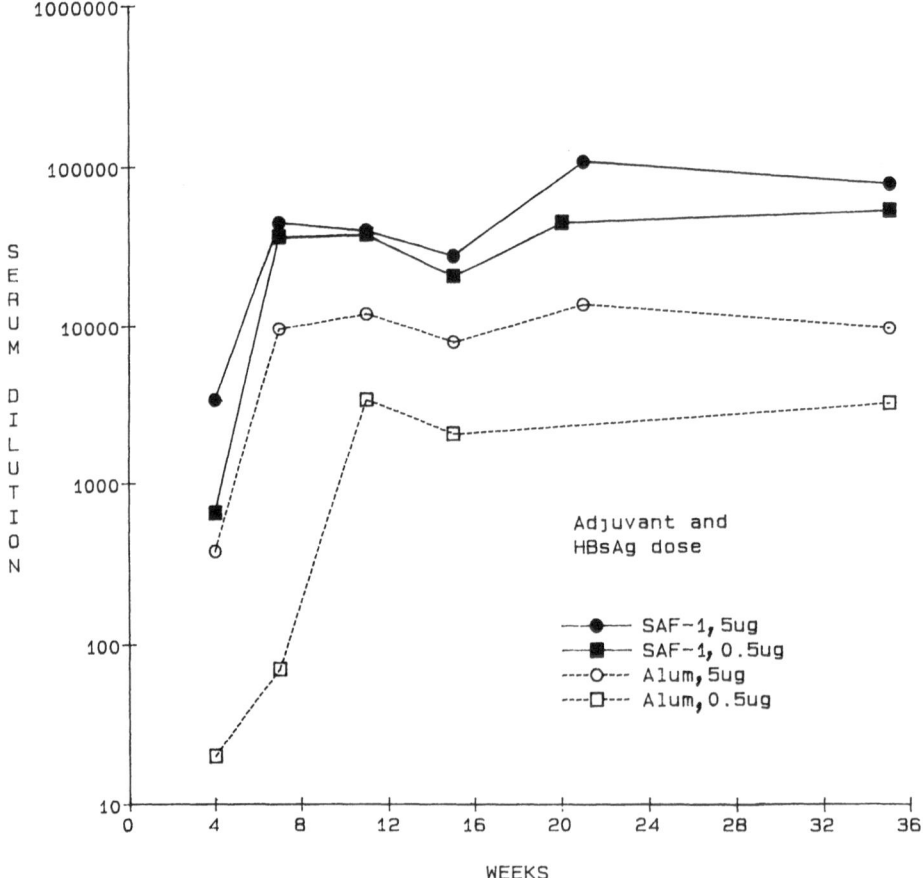

Fig. 1. Time course of antibody titres in pooled sera of guinea pigs immunized at 0 and 4 weeks with HBsAg in SAF or on alum.

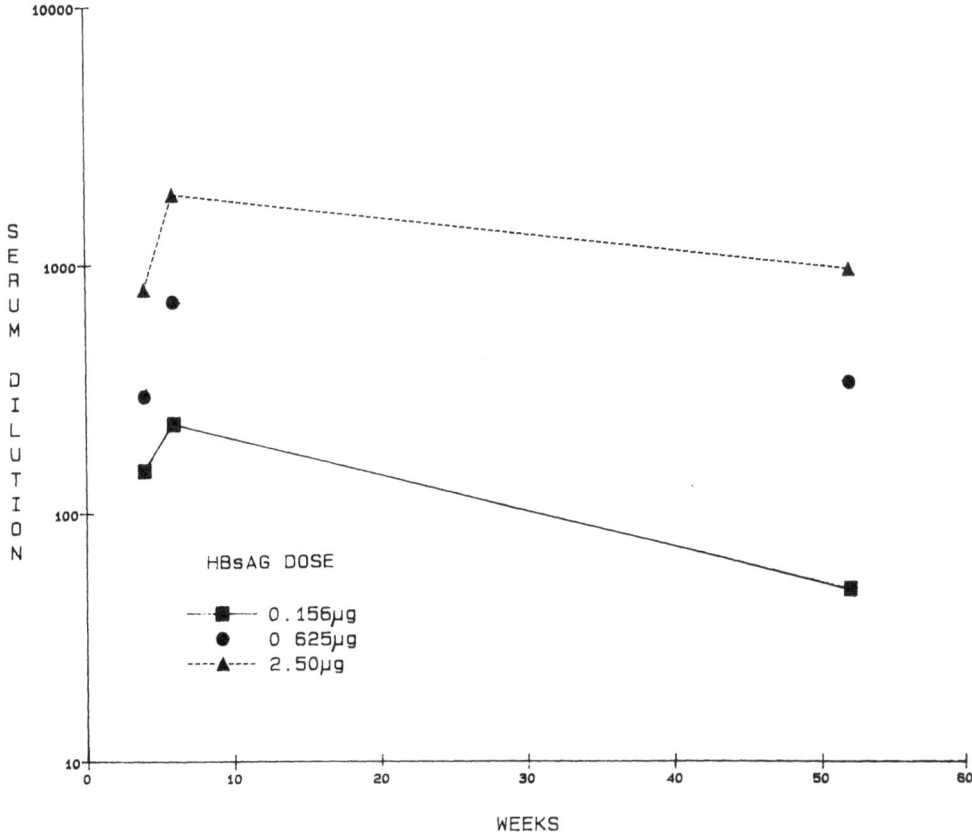

Fig. 2.   Time course of antibody titres in pooled sera of Balb/c mice after
          one injection of HBsAg in SAF.

fied by vigorous vortexing. Larger volumes are more conveniently prepared
using a mechanical mixer. In a correctly prepared emulsion there is either
no creaming (separation of lipid and aqueous phases) or creaming occurs very
slowly.

A solution containing 2X concentrations of antigen and threonyl-MDP
([Thr^1]-MDP) is prepared in PBS. The 2X antigen-[Thr^1]-MDP solution is then
added to an equal volume of the 2X emulsion and mixed gently. Since the
emulsion is an oil-in-water type, addition of the aqueous antigen-[Thr^1]-MDP
solution does not disturb it. The resulting vaccine should have the appearance and consistency of milk, easily injected through a fine guage needle.

(b) HBsAg immunization. Groups of 8 female Hartley strain guinea pigs
were immunized subcutaneously (sc) using 5.0 or 0.5 μg of hepatitis B surface antigen either adsorbed to alum or in SAF with 50 μg of [Thr^1]-MDP.
The aqueous and adsorbed HBsAg were kindly provided by Dr. Dale Lehman of
Merck, Sharp and Dohme Research Laboratories. Animals were injected at 0
and 4 weeks, and then bled by cardiac puncture at intervals up to 35 weeks.
Groups of 8-10 Balb/c and B10.M mice were immunized sc with 0.01 to 20 μg of
HBsAg either in SAF (with 50 ug [Thr^1]-MDP or adsorbed to alum. They were
boosted at week 3 with the same preparations used for the primary injections. Animals were bled via the retro-orbital plexus at intervals up to 25
weeks.

Sera from both guinea pigs and mice were assayed using ELISA techniques. Duplicate serial dilutions of sera were made in wash buffer (PBS

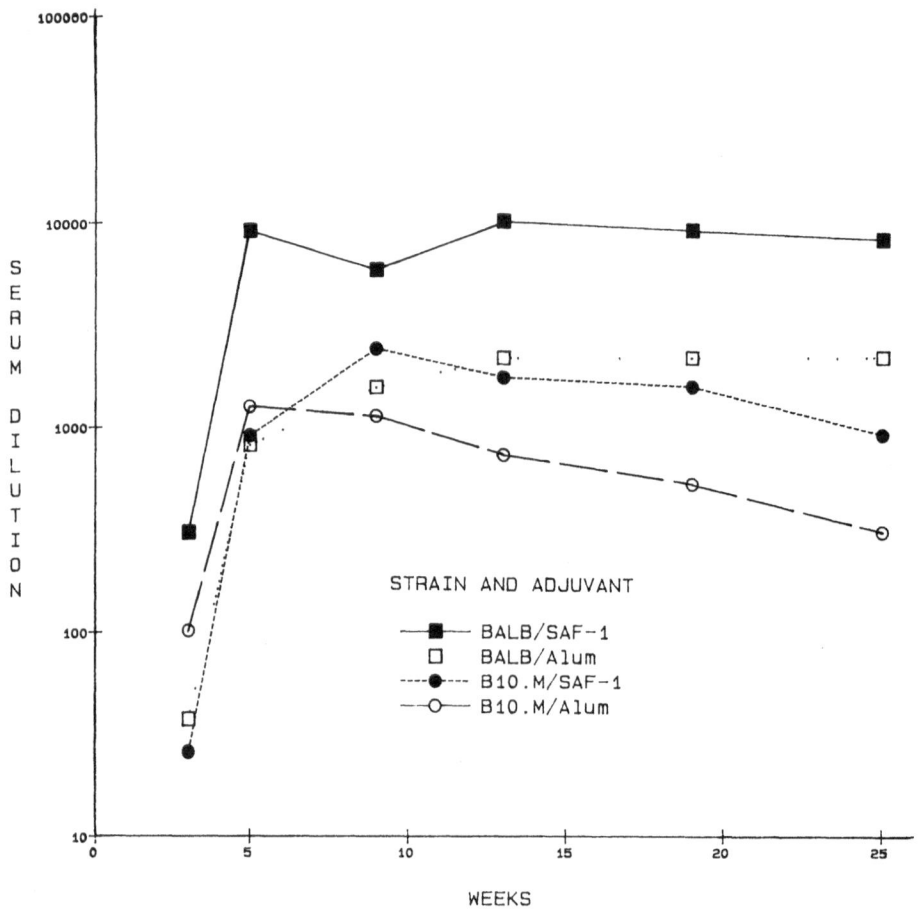

STRAIN AND ADJUVANT
■ BALB/SAF-1
□ BALB/Alum
● B10.M/SAF-1
○ B10.M/Alum

WEEKS

Fig. 3. Time course of antibody titres in pooled sera of Balb/c and B10.M mice immunized at 0 and 3 weeks with 1 µg of HBsAg in SAF or on alum.

with 0.05% Tween 20, 0.1% BSA and 0.01% thimerosal) and added to 96-well ELISA plates which had been coated with a solution of 2 µg/ml of HBsAg in pH 9.6 bicarbonate buffer. Following overnight incubation at 4-8°C, the plates were developed at room temperature with peroxidase labeled rabbit anti-guinea pig IgG (H + L) or peroxidase labeled rabbit anti-mouse Ig (G + A + M). The enzyme substrate used was o-phenylenediamine in 0.1 M citrate buffer, pH 4.5 with 0.03% hydrogen peroxide. The reaction was stopped using 5% sodium dodecyl sulfate. The titre was defined as the reciprocal of the serum dilution which resulted in an optical density value of 0.5 at 450 nM.

(c) <u>Influenza virus HA immunization</u>. Similar methods were used to immunize guinea pigs and Balb/c mice with the hemagglutinin of influenza B virus (B/USSR/100/83). This antigen was kindly provided by Ian Furminger, Evans Medical Ltd., England. The guinea pigs used were female Hartley strain animals, 350-450 g at the time of primary immunization. Secondary immunization was at week 4. We used female Balb/c mice of different ages, i.e. 3-4 weeks, 7 weeks or $13\frac{1}{2}$ months at the time of primary vaccination. The animals were boosted at week 3. The animals were bled at intervals up to week 35 (guinea pigs) or week 25 (mice). Antibody titres were determined using ELISA methods similar to those described above for the hepatitis experiments, except that the plates were coated with 0.5 µg/ml of B/USSR hemagglutinin.

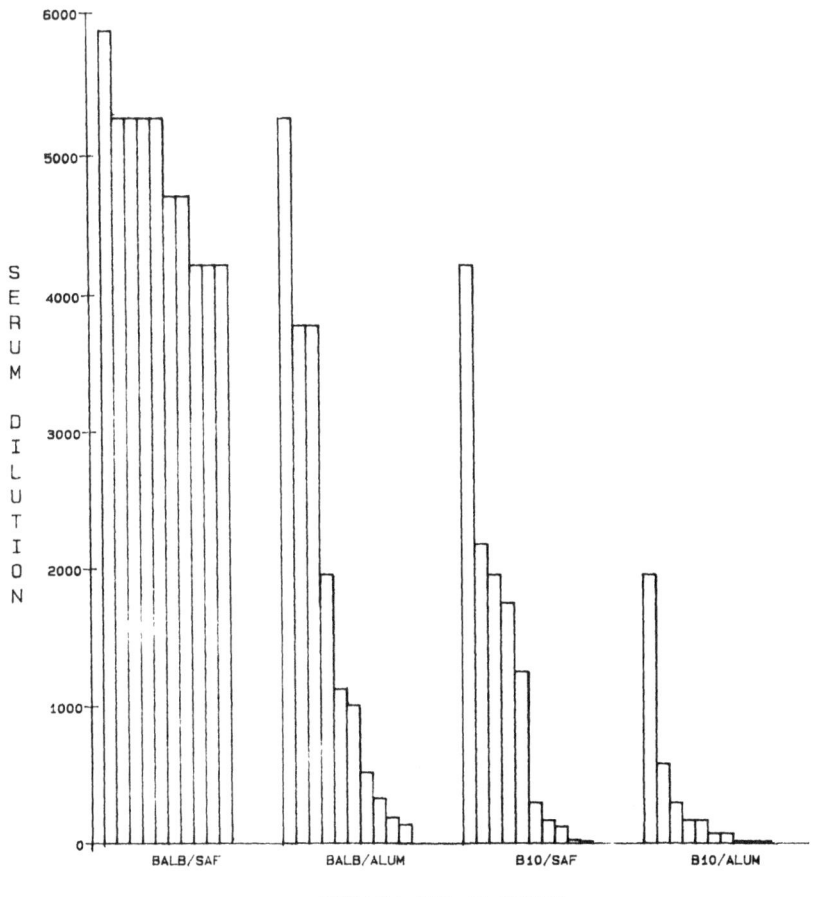

STRAIN AND ADJUVANT

Fig. 4. Anti-HBsAg titres of individual Balb/c and B10.M mice 25 weeks
after the primary immunization with 1 µg HBsAg in SAF or on alum.

RESULTS

1) Hepatitis

(a) Guinea pigs. Figure 1 shows the time course of the antibody titres
determined by ELISA, using pooled sera. The anti-HBsAg titres of sera from
the animals immunized using SAF were much greater than those of animals
given the alum-adjuvanted vaccines. Sera from individual animals in all
groups were assayed separately for the 11 week bleed. The mean titres for
the groups immunized using SAF were significantly greater than those of the
corresponding groups immunized using alum. The mean titres were very similar to the titres found for the pooled sera from these groups.

Over the course of the experiments, titres for the group given 5.0 µg
HBsAg in SAF were 4-9 fold higher than the group given 5.0 µg HBsAg on alum.
The difference in titre was even more striking at the 0.5 ug dose with the
SAF group having 10 to >450-fold higher titres than the alum group.

(b) Mice. Figure 2 shows the results of a time course experiment in
which Balb/c mice were given only one vaccination of HBsAg in SAF. There
was a clear dose response and, at the two higher doses in particular, the
antibody levels fell only slightly between weeks 6 and 52 post-vaccination.

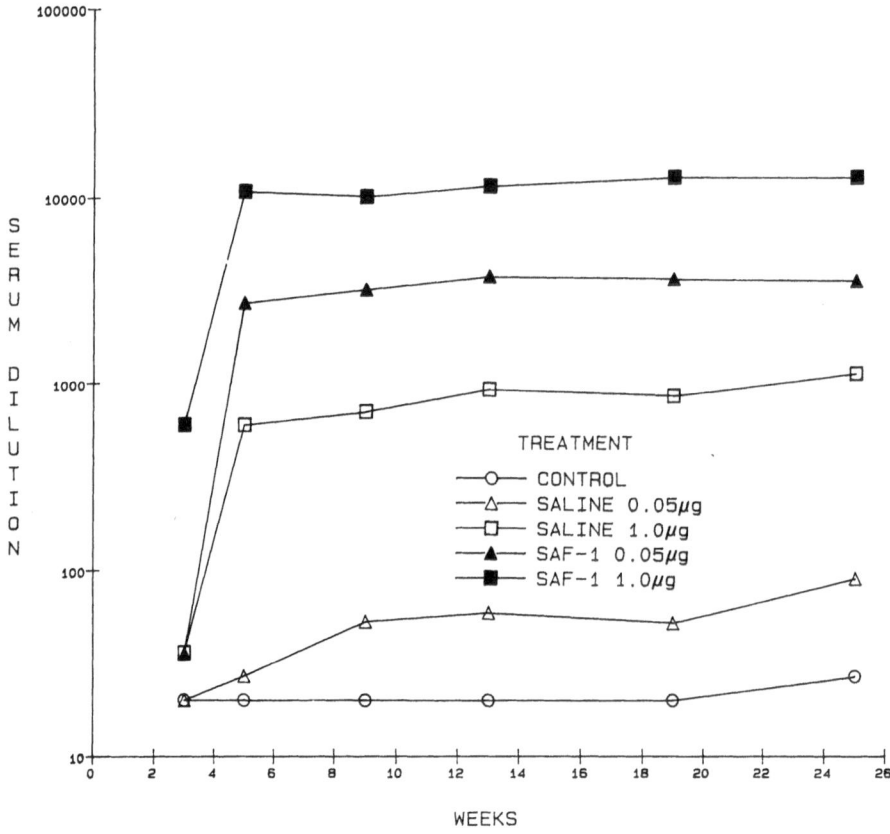

Fig. 5. Time course of anti-influenza B virus HA titres in pooled sera of young mice immunized at 0 and 3 weeks with HA in PBS or SAF.

It has been reported that Balb/c and B10.M (H-2^f) mice are high and non-responders respectively following intraperitoneal (ip) injection of HBsAg in complete Freund's adjuvant (Milich and Chisari, 1982). We found that both Balb/c and B10.M mice responded to sc injections of 20 μg or 1 μg of HBsAg in SAF or adsorbed to alum. In both these strains, the antibody response to HBsAg in SAF was greater than the response to HBsAg on alum, as shown in Fig. 3 which shows ELISA data obtained with pooled sera of mice vaccinated with 1 μg of HBsAg. The Balb/c mice had higher titres than the B10.M mice at all times when SAF was used, and when alum was used Balb/c titres were higher than B10.M titres from week 9 to 25. Individual sera were assayed at week 25, when mice were killed and bled out. Figure 4 shows that Balb/c mice given HBsAg in SAF had higher and more consistent titres than Balb/c mice given HBsAg on alum. Among the B10.M mice, 5 of 10 given HBsAg on SAF had moderate or high titres, and only 2 were non-responders, while only one of 10 B10.M mice vaccinated using alum had a titre over 1,000, 6 mice had low responses and 3 were non-responders.

2) Influenza vaccine

We found that immunologically mature mice and guinea pigs responded much better to influenza B virus HA in SAF than they did to HA in saline (data not shown). We also immunized young immunologically immature mice (3 weeks old) and aging mice (13½ months old) with two doses of influenza virus HA in SAF or saline. As shown in Figs. 5 and 6, both the young and the old mice had much higher titres following immunization with HA in SAF than they did after immunization with HA in saline. Furthermore, we again found much

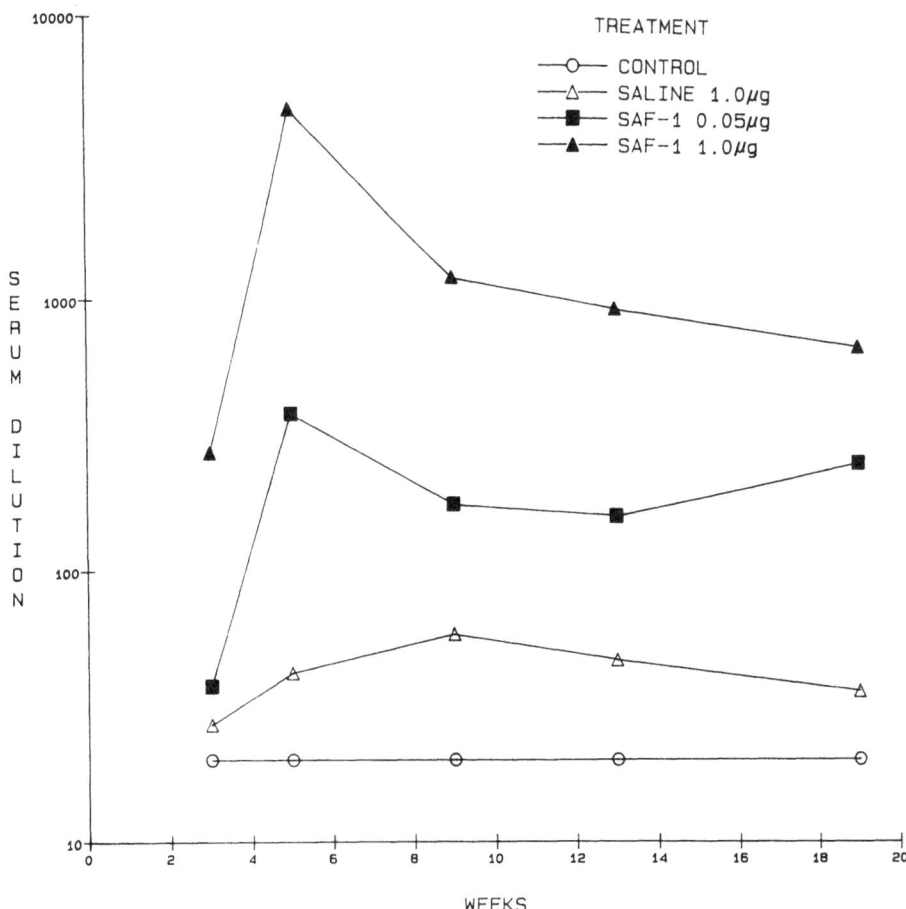

Fig. 6. Time course of anti-influenza B virus HA titres in pooled sera of old mice immunized at 0 and 3 weeks with HA in PBS or SAF.

more consistent responses among the mice given the vaccines formulated with SAF compared to those given the saline formulations. For example, in two experiments, all 11 old mice given HA in SAF had antibody titres detectable by ELISA, but only 2 of 10 mice given HA in saline had detectable titres 19 weeks after vaccination.

DISCUSSION

We have found that vaccines formulated with SAF are significantly more efficacious than those formulated with alum as adjuvant, as is the case with HBsAg, or those with no adjuvant at all, as is the case with the currently a available influenza vaccines. No side effects have been observed in the animals given these vaccines. The formulation is easy to use, and since the emulsion is made prior to antigen addition, there is no risk of antigen denaturation during vaccine preparation. This is in contrast to vaccines prepared by the emulsification of aqueous antigen in the oil phase, when the sheer forces can denature antigens.

If our findings in experimental animals can be extrapolated to humans it should be possible to elicit protective immunity with two small doses of HBsAg and a single low dose of influenza HA formulated with SAF. The adjuvant should improve immune responses especially in newborn children (relevant for HBV vaccination in developing countries) and persons over 65 (relevant for influenza vaccination).

REFERENCES

Alling, D.W., Blackwelder, W.C. and Stuart-Harris, C.H., 1981, A study of excess mortality during influenza epidemics in the United States, 1968-1976, Am.J.Epidem., 113:30.

Arden, N.H., Patriarca, P.A. and Kendal, A.P., 1986, Experiences in the use and efficacy of inactivated influenza vaccine in nursing homes, in: "Options for the Control of Influenza", A.P. Kendal and P.A. Patriarca, eds., Alan R. Liss, Inc.

Beasely, R.P., Lin, C.C., Hwang, L. and Chien, C., 1981, Hepatocellular carcinoma and hepatitis B virus: a prospective study of 22,707 men in Taiwan, Lancet, ii:1129.

Beasely, R.P., Hwang, L.Y., Lee, G.C., Lin, C.C., Roan, C.H., Hwang, F.Y. and Chen, C.L., 1983, Prevention of perinatally transmitted hepatitis B virus infections with hepatitis B immune globulin and hepatitis B vaccine, Lancet ii:1099.

Byars, N.E. and Allison, A.C., 1987, Adjuvant formulation for use in vaccines to elicit both cell-mediated and humoral immunity, Vaccine, 5:223.

Byars, N.E., Allison, A.C., Harmon, M.W. and Kendal, A.P., Enhancement of antibody responses to influenza B virus hemagglutinin by use of a new adjuvant formulation, (Vaccine, submitted).

Hilleman, M.R., 1985, Newer directions in vaccine development and utilization, J.Infect.Dis., 151:407.

Letvin, N.L., Daniel, M.D., King, N.W., Kannagi, M., Chalifoux, L.V., Sehgal, P.K., Desrosiers, R.C., Arthur, L.O. and Allison, A.C., 1987, AIDS-like disease in macaque monkey induced by simian immunodeficiency virus: A vaccine trial, in: "Vaccines 87", R.M. Chanock, R.A. Lerner, R.A. Brown and H. Ginsberg, eds., Cold Spring Harbor Laboratory.

Lui, K-J. and Kendal, A.P., 1987, The impact of influenza epidemics on mortality in the United States from October 1972 to May 1985, Am.J.Pub.Health, 77:712.

Marx, P.A., Pedersen, N.C., Lerche, N.W., Osborn, K.G., Lowenstine, L.J., Lackner, A.A., Maul, D.H., Kwang, H-S., Kluge, J.D., Zaiss, C.P., Sharpe, V., Spinner, A.P., Allison, A.C. and Gardner, M.B., 1986, Prevention of simian acquired immune deficiency syndrome with a formalin-inactivated type D retrovirus vaccine, J.Virol., 60:431.

Milich, D.R. and Chisari, F.V., 1982, Genetic regulation of the immune response to hepatitis B surface antigen (HBsAg). I. H-2 restriction of the murine humoral response to the a and d determinants of HBsAg, J.Immunol., 129:320.

Robey, W.G., Arthur, L.O., Matthews, T.J., Langlois, A., Copeland, T.D., Lerche, N.W., Oroszolan, S., Bolognesi, D., Gilden, R.V. and Fischinger, P.J., 1986, Prospect for prevention of human immunodeficiency virus infection: Purified 120-kDa envelope glycoprotein induces neutralizing antibody, PNAS, 83:7023.

Szmuness, W., 1978, Hepatocellular carcinoma and the hepatitis B virus: Evidence for a causal association, Prog.Med.Virol., 24:40.

IMMUNOSTIMULATING COMPLEX (ISCOM)

B. Morein[1], K. Lövgren[2] and S. Höglund[3]

[1]Swedish University of Agricultural Sciences, College of
Veterinary Medicine, Department of Microbiology, Section of
Virology, Biomedicum, Box 585, S-751 23 Uppsala, Sweden
[2]Department of Virology, National Veterinary Institute
Biomedicum Box 585, S-751 23 Uppsala, Sweden
[3]Institute of Biochemistry, Biomedicum, Box 576, S-751 23
Uppsala, Sweden

INTRODUCTION

For almost two centuries vaccines have been based on whole micro-organisms, killed (inactivated) or attenuated. With increasing knowledge of the molecular composition of pathogenic micro-organisms and the function of different molecules, protective antigens also became identified. In spite of the fact that some of the early killed whole micro-organism vaccines proved to be effective there are several reasons why vaccines in the future should be formulated based on defined antigens, e.g.: (i) To avoid hazards due to toxicity or to genetic material which may cause replication, or as regards retroviruses the integration of viral genes into the infected host-cell genome. (ii) To limit the number of antigens in a vaccine in order to decrease the risk for induction autoimmune or allergic reactions. (iii) Further, reorganization of the prospective protective antigens may break a strategy for survival of the pathogen in the "hostile" immunological environment of the infected host.

The first experimental subunit vaccines were made mainly to contain the envelope proteins of influenza viruses. Although non-toxic, they turned out to have very low immunogenicity and never came out for commercial use (for references see Morein & Simons, 1985). Later work showed that isolated antigens, e.g., the surface proteins of Semliki forest virus in monomeric form or in non-defined aggregations had low immunogenicity (Morein et al, 1978). To obtain an immune response equivalent to that obtained with a whole killed virus it has been shown in several works that the antigens should be presented in defined multimeric constructions, e.g. in the shape of micelles or integrated into liposomes (for references, see Morein & Simons, 1985). In spite of the fact that liposomes with integrated envelope protein of virus, so-called virosomes (Almeida et al, 1975 and Helenius et al, 1977) like micelles are highly immunogenic, it became evident that an optimal physical presentation as such does not necessarily suffice to obtain optimal immunogenicity and protective immunity (Morein et al, 1983). In the referred case, the micelle preparation had to be supplemented with an oil adjuvant, to protect against pneumonia induced by parainfluenza-3 virus. This adjuvant caused local as well as systemic reactions.

The Iscom and Iscom Matrix

The iscom was consructed to combine the physical presentation of a sub-
microscopic particle with a built-in adjuvant. The choice of adjuvant was
Quil A because it has proved to be an effective adjuvant and it has amphipathic properties forming micelles at a concentration of 0.3 mg per liter
(Dalsgaard, 1978).

In the iscom the antigen is attached to a matrix - the iscom matrix -
by hydrophobic interaction. This means that at least one hydrophobic domain
is required within the antigen.

The unique component of the iscom matrix is Quil A, a triterpenoid with
two carbohydrate chains. The other component, which cannot be substituted
or omitted is cholesterol (Lövgren and Morein, 1988). Quil A and cholesterol can, alone in a one to one molar ratio, form the cage-like structure
of the iscom matrix. To include a protein it has been necessary to include
also a lipid less rigid than cholesterol, e.g. phosphatidyl choline. Excess
of lipid is not included in the iscom matrix or in the iscom since the excess lipid forms vesicles in a water solution. The final formulation of an
iscom with an optimal amount of protein is a molar ratio of Quil A, cholesterol, phosphatidyl choline and protein, 1:1:1:1. However, less protein can
be included. Generally, envelope protein of viruses can be included according to the above formulation. Other hydrophobic molecules like outer membrane proteins (OMP) of bacteria are more difficult to handle and may require special solubilization systems. Increased amounts of lipids might,
therefore, be required in the system to prevent uncontrolled aggregation of
the OMP. Since Quil A binds to the lipids, the amount of Quil A will increase proportionally with the amount of lipid, resulting in increased
amounts of Quil A integrated in the iscoms in relation to protein.

Irrespective of the proteins incorporated into the iscom, it has a
cage-like shape with a diameter of approximately 35 nm. In the shell of the
iscom, subunits of about 12 nm in size are assembled (Morein et al, 1984).
Recently Höglund et al (submitted for publication) showed on freeze-dried
specimens that the iscom appears isomorphic and that the uranyl acetate
stain used readily penetrated into the shell of the iscom particle (Fig. 1)
suggesting a hollow structure (Fig. 1c). The stability demonstrated by the
iscoms might be explained by an energetically favourable symmetrical assembly of the particles.

Preparation of Iscoms

The iscom was first prepared by a centrifugation method (Morein et al,
1984). The procedure involves solubilization of the virus to obtain viral
proteins in a monomeric form. Preferably a non-ionic detergent is used,
which will interact mainly with the hydrophobic region of the protein leaving the hydrophilic epitopes unaffected, i.e. where we are likely to find
the protective antigenic determinants. The solubilized virus (e.g. 2 mg in
200 µl PBS) is layered onto a gradient consisting of 20 to 50% sucrose in
PBS (or any other suitable buffer) containing 0.1% Quil A on top of which
200 µl 10% sucrose containing 0.1% Triton X-100 is layered (see Fig. 2).
The gradient is centrifuged for 18 to 20 hrs at 20°C at 40,000 r.p.m. using
a Kontron SW-40 rotor. The gradient is collected in, for instance, 200 µl
fractions. The envelope proteins can be detected by ELISA using suitable
antibody. It is convenient to use radiolabelled envelope protein e.g., by
the technique of Luukkonen et al, 1977), which makes it easy to identify and
trace the envelope proteins in the gradient. The selected fractions are
collected and analysed by electron microscopy for identification of the
iscom structure (Fig. 1), pooled and dialysed extensively against a suitable
buffer. The iscoms can then be isolated by sedimenting the dialysate for 18

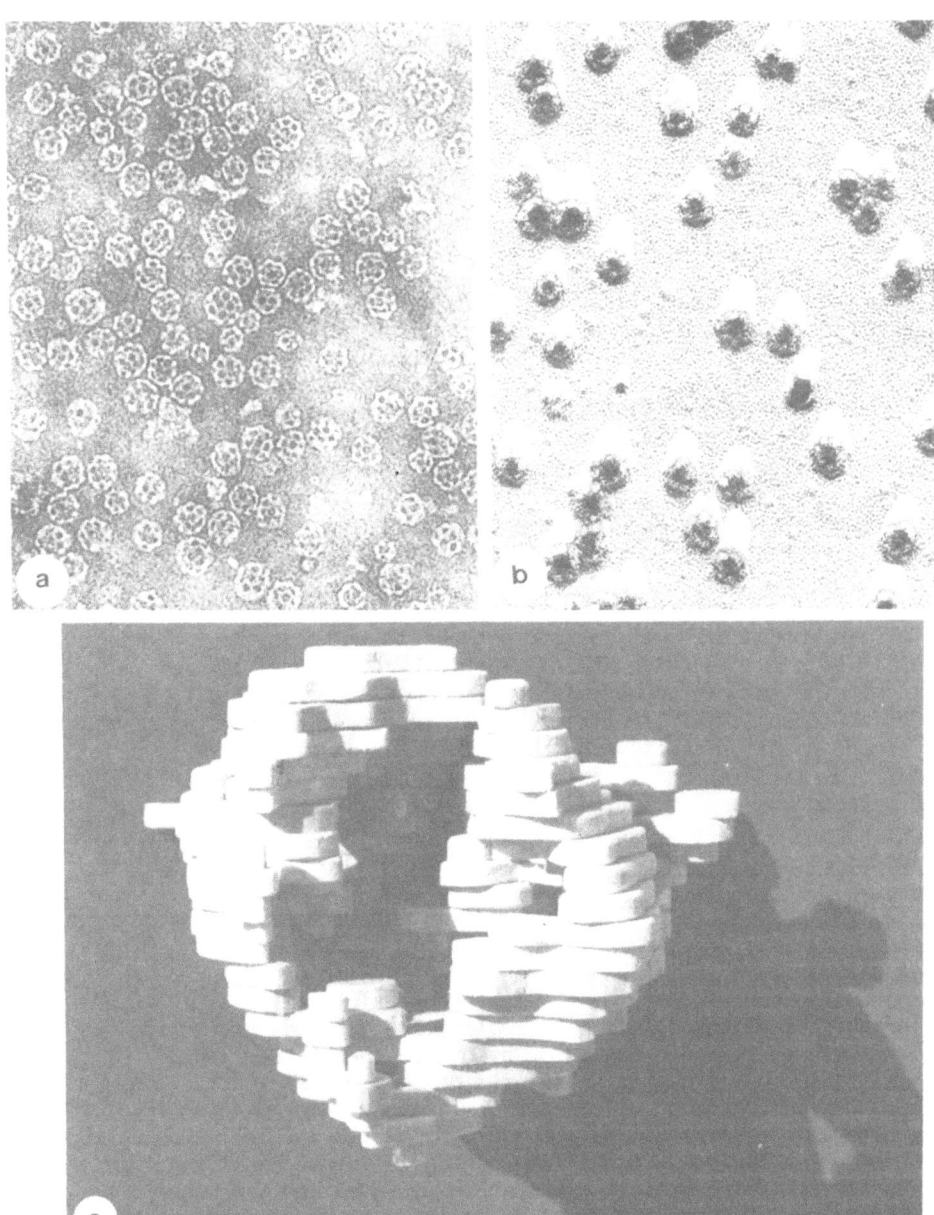

Fig. 1.   HIV proteins incorporated into iscom particles which demonstrate a
cage-like shape, readily penetrated by a heavy metal salt solution.

(a) Iscoms prepared by negative staining.
(b) Negatively stained iscoms were lyophilized, excess stain was
    removed followed by shadow casting. An isomorphic shape of the
    particles is shown (x 160,000). (For reference see Höglund et
    al, submitted.)
(c) Electronmicrographs which were obtained from sectioned iscom
    particles were analysed by electronmicroscopic tomography re-
    sulting in reconstruction of iscoms. This view clearly shows a
    hollow iscom particle. (By courtesy of Xiao-dong Su and Lars-
    Göran Öferstedt.)

hrs at 40,000 r.p.m. through a cushion of 20% sucrose. If a thin layer of detergent in 10% sucrose is included on the top of the cushion, virtually all free lipid and Quil A will be removed. The above method will yield a preparation of low toxicity. However, to quantitate the amount of Quil A and detergent in a preparation, a rocket electrophoresis (rocket haemolysis) assay was designed (Sundquist et al, 1983). This test will not detect non-ionic detergents. In Fig. 1, a scheme illustrating the preparation of iscoms in comparison with protein micelles is presented.

Characterization of Iscoms

The iscom is characterized by its morphology, its s-value being higher than the monomeric form of the protein included but lower than the micelle form of the protein. The characteristic morphology is identified by electronmicroscopy (Fig. 1). The s-value can be determined in a sucrose gradient according to McEwen (1967), using bovine serum albumin (4s) and the micelle form of the protein (about 30s) or thyroglobulin (19s) as standards. Generally, the iscoms have been found to have an s-value of 19s, while the iscom matrix in the same system sediment with s-values varying from 14-20s, depending on the amount and the type of lipids that are incorporated. Pure iscom matrix consisting only of Quil A and cholesterol sediments with an s-value of 20. For the determination of the protein content any method for quantification of protein can be used, for instance, the method of Bradford, 1976. More adequate is, of course, a specific immunological quantification system, e.g. a quantitative ELISA. Further protein analysis may include polyacrylamide gel elctrophoresis and immuno (Western) blotting techniques. In case monoclonal antibodies recognizing important epitopes are available, they can be used for quality control. This was done by Merza et al (in press) to verify that neutralizing epitopes of gp 51 from bovine leukemia virus were not destroyed during integration into iscom.

The Immunogenicity of Iscoms

The immunogenicity includes the ability of iscoms to induce antibody mediated immunity (AMI) and cell mediated immunity (CMI) (for a general review see Morein et al, 1987 and Hoglund et al, in press). But for vaccine purposes, the efficacy measured by protection, if possible in the relevant host, is what counts. Generally, the antibody titres induced by iscoms have been ten-fold higher or even more than titres induced by the same antigen in a killed micro-organism or in a micelle form. The same antigen in a monomeric or undefined form induces even lower antibody titres if any (Morein et al, 1978; Morein and Simons, 1985). The serum antibody response evoked by iscoms containing haemagglutinin of influenza virus given subcutaneously is a classical one starting with an IgM response followed by an IgG response (Lövgren, 1988). The antibody response is distributed in all isotypes of IgG. In general the iscoms induced higher antibody titres in all isotypes than the same antigen in the micelle form (Fig. 2). Although the micelles generally induced an antibody response that is most prominent within the IgG2a and IgG2b subclasses, the route of administration seems to play a major role in determination of the IgG subclass profile. Subcutaneous administration of both iscoms and micelles induced considerably more IgG1 than the intranasal route. For replicating virus the response is mainly found in IgG2b (Coutelier et al, 1987). The intranasal administration of envelope proteins of influenza virus in iscom induced antibody titres of similar magnitude to that of subcutaneous administration. These results indicate that the biological activities (Sundquist et al, 1988) of the envelope proteins are well preserved and expressed, allowing a biologically active penetration through the mucus. As shown by Jones et al (1988), the local application, twice, of 5 ug influenza virus iscoms induced a cytotoxic T cell response and these iscoms also efficiently boostered the cytotoxic T cell response. This T cell response had a broader specificity than

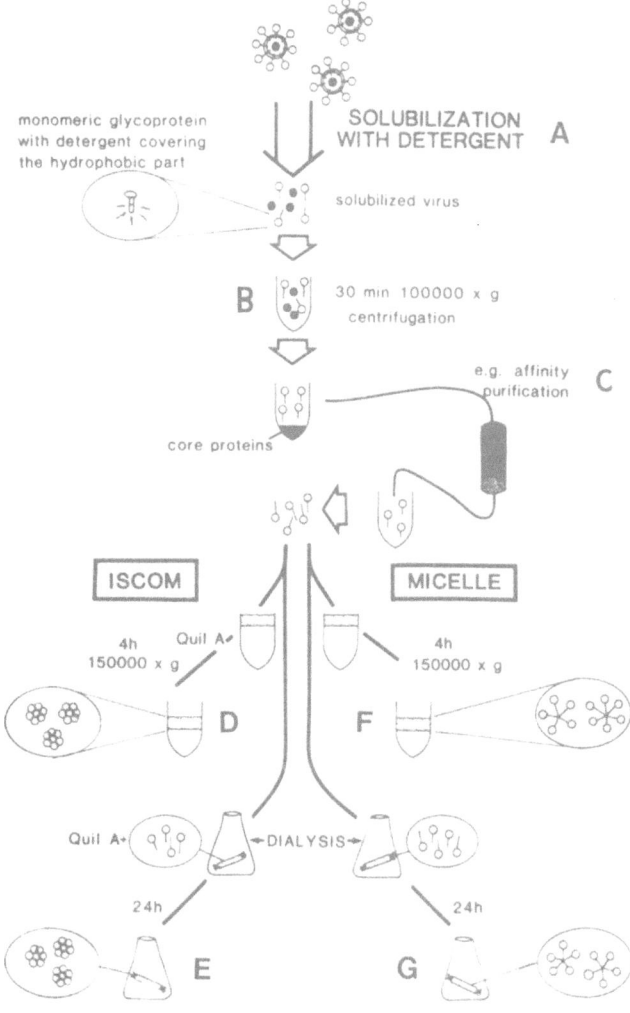

Fig. 2. Preparation of iscoms and micelles from viral envelope glycoproteins.

A. Solubilization of the viral membrane by use of a detergent.
B. Solubilized membrane components are separated from the viral
   core by ultracentrifugation.
C. The solubilized proteins can be further purified, e.g. by
   affinity chromatography.
D. Extracted envelope proteins are centrifuged through a sucrose
   layer containing detergent, into a Quil A-containing zone. In
   this zone, in the absence of detergent, amphipatic proteins
   associate with Quil A and iscoms are formed.
E. Iscoms are formed from the mixture of envelope proteins and Quil
   A as the detergent is removed by dialysis.
F. Extracted envelope proteins are centrifuged through a sucrose
   layer containing detergent into a detergent-free zone, in which
   micelles are formed.
G. Micelles are formed by the envelope proteins as the detergent is
   removed by dialysis.

Table 1. Examples of Viruses to which Neutralizing Antibodies have been Evoked by Iscom Preparations

VIRUS	ANTIGEN	ANIMAL
Influenza	Haemagglutinin (H) Neuraminidase (N)	mouse, guinea-pig, horse, swine
RS-virus	Fusion (F)-protein	guinea-pig
Rubella	El-glycoprotein and nucleoprotein	rabbit
Bovine herpes virus type 1	Envelope protein	cattle
Pseudorabies	Envelope protein	lamb
Pseudorabies	Envelope protein	swine
Rabies	G-protein	mouse
Canine distemper	Envelope protein	dog
Measles	H and F	monkey, rat
HIV	gp120	mouse
HIV	gp120	monkey

Table 2. Protective Immunity Induced by Iscoms Containing Various Microbial Antigens

ANTIGEN	ANIMAL	DISEASE
Haemagglutinin (H) Neuraminidase (N) influenza virus	mice	pneumonia
Haemagglutinin measles	mice	encephalitis
Fusion protein measles	mice	encephalitis
Toxoplasma surface antigens	mice	lethal infection
Bovine virus diahorrea virus envelope protein	sheep	abortion
gp70 Feline leukemia virus	cat	tumour

Table 2. (Cont'd.)

ANTIGEN	ANIMAL	DISEASE
gp340 Epstein-Barr virus	tamarin monkey	tumour
Envelope proteins pseudorabies virus	sheep	lethal infection
envelope proteins Bovine herpes virus, type 1	cattle	pneumonia, fever
Rabies virus	mouse	lethal infection
Canine distemper virus	dog	viremia, encephalitis

the antibodies, since it not only lysed target cells infected with the homologous H1N1 subserotype but also cells infected with the heterologous H3N2 subserotype. In protection experiments we have seen that one ug of the envelope proteins included in iscoms induced protective immunity both by the subcutaneous and the intranasal routes (Lövgren, manuscript in preparation).

The in vitro test which is often considered to reflect protective immunity, is the neutralization test. Iscoms prepared with different envelope proteins from various viruses were found to induce virus neutralization antibodies (Table 1) including rubella, respiratory syncytial virus (Trudel et al, in press), Epstein-Barr virus (EBV) (Morgan et al, 1988) or measles virus (De Vries et al, 1988), all of which have not earlier been successfully formulated as subunit vaccines. Protective immunity measured by challenge infection following the immunization is of course the most relevant test. In Table 2 experimental iscom vaccines are listed which have induced protection in various challenge systems. This includes protection against lethal challenge infection with rabies, abortion caused by bovine virus, diarrhoea virus infection in lambs, or the development of tumor caused by Epstein-Barr virus in cotton-top tamarin monkeys.

Oligopeptides or other small molecules in general show properties like haptens and are therefore dependent on a carrier molecule. Such a carrier, in this case haemagglutinin of influenza virus, can be integrated into iscoms as described by Lövgren et al (1987). Such an iscom-carrier molecule made the integrated small molecule strongly immunogenic (e.g. biotin or the β-chain of human chorionic gonadotrophin).

Toxicity of Iscoms

Most, if not all, adjuvants are toxic given in sufficiently high doses. The concept with the iscom is to avoid high doses of adjuvant by binding the adjuvant to the antigen in one particle where they are presented in optimal forms as multimers. Therefore, the dose of Quil A can be reduced several hundred-fold. In mice between 10 to 50 µg of Quil A administered parenterally start to get toxic. Toxicological studies in mice have shown that an experimental measles virus iscom vaccine was well tolerated in mice (Speijers et al, 1987).

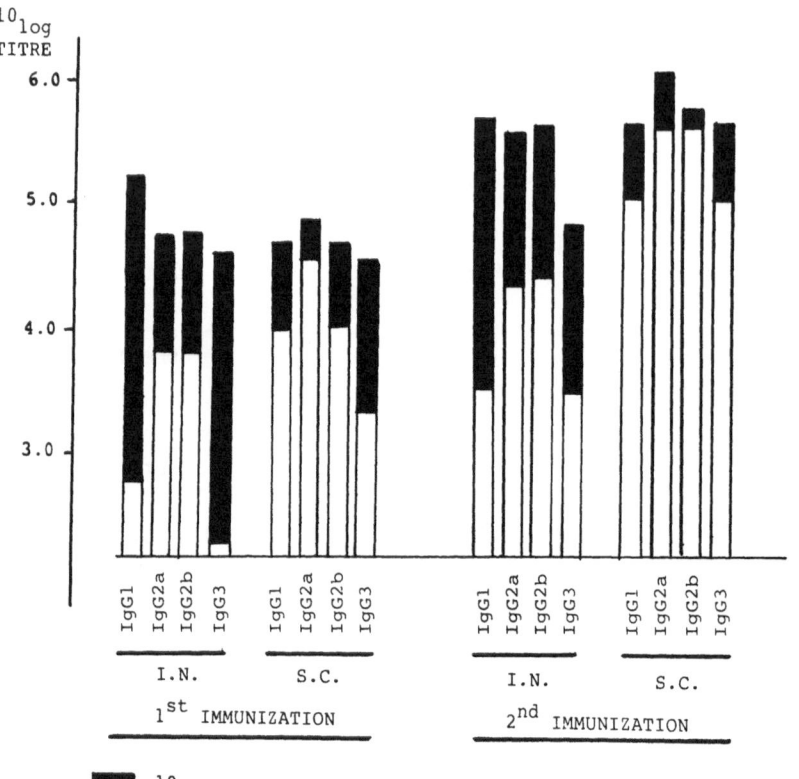

Fig. 3. The serum antibody response in mice distributed into IgG isotypes after intranasal (I.N.) or subcutaneous (S.C.) administration of iscoms or micelles. The antibody response was tested 7 weeks after the first immunization when the mice were boostered and 2 weeks after the booster.

REFERENCES

Almeida, J.D., Brand, D.C., Edwards, D.C. and Heath, T.D., 1975, Formation of virosomes from influenza subunits and liposomes, Lancet 2:899.

Bradford, M.M., 1976, A rapid and sensitive method for the quantification of microgram quantities of protein utilizing the principal of protein dye binding, Analyt.Biochem., 72:249.

Coutelier, J-P., van der Logt, J.T.M., Heessen, F.W.A., Warnier, G. and van Snick, J., 1987, IgG2a restriction of murine antibodies elicited by viral antigens, J.Exp.Med., 165:64.

Dalsgaard, K., 1978, A study of the isolation and characterization of the saponin Quil A. Evaluation of its adjuvant activity with special reference to the application in the vaccination of cattle against foot and mouth disease, Acta Vet.Scand.Suppl., 69:1.

De Vries, P., Van Binnendijk, R.S., Van der Marel, P., Van Wesel, A.L., Voorma, H.O., Sundquist, B., Uyt de Haag, F.G.C.M. and Osterhaus, A.D.M.E., 1988, Measles virus fusion protein presented in an immune-stimulating complex (iscom) induces haemolysis-inhibiting and fusion inhibiting antibodies, virus-specific T cells and protection in mice, J.Gen.Virol., 69:549.

Helenius, A., Fries, E. and Kartenbeck, J., 1977, Reconstitution of Semliki Forest virus membrane, J.Cell.Biol., 75:866.

Höglund, S., Dalsgaard, K., Lövgren, K., Sundquist, B., Osterhaus, A. and
    Morein, B., Iscoms and immunostimulation with viral antigens,
    Subcell.Biochem., in press.
Höglund, S., Ozel, M., Gelderblom, H., Akerblom, L., Villacres, M. and
    Morein, B., 1988, A construct of iscom of HIV antigens: structural
    and immunological function, Proc.Int.Conf.AIDS, Stockholm.
Jones, P.D., Tha Hla, R., Morein, B., Lövgren, K. and Ada, G.L., 1988,
    Cellular immune response in the murine lung to local immunization
    with influenza A virus glycoproteins in micelles and iscom, Scand.J.
    Immunol., 27:645.
Luukkonen, A., Gahmberg, C.G. and Renkonen, O., 1977, Surface labelling of
    Semliki forest virus glycoprotein using galactose oxidase, Virology,
    76:55.
Lövgren, K., Lindmark, J., Pipkorn, R. and Morein, B., 1987, Antigenic
    presentation of small molecules and peptides conjugated to a pre-
    formed iscom as carrier, J.Immunol.Meth., 98:137.
Lövgren, K. and Morein, B., 1988, The requirement of lipids for the for-
    mation of immunostimulating complexes (iscoms), Biotechnol.Appl.
    Biochem., 10:161.
Lövgren, K., 1988, The serum antibody response distributed in subclasses and
    isotypes after intranasal and subcutaneous immunization with influ-
    enza virus immunostimulating complexes, Scand.J.Immunol., 27:241.
McEwen, C.R., 1967, Tables for estimating sedimentation through linear con-
    centration gradients of sucrose solution, Anal.Biochem., 20:114.
Merza, M.S., Linne, T., Hoglund, S., Morein, B., Portetelle, D. and Burny,
    A., Bovine leukemia virus iscoms: biochemical characterization,
    Vaccine, in press.
Morein, B., Helenius, A., Simons, K., Pettersson, R., Kääriäinen, L. and
    Schirrmacher, V., 1978, Effective subunit vaccines against enveloped
    animal virus, Nature, 276:715.
Morein, B., Sharp, M., Sundquist, B. and Simons, K., 1983, Protein subunit
    vaccines of Parainfluenza type 3 virus: Immunogenic effect in lambs
    and mice, J.gen.Virol., 64:1557.
Morein, B., Sundquist, B., Höglund, S., Dalsgaard, K. and Osterhaus, A.,
    1984, Iscom, a novel structure for antigenic presentation of membrane
    proteins from enveloped viruses, Nature, 308:457.
Morein, B. and Simons, K., 1985, Subunit vaccines against enveloped viruses:
    virosomes, micelles and other protein complexes, Vaccine, 3:83.
Morein, B., Lövgren, K., Höglund, S. and Sundquist, B., 1987, The iscom: an
    immunostimulating complex, Immunol.Today, 8:333.
Morgan, A.J., Finerty, S., Lövgren, K., Scullion, F.T. and Morein, B.,
    1988, Prevention of Epstein-Barr (EB) virus induced lymphoma in
    cottontop tamarins by vaccination with the EB virus envelope glyco-
    protein gp340 incorporated into immune-stimulating complexes, J.gen.
    Virol., 69:2093.
Speijers, G.J.A., Danse, L.H.J.C., Beuvery, J.J.T.W., Strik, A. and Vos,
    J.G., 1987, Local reactions of the Saponin Quil A and a Quil A con-
    taining iscom measles vaccine after intramuscular injection of rats:
    a comparison with the effect of DPT-polio vaccine, Fund.Appl.
    Toxicol., 10:425.
Sundquist, B., Lövgren, K., Höglund, S. and Morein, B., 1988, Influenza vir-
    us iscoms: Biochemical characterization, Vaccine, 6:44.
Sundquist, B., Dalsgaard, K. and Morein, B., 1983, Assay of detergents by
    rocket electrophoresis in agarose gels containing red blood cells:
    "rocket hemolysis", Biochem.Biophys.Res.Comm., 114:699.
Trudel, M., Nadon, F., Séguin, C., Simard, C. and Lussier, G., Experimental
    polyvalent iscoms subunit vaccine based on the fusion protein induces
    antibodies that neutralize human and bovine respiratory syncytial
    virus, Vaccine, in press.

RECOMBINANT SUB-UNIT AND PEPTIDE VIRAL VACCINES

F. Brown

Wellcome Biotechnology
Langley Court
Beckenham, Kent BR3 3BS, UK

INTRODUCTION

Vaccination has been such a major factor in the control of many virus diseases during the last five decades that it is often forgotten how successful vaccines have been, not only in human medicine but also in veterinary medicine. Diseases such as poliomyelitis, measles, mumps, German measles and yellow fever are now rare events in developed countries and smallpox has been eradicated from all countries of the world. In the veterinary field, foot-and-mouth disease, Newcastle disease, Marek's disease and rinderpest are now largely controlled by vaccination.

All these vaccines are either killed or attenuated. Killed vaccines, which are prepared by inactivating large amounts of virus which has been grown in animals, eggs or tissue culture cells, usually require two or more injections at suitably spaced intervals to elicit the protective immune response. For attenuated vaccines, the objective is to select a strain of virus which multiplies sufficiently well to elicit an immune response without causing clinical disease. In the instances of smallpox (in man) and Marek's disease (in chickens) the vaccine strains are naturally occurring viruses isolated from calves and turkeys respectively.

Why then, if these vaccines are so successful, are we concerned with producing different vaccines? There are several reasons.

1. The current vaccines have certain disadvantages which are summarised in Tables 1 and 2.

2. There are diseases for which no classical vaccine is available, either because the virus cannot be grown in sufficient amounts to allow a killed vaccine to be produced or because it has not been possible to isolate an attenuated strain which is sufficiently innocuous but which will still induce a protective immune response.

3. The advantage of a product which would be defined in precise chemical terms; progress in molecular biology now enables this objective to be realised.

4. The opportunity to produce a vaccine which is stable and would not require storage and transport at refrigeration temperatures.

Table 1. Disadvantages with Live Attenuated Vaccines

1. Possible presence of adventitious agents in the cells and medium used for growth

2. Reversion to virulence which causes a small but significant number of clinical cases each year

3. Refrigerator temperatures are required for storage and transport

4. Limited shelf life

Table 2. Disadvantages with Killed Vaccines

1. Hazards to personnel working with large amounts of human pathogens (e.g rabies virus)

2. Hazard to environment when working with large amounts of virus which will infect livestock (e.g. foot-and-mouth disease virus)

3. Need to ensure complete inactivation of infectivity

4. Presence of considerable amounts of cellular material, leading to side-effects (e.g. rabies vaccine produced in the brains of sheep and goats can produce neurological problems in man; foot-and-mouth disease vaccine produced in tissue culture cells can cause hypersensitivity and anaphylaxis)

5. More than one injection usually required

6. Refrigerator temperatures are required for storage and transport

7. Limited shelf-life

5. The desirability of moving away from products which are based on empiricism.

Viruses are the simplest micro-organisms in terms of structure and have provided the best models for exploring the concept of molecular vaccines. In this paper the methods which have been used to achieve these objectives are described.

BASIC CONCEPTS

The impact of molecular biology on the study of viruses has been profound. The realisation that viruses are molecules, stemming from the crystallisation of tobacco mosaic virus and the studies which showed that they contained only nucleic acid and protein as essential constituents led to the intensive chemical analysis of viruses causing a variety of diseases. These studies led to a physico-chemical description of many different viruses which forms the current basis of virus classification (Matthews, 1982). This chemical approach led in turn to an appraisal of the structural entities which confer immunity and to the conclusion that immunogenic activity is usually associated with only one of the structural proteins. This conclusion was reached by dissecting several lipid-containing viruses into subunits, separating these by high-speed centrifugation and other techniques and measuring their ability to elicit neutralizing antibody. Dissection of lipid-containing viruses was accomplished under mild conditions by dissolving the envelope of the viruses in a lipid solvent or mild detergent, thus allowing the biological activity of the sub-units to be retained. The conclusion to emerge was that the immunogenic activity of these viruses was associated with the spike projections clearly observable in electron micrographs. There was indisputable evidence, however, that the activity was two or more orders of magnitude lower than that of the intact virus particle (see review by Brown, 1984).

This response can be enhanced, however, by presenting the sub-units in liposomes or as immunostimulating complexes (ISCOMS; Morein and Simons, 1985). However, the physical chemistry underlying the preparation of liposomes and ISCOMS is rather rudimentary but the results obtained with the immunogenic proteins of measles, influenza and rabies viruses, particularly when presented as ISCOMS, are very encouraging and the results of the exploration of this technology at a more fundamental level are eagerly awaited. The important points to emerge are that individual proteins can induce the production of protective neutralizing antibodies but that the way in which the antigens are presented is important if acceptable immune responses are to be obtained. In other words, the configuration of the protein is critical in obtaining a good immune response.

BIOENGINEERED ANTIGENS

Studies on the replication of viruses have provided considerable information on their genetic structure and identified the genes coding for many of the virus-induced proteins found in infected cells, including those that code for the capsid proteins of the virus particles. Mapping experiments have identified the location of these genes on the nucleic acid of the virus and allowed them to be expressed in suitable systems. The protein VP1 of foot-and-mouth disease virus was expressed in E. coli cells by Kleid et al (1981) and shown to be immunogenic. As expected, its activity was of the same order as that of the protein isolated from virus particles. However, the surface protein of hepatitis B virus, expressed in yeast cells (McAleer et al, 1984), proved to be highly immunogenic, presumably because it assembled into particles similar to the immunogenic particles found in the plasma of infected individuals. This yeast expressed product is the first genetically engineered viral protein which has become a commercially viable product.

It is clear that the configuration of the protein is vitally important in evoking a protective immune response. With the hepatitis B surface protein it was fortuitous that the expressed product assembled into a configuration which evoked such a response. In the immediate future it would appear that formulation into ISCOMS provides the most direct route for the assessment of engineered proteins as immunogens.

PEPTIDES

The possibility of using peptides as vaccines was first envisaged more than 20 years ago when Anderer (1963) showed that a short fragment of the protein of tobacco mosaic virus would elicit the formation of antibody which neutralized the virus. This observation, which was followed up by Anderer and Schlumberger (1965) showing that the synthetic peptide corresponding in sequence to the fragment would also elicit neutralizing antibody, has provided the basis for the work on peptide vaccines which has ensued since the sequences of immunogenic proteins became available.

Advantages

The advantages of a peptide vaccine are summarised in Table 3. Clearly the degree of the advantage of any item on the list will vary with the particular circumstances but the problem of stability at ambient temperatures cannot be over-emphasised, particularly in those situations where it is difficult to maintain a cold chain. Moreover, delayed release mechanisms which are used extensively in drug delivery could be used for a dry vaccine such as a peptide. Consequently the opportunity exists for boosting the immune response without the need for further injections. Such a system would have considerable advantages for all vaccines where booster inoculations are required; this would apply particularly in the developing countries where transport can often pose difficulties.

Identification of Immunogenic Sites

Several methods have been used for identifying immunogenic sites (Table 4). The direct approach (Method 1) in which the immunogenic protein is cleaved, either by enzymes or at specific sites by chemical agents such as cyanogen bromide, is time consuming and the recovery of the fragments is often far from quantitative. Moreover, the cleavage procedures used may destroy the active site. Nevertheless it has led to the identification of active sites in several proteins. Of the others, probably the most successful predictions have come from comparisons of the amino acid sequences of antigenic variants (Method 4). These variants can either be naturally occurring (Bittle et al, 1982) or produced in the laboratory by growing viruses in the presence of neutralizing monoclonal antibodies so that escape mutants are isolated (Minor et al, 1986). However, the more direct methods described by Nunberg et al (1984; Method 5) and Geysen et al (1984; Method 6) hold out much promise. In Nunberg's method, the cDNA corresponding to the gene coding for the immunogenic protein of feline leukemia virus was hydrolysed with DNAse and the individual fragments expressed in λ phage. The phage library was then screened with neutralizing monoclonal antibody to identify those clones expressing the specific antigenic determinants. Sequencing the DNA fragment of the immuno-reactive phage then allowed the sequence of the antigenic region to be derived. The approach by Geysen et al (1984) was even more direct. Working with the immunogenic protein of foot-and-mouth disease virus (FMDV), overlapping hexapeptides corresponding to amino acids 1-6, 2-7, 3-8.....208-213 were synthesized on plastic sticks and then allowed to react with neutralizing antibody. Those peptides with which the antibody had reacted were detected by further reaction with anti-species antibody. The method, now known as the 'Pepscan' method, has the advantage of rapidity and its potential is high provided that the config-

Table 3. Advantages of a Peptide Vaccine

---

1. Product chemically defined

2. Stable indefinitely at ambient temperatures

3. No infectious agent present - hence no problems with innocuity

4. No large-scale production plant required

5. No downstream processing required

6. Can be designed to stimulate appropriate immune responses

7. Provides opportunity to use delayed release mechanisms

Table 4. Identification of Immunogenic Sites on Proteins

1   Measure biological activity of fragments.

2   Identify regions accessible to water.

3. Interpret the secondary structure.

    e g   amphipathic helical region, β-turns, flexibility

4   Relate antigenic variation to amino acid sequence variation in naturally occurring and laboratory derived variants

5   Measure the antigenic activity of proteins expressed from fragments of the coding gene

6   Map synthetic fragments of the protein with neutralizing antibody.

---

uration of the peptides on the plastic sticks is sufficiently similar to that of the naturally occurring sequence on the virus particle.

Several potentially immunogenic peptides have now been assessed for activity. Of these, most work has been done with a linear epitope from protein VP1 of FMDV. These studies will be described in some detail because they provide an example of the problems which remain to be solved once an immunogenic site has been identified and shown to elicit neutralizing antibody.

The Foot-and-Mouth Disease Virus Peptide

FMDV is a spherical particle, 30 nm in diameter and consists of one molecule of ssRNA, mol wt 2.6×10^6 and 60 copies of each of four proteins VP1-VP4. The three proteins VP1-VP3 have mol wt of 24×10^3 and VP4 has a mol wt of 10×10^3. The virus has two distinctive properties which are particularly significant when considering its immunogenic activity: (1) its vulnerability to environments below pH7, when it is disrupted into the infectious RNA, an aggregate of VP4 and a pentameric 12S unit comprising VP1-VP3; the disrupted virus has low immunogenic activity and (2) its vulnerability to proteolytic enzyme cleavage, resulting in a dramatic loss of infectivity due to impairment of cell attachment. With some strains of virus, the cleavage also leads to considerable loss of immunogenicity.

The fact that the only measurable physical alteration ensuing from the enzymic hydrolysis was the cleavage of VP1 pointed to the major importance of this protein in the protective immune response. This observation led to the anticipation that an immune response could be obtained with this protein alone and in 1973 Laporte and his colleagues showed that the separated protein induced neutralizing antibody in pigs although there was no response to VP2 and VP3. However, the amount of VP1 required to elicit a protective level of response was much greater than the amount of inactivated virus particles normally used in vaccination. Thus about 500 μg of VP1 were required, as two inoculations, to achieve protection compared with about 10 μg of virus particles as one inoculation. The greater amount required could be ascribed to the difference in configuration of the isolated VP1 when it was released from the virus particle. A similar argument would account for the low activity of the 12S particle. However, there remains the possibility that the immunogenic site is a complex structure of the surface proteins which is disrupted when the virus is converted to 12S particles or VP1 is cleaved <u>in situ</u> with proteolytic enzymes. Mapping with neutralizing monoclonal antibodies indicates that there is more than one protein involved

(Xie et al, 1987) but the interpretation of these results awaits the three-dimensional structure of the virus.

The apparent immunodominance of VP1 focused attention on this protein and led to studies to locate immunogenic sites. Using the approach described by Anderer in 1963 for tobacco mosaic virus and extended by Sela and his colleagues with MS2 bacteriophage (Langebeheim et al, 1976), Strohmaier and his colleagues (1982) predicted that amino acid sequences 146-154 and 200-213 would be contained in an immunogenic site. Using a different approach Bittle and his colleagues (1982), reasoning that antigenic variation would be reflected in amino acid variation, compared the amino acid sequences of viruses belonging to different serotypes. Three regions of amino acid variability were found, at positions 41-60, 138-160 and 194-205, with the greatest variation being at 138-160. Synthetic peptides corresponding to the 138-160 region from three serotypes elicited levels of neutralizing antibodies which would protect guinea pigs against infection after a single inoculation. Moreover, this protection could be obtained with only one inoculation of the peptide. Other regions of the protein gave either no neutralizing antibody response or very low levels. The second site predicted by Strohmaier et al (1982), namely 200-213, gave low levels of neutralizing antibody and on a weight basis possessed only about 1% of the activity of the 138-160 sequence.

Supportive evidence that the 138-160 region contained the immunodominant site was provided by experiments with naturally occurring antigenic variants of a virus belonging to serotype A, sub-type 12. Rowlands et al (1983) showed that the variants, differing only at positions 148 and 153 in the protein VP1 could be distinguished readily by cross-neutralization tests. Moreover, antisera produced by injection of the corresponding peptides distinguished between the variants in cross-neutralization and immunoprecipitation tests.

These observations inevitably led to the conclusion that a vaccine could be produced with peptides corresponding to this region of VP1 as their basis. The first experiments were made with the peptide coupled to a protein such as keyhole limpet haemocyanin or bovine serum albumin. Protection of guinea pigs could be achieved with a single inoculation of 30 μg when Freund's incomplete adjuvant was used. However, the response in cattle and pigs was lower and in general a single inoculation did not afford protection. The different responses in the three host species indicated that the way in which the peptide was presented to the recipient could be crucial and led to an investigation of this aspect of immunization.

Attachment of the Peptide to the N Terminus of β-galactosidase

The ordered coupling of peptides to proteins has not been investigated in any detailed manner. The literature reveals that several methods have been used, ranging from crude uncontrolled linking with glutaraldehyde to the more specific linking via an added cysteine at one of the termini. Francis et al (1987a) found that the peptide alone was immunogenic in guinea pigs and that coupling it to different carrier protein molecules did not lead to an enhanced response. Indeed in some instances it was found that the response was lower. Consequently a more defined attachment of the peptide to a carrier protein has been investigated to determine whether it would lead to better antibody responses.

In the first experiments, the peptide sequence was fused to β-galactosidase at its N terminus by expressing in E. coli cells the gene coding for the peptide ligated to the gene coding for β-galactosidase. This construction has the potential added advantage that the antigenic sites on β-galactosidase which are recognised by helper T cells are known. However Winther

and his colleagues (1986) showed that this fusion protein was no more immunogenic for mice or guinea pigs than the chemically linked peptide.
Broekhuijsen et al (1986) reported similar results.

However, it has now been shown (Table 5) that proteins consisting of
two or four copies of the peptide linked to β-galactosidase at the N terminus have a much greater immunogenic activity than the single copy construct (Broekhuijsen et al, 1987). Moreover, 40 µg of the peptide in the
four copy construct, given as a single inoculation, protected pigs against
challenge inoculation with 60000 ID_{50} of virus.

The reason for the much greater activity of the two and four copy constructs is not known. It could be the repetition of the antigenic determinant which is important and a detailed structural study to compare the configuration of the peptide sequence on the one, two and four residue constructions could be rewarding.

Expression of the Peptide as Part of the Hepatitis B Core Particle

The response to the peptide sequence when it forms part of the virus
particle is very high. As little as 1 µg of virus, which contains 0.02 µg
of the 141-160 sequence on the 60 copies of VP1, will provide protection
against challenge infection. Consequently Clarke et al (1987) have investigated whether multiple copies of the peptide, presented on a particle of the
same size as a picornavirus, will evoke a similar high response. The core
protein of hepatitis B virus self assembles into particles which are 27 nm
in diameter and contain several hundred copies of the molecule. Moreover,
Peter Highfield (personal communication) had shown that hybrid proteins in
which foreign amino acid sequences are expressed at the N terminus of the
hepatitis B core protein will also self assemble into particles. On the
basis of this information, Clarke et al (1987) expressed the DNA coding for
the FMDV peptide, six amino acids of the pre-core particle and the entire
core protein. Although this construct piosoned the E. coli system, it was

Table 5. Anti-peptide Antibody, Neutralizing Antibody and Protective Immune
Response of Guinea Pigs Inoculated with β-galactosidase Fusion
Proteins Containing One, Two or Four Copies of the FMDV Peptide

Sample	Dose as µg of peptide	Anti-peptide antibody (log_{10})		Neutralizing antibody (log_{10})		Protection at 56 days
		28 days	56 days	28 days	56 days	
One copy	5	1.2	1.8	<0.6	<0 6	NP
Two copies	10	3.8	3.9	2.0	1.6	P
Four copies	20	3.3	3.5	1 5	1.6	P
Synthetic peptide 137-160 Cys	5	-	-	<0.6	0.6	NP

P = Protected, NP = Not protected

found that it could be expressed in vaccinia virus. Moreover, the expressed hybrid protein assembled into particles which reacted with antisera prepared against hepatitis B core particles, the FMDV peptide and, crucially, the FMDV particle. The particles were separated from the vaccinia virus and tested for their immunogenic activity in guinea pigs. One injection of as little as 2 µg of the particles, containing the equivalent of 0.2 µg of peptide, elicited very high neutralizing antibody levels which protected against challenge infection. This response is by far the best that has been obtained with a single inoculation of the peptide and is not very much lower than that obtained with virus particles (Table 6).

The excellent response to the peptide when it is expressed in this form may be explained in purely physical terms. However, other factors may be playing a role in the enhanced response since Milich and his colleagues (1986) have shown that the hepatitis B core protein can induce antibody responses via both T-cell dependent and T-cell independent pathways. Moreover, the core specific helper T cells can help B cells produce antibody against the envelope antigens of hepatitis B virus as well as the core proteins, even though these antigens are on different molecules. These immunological properties may account for the excellent response to the FMDV peptide when it is presented as part of the core particle. It is clear that the exploration of this method of presentation may provide important clues for the practical application of peptide vaccines.

The Role of T Cell Epitopes

As mentioned above, the 141-160 peptide region of VP1 elicits a neutralizing antibody response which protects guinea pigs against experimental infection, but the response in pigs and cattle is lower although the antibodies which are elicited are qualitatively similar in all three species. These observations pointed to a genetic restriction in the response. It had been appreciated that to realise their full potential as vaccines it is necessary for the peptides to combine with helper T cell receptors and Ia

Table 6. Comparative Immunogenicity in Guinea Pigs of Inactivated FMDV Particles, Serotype O_1 and the Hepatitis B Core Peptide Fusion

Test Sample	Dose (µg)		Days post inoculation		
	Total antigen	FMDV VP1 142-160 sequence	0	28	42
Inactivated FMDV particles	1	0.02	<0.6*	1.6	1.7
HBcAg/142-160 fusion protein	2	0.2	<0.6	2.1	2.9

*$\log_{10}$ SN_{50} against 100 $TCID_{50}$ of virus

antigens in addition to antibody binding sites. It has been shown recently that non-immunogenic B cell epitopes of the malaria circumsporozoite (Good et al, 1987) and hepatitis B surface antigen (Leclerc et al, 1987) can be made immunogenic either by conjugation with a 'natural' T cell epitope or even by co-polymerisation with a 'foreign' T-cell epitope. Similarly, Francis et al (1987b) have shown that non-responsiveness in mice to the FMDV peptide can be overcome by adding foreign T cell sites. In the experiments with the FMDV peptide, the authors found that only two of six strains of congenic B10 mice, differing only at the locus of the H-2 complex, responded to the free peptide in incomplete Freund's adjuvant. These were the H-2^k and H-2^r strains. Among the non-responding mice was the B10.D2 strain, which belongs to the H-2^d haplotype and for which several T-helper cell epitopes have been described. Three of the epitopes, one from ovalbumin (amino acids 323-339; OVA) and two from sperm myoglobin (amino acids 132-148; SWM1 and amino acids 105-121; SWM2) were added to the C terminus of the 141-160 FMDV peptide. As a control, a fourth peptide consisting of the 141-160 peptide and the 161-177 sequence was used. Each peptide also had a non-natural cysteine residue added to the C-terminus for increased immunogenicity.

Neither the B10.D2 or Balb/c strains of mice (both H-2^d haplotype) produced any antibody response to the 141-160 + 161-177 peptide but each of the 141-160 + OVA, 141-160 + SWM1 and 141-160 + SWM2 peptides elicited a good response. Unexpectedly, the 141-160 + SWM2 peptide did not elicit any neutralizing antibody in either strain after one injection, although the Balb/c mice produced neutralizing antibody after two injections. This suggests that T-helper cell epitopes could control antibody production of specific B cell clones.

These results provide further evidence that genetically controlled non-responsiveness to the 141-160 peptide can be overcome by adding foreign T cell epitopes and that functional (i.e. neutralizing) antibody can be produced. The significance of this result for the vaccination of cattle with the 141-160 peptide is that the poor response to the uncoupled sequence may be overcome by presenting it with a suitable T-helper cell epitope.

CONCLUSIONS

Reductionism, in the form of molecular biology, is driving research in many areas of biology. The results, in terms of a better understanding of a large number of diseases, have been spectacular. In no area have the advances been more rapid than in virology and in the basic understanding of immunity to virus diseases. The relatively simple structure of virus particles, compared with other micro-organisms, has allowed them to be studied more readily at the molecular level and foot-and-mouth disease virus has provided an excellent model for this type of work. Not only does it cause an important disease but its molecular structure has been worked out in great detail.

The clear message which has emerged from this approach to the designing of new vaccines is that it should be possible to construct molecules which will provide immunity comparable to that elicited by conventional vaccines. Moreover, it should also be possible, as we learn more about the structure of the B and T cell epitopes of a particular virus, to design vaccines which can overcome problems such a antigenic variation against which the current vaccines do not afford protection.

Although it is probably too radical at present to state, as Jim Watson is alleged to have said "There are only atoms, everything else is merely social work", it seems to me that experiments aimed at gaining an understanding of immunity to virus diseases at the molecular level will eventually bear a rich harvest.

REFERENCES

Anderer, F.A., 1963, Versuche zur bestimmung der serologisch terminaten gruppen des tobakmosaikvirus, Z.Naturf.B., 188:1010.

Anderer, F.A. and Schlumberger, H.D., 1965, Properties of different artificial antigens immunologically related to tobacco mosaic virus, Biochim.Biophys.Acta, 97:503.

Bittle, J.L., Houghten, R.A., Alexander, H., Shinnick, T.M., Sutcliffe, J.G., Lerner, R.A., Rowlands, D.J. and Brown, F., 1982, Protection against foot-and-mouth disease by immunization with a chemically synthesised peptide predicted from the viral nucleotide sequence, Nature, 298:30.

Broekhuijsen, M.P., Blom, T., Kottenhagen, M., Pouwels, P.H., Meloen, R.H., Barteling, S.J. and Enger-Valk, B.E., 1986, Synthesis of fusion proteins containing antigenic determinants of foot-and-mouth disease virus, Vaccine, 4:119.

Broekhuijsen, M.P., Van Rijn, J.M.M., Blom, A.J.M., Pouwels, P.H., Enger Valk, B.E., Brown, F. and Francis, M.J., 1987, Fusion proteins with multiple copies of the major antigenic determinants of foot-and-mouth disease virus protect both the natural host and laboratory animals, J.gen.Virol., 68:3137.

Brown, F., 1984, Synthetic viral vaccines, Ann.Rev.Microbiol., 3:221.

Clarke, B.E., Newton, S.E., Carroll, A.R., Francis, M.J., Appleyard, G., Syred, A.D., Highfield, P.E., Rowlands, D.J. and Brown, F., 1987, Improved immunogenicity of a peptide epitope after fusion to hepatitis B core protein, Nature, 330:381.

Francis, M.J., Fry, C.M., Rowlands, D.J., Bittle, J.L., Houghten, R.A., Lerner, R.A. and Brown, F., 1987a, Immune response to uncoupled peptides of foot-and-mouth disease virus, Immunology, 61:1.

Francis, M.J., Hasting, G.A., Syred, A.D., McGinn, B., Brown, F. and Rowlands, D.J., 1987b, Non-responsiveness to a foot-and-mouth disease virus peptide overcome by addition of foreign helper T-cell determinants, Nature, 330:168.

Geysen, H.M., Meloen, R.H. and Barteling, S.J., 1984, Use of peptide synthesis to probe viral antigens for epitopes to a resolution of a single amino acid, Proc.Natl.Acad.Sci.USA, 81:3998.

Good, M.F., Maloy, W.K., Lunde, M.N., Margalit, H., Cornette, J.L., Smith, G.L., Moss, B., Miller, L.H. and Berzofsky, J.A., 1987, Construction of synthetic immunogen: use of new T-helper epitope on malaria circumsporozoite protein, Science, 235:1059.

Kleid, D.G., Yansura, D., Small, B., Dowbenko, D., Moore, D.M., Grubman, M.J., McKercher, P.D., Morgan, D.O., Robertson, B.H. and Bachrach, H.L., 1981, Cloned viral protein vaccine for foot-and-mouth disease; Responses in cattle and swine, Science, Wash., 214:1125.

Langebeheim, H., Arnon, R. and Sela, M., 1976, Antiviral effect on MS2 coliphage obtained with a synthetic antigen, Proc.Natl.Acad.Sci.USA, 73:4636.

Leclerc, C., Przewlocki, G., Schutze, M. and Chedid, L., 1987, A synthetic vaccine constructed by copolymerization of B and T cell determinants, Eur.J.Immunol., 17:269.

McAleer, W.J., Buynak, E.B., Maigetter, R.Z., Wampler, D.E., Miller, W.J. and Hilleman, M.R., 1984, Human hepatitis B vaccine from recombinant yeast, Nature, 307:178.

Matthews, R.E.F., 1982, Classification and nomenclature of viruses, 4th Report, Intervirology, 17:1.

Milich, D.R. and McLachlan, A., 1986, The nucleocapsid of hepatitis B virus is both a T-cell-independent and a T-cell-dependent antigen, <u>Science</u>, 234:1398.

Minor, P.D., Ferguson, M., Evans, D.M.A., Almond, J.W. and Icenogle, J.P., 1986, Antigenic structure of polioviruses of serotypes 1, 2 and 3, <u>J.gen.Virol</u>., 67:1283.

Morein, B. and Simons, K., 1985, Subunit vaccines against enveloped viruses; virosomes, micelles and other protein complexes, <u>Vaccine</u>, 2:83.

Nunberg, J.H., Rodgers, G., Gilbert, J.H. and Snead, R.M., 1984, Method to map antigenic determinants recognised by monoclonal antibodies: Localisation of a determinant of virus neutralization on the feline leukemia virus envelope gp70, <u>Proc.Natl.Acad.Sci.USA</u>, 81:3675.

Rowlands, D.J., Clarke, B.E., Carroll, A.R., Brown, F., Nicholson, B.H., Bittle, J.L., Houghten, R.A. and Lerner, R.A., 1983, Chemical basis of antigenic variations in foot-and-mouth disease virus, <u>Nature</u>, 306:694.

Strohmaier, K., Franze, R. and Adam, K-H., 1982, Localisation and characterisation of the antigenic portion of the foot-and-mouth disease virus protein, <u>J.gen.Virol</u>., 59:295.

Winther, M.D., Allen, G., Bomford, R.H. and Brown, F., 1986, Bacterially expressed antigenic peptide from foot-and-mouth disease virus capsid elicits variable immunologic responses in animals, <u>J.Immunol</u>., 136:1835.

Xie, Q-C., McCahon, D., Crowther, J.R., Belsham, G.J. and McCullough, K.C., 1987, Neutralization of foot-and-mouth disease can be mediated through any of at least three separate antigenic sites, <u>J.gen.Virol</u>., 68:1637.

SYNTHETIC ANTIGENS AND VACCINES

Ruth Arnon

Department of Chemical Immunology
The Weizmann Institute of Science
Rehovot, 76100, Israel

Early studies in our laboratory have demonstrated that synthetic antigens containing an immunoreactive region of a protein can give rise to a
specific, and often conformation-dependent, immune response towards the
intact native protein (Arnon et al, 1971). When the protein in question is
a component of a virus e.g. the coat protein of MS-2 coliphage, the antibodies induced by a synthetic fragment were capable of neutralizing the viability of the phage (Langbeheim et al, 1976). These findings paved the way
for the study of synthetic vaccines. We have employed this approach for the
study of three systems - the influenza virus and the bacterial toxins of
cholera and shigella. The results achieved to date as well as the future
prospects of these three synthetic vaccines are discussed below.

I. SYNTHETIC VACCINE AGAINST INFLUENZA VIRUS

Influenza virus contains two major membranal proteins - hemagglutinin
and neuraminidase. The hemagglutinin (HA) occurs as trimeric spikes pro-
jecting from the viral proteolipid envelope. It is quantitatively the most
important glycoprotein in the viral surface and is the major antigenic protein, against which neutralizing antibodies are directed. In view of these
properties it was considered a good candidate for studying the synthetic
approach to vaccination.

The influenza virus provides a very suitable model for studying the
synthetic approach to vaccination, for several reasons: a) detailed information is available on the structure and function of this virus, as well as
on its serological specificities and genetic variations; b) various reliable
assays of the virus are available for evaluating the effect of the immune
response on the different viral functions; c) sufficient information is
available on the amino acid sequence of the influenza hemagglutinin of many
viral strains, as well as on its three dimensional structure and immunochemical properties, to allow the synthesis of peptide fragments that might
carry some of its antigenic determinants.

Another advantage of this system is that once such fragments have been
synthesized, the immune response they elicit can be assessed on four different levels: 1. the immunochemical reaction - namely, the capacity of the
elicited antibodies to interact with the peptide as such and to cross-react
with the intact virus; 2. the inhibitory effect on the biological activity

of the hemagglutinin; 3. the _in vitro_ neutralization of the virus, as expressed by reduction of virus plaque formation in tissue-cultured cell monolayers; and finally, 4. the most crucial criterion - the _in vivo_ protection of animals. Since mice are susceptible to the same viral strains that are infectious in man, they provide an adequate animal model for evaluating the decrease in the incidence and/or severity of infection after active immunization with the synthetic antigens.

The first peptide we synthesized, before the three-dimensional structure was known, consisted of 18 amino acid residues corresponding to the sequence 91-108 of the HA molecule. This region, which is common to at least 12 H3 strains, was computer-predicted to be immunologically reactive. Indeed, a conjugate of this peptide with tetanus toxoid elicited in both rabbits and mice antibodies that reacted immunochemically with the synthetic peptide, as well as with the intact influenza virus of several strains of type A.

These antibodies were capable of inhibiting the capacity of the HA of the relevant strains to agglutinate chicken red blood cells. They also interfered with the _in vitro_ growth of the virus in tissue culture, causing up to 60% reduction in viral plaque formation. Furthermore, as shown in Table 1, mice immunized with the peptide-toxoid conjugate were partially protected against further challenge infection with the virus (Müller et al, 1982).

One of the crucial factors concerning the influenza vaccine is the tremendous genetic variations among the various virus strains - shifts and drifts - and their reflection on the seriological specificities (Laver and Air, 1979). Detailed studies of amino acid sequences and X-ray diffraction (Wiley et al, 1981) resulted in the location of four antigenic sites, designated A through D, on the HA molecule. Amino acid substitutions at these sites caused by mutation led to the development of new virus strains with changed immunogenic properties, whereas other regions were shown to be "constant".

The peptide 91-108 was deliberately chosen to be part of a conserved sequence, and although it does not overlap with any of the four proposed antigenic determinants of the native HA, it is adjacent in the three-dimensional structure to the antigenic site D. This could provide an explanation for the partial protective effect achieved by immunization with this peptide. Furthermore, because it is a conserved region, namely the same sequence is present in the HA of several influenza strains, immunization with the same conjugate elicited protection against more than one H3 strain (Table 1) - thus indicating that the synthetic approach might lead to multivalent vaccines for cross-strain protection.

As evident from the three-dimensional structure of the influenza HA (Wiley et al, 1981), the region 140-146, which forms antigenic site A, is a "loop" of seven amino acid residues unusually protruding from the surface of the molecule. It is also exposed in the trimeric structure assumed by the HA in the spikes on the virus surface. One would expect that peptides corresponding to this region, where "natural" immunogenic determinants of the HA molecule are located, would elicit a better and more specific immune response against the virus. Yet Jackson et al (1982), using a synthetic peptide with the sequence 123-151 that includes the loop region, for immunization of rabbits, did not find significant binding between the antipeptide antibodies and the X-31 virus.

In our studies (Shapira et al, 1984), four peptides of this region have been synthesized. Two of them corresponded to the sequence 139-146, with either Gly or Asp at position 144. The third peptide corresponded to the

Table 1. Protection of Mice against Challenge Infection with Influenza Virus

Infectious Strain	Group	Incidence of infection at 10^{-2} dilution into egg	EID[***]
A/TEX/77	Immunized*	19/36 (52%)	$10^{-1.98}$
	Control**	21/23 (91%)	$10^{-3.56}$
A/PC/75	Immunized	4/18 (22%)	$10^{-0.61}$
	Control	8/19 (42%)	$10^{-1.53}$
A/ENG/42/72	Immunized	8/18 (44%)	$10^{-1.61}$
	Control	15/21 (71%)	$10^{-2.9}$
A/PR/34(H1)	Immunized	18/19 (95%)	$10^{-3.47}$
	Control	20/21 (95%)	$10^{-3.95}$

  *Immunized with a conjugate of the peptide 91-108 and tetanus toxoid in CFA.
 **Control groups were injected with the tetanus toxoid alone in CFA.
***Egg-infectious dose.

sequence 147-164, and the fourth included both regions and corresponded to the sequence 138-164. Conjugates of these peptides with tetanus toxoid (TT) elicited in rabbits high antibody titer against the respective homologous peptides, with a significant extent of cross-reactivity among them. They differed, however, in their cross-reactivity with the intact H3 influenza virus as indicated by radioimmuno-binding assays. Thus, antibodies induced by the two longer peptides, 147-164 and 138-164, showed significant binding with the intact virus, whereas the extent of virus binding of the antisera against the two short octapeptides 139-146 was essentially insignificant. In contrast, antibodies raised against the intact virus (A/Mem/102/72) or against the isolated hemagglutinin were capable of recognizing the synthetic "loop" octapeptides, but did not react at all with the region 147-164.

The role of the size of the peptide fragment in this system was even more emphasized when evaluating the interference of the antibodies with the biological activity of the virus. In this reaction the only peptide which proved effective was the 138-164 fragment. Moreover, immunization of mice with the tetanus toxoid conjugate of this peptide (but not with other conjugates) resulted in their partial protection against infection with the A/Eng/42/72 strain, with which the sequence of the synthetic peptide corresponds.

These results indicate that the loop region 140-146 of the influenza hemagglutinin, although constituting a major naturally occurring antigenic determinant of the intact virus, is by itself too short to fold into a loop. However, when forming a part of the longer peptide 138-164, the folding to the right conformation is facilitated, to yield a protective epitope.

Preliminary experiments from our laboratory show that a third region in the influenza hemagglutinin molecule is also influential in the immunological reactivity and this is the region 181-200, which, according to Wiley et al (1981), constitutes the antigenic site B. Antibodies raised by a conjugate of this peptide were cross-reactive with the intact virus and were capable of inhibiting its biological activity and viability both _in vitro_ and _in vivo_ (Shapira et al, 1984).

II. SYNTHETIC ANTI-INFLUENZA VACCINE WITH BUILT-IN ADJUVANTICITY

All the results reported above were achieved by immunization in complete Freund's adjuvant (CFA) for augmenting the immune reactivity. This adjuvant, which consists of a water-in-oil emulsion containing killed mycobacteria, is a very effective adjuvant evoking high level and long lasting immunity. However, it is not suitable for human use, since it induces local reactions and granulomas, inflammation and fever, due to both the low metabolizable mineral oil and the mycobacteria. The choice of adjuvant for vaccine preparation is of crucial importance, particularly when a synthetic vaccine is being considered, since the synthetic materials are usually water soluble and therefore less immunogenic than particulate substances.

It has been demonstrated that the mycobacteria in Freund's adjuvant can be replaced with less damaging substances. The minimal structure that can substitute for mycobacteria was identified as N-acetyl-muramyl-L-alanyl-D-isoglutamine, denoted MDP for muramyl dipeptide (Ellouz et al, 1974). This material and some of its synthetic analogs have already been used in combination with synthetic antigens (Arnon et al, 1980; Audibert et al, 1982; Carelli et al, 1982).

We have employed MDP in combination with the 91-108 peptide of hemagglutinin for induction of anti-influenza response. Our findings showed that this adjuvant was similar to CFA in the induction of antibodies specific towards the peptide (Shapira et al, 1985). As for the cross-reactivity with the intact virus, although a marked difference was observed between the titer of antibodies raised in the presence of CFA and the MDP, we found that both adjuvants were capable of inducing protective immunity against a viral challenge. This impliles a probable participation of cellular immunity induced by the synthetic peptide and is in accord with several recent publications which emphasize the role of cellular immunity in anti-viral protection (Ada et al, 1981).

Incidentally, MDP was efficient only when coupled covalently to the conjugate (91-108) TT. However, in that form it led to the highest protection rate against _in vivo_ viral challenge, even higher than that induced in the presence of CFA. It should be emphasized that the conjugate of the synthetic peptide and MDP with the tetanus toxoid is water soluble and was administered in a physiological aqueous solution. Hence, this constitutes a synthetic vaccine with built-in adjuvanticity, which should be suitable for use in humans.

III. BACTERIAL TOXINS

The synthetic approach to vaccination can be employed also in the case of bacterial toxins. This has been demonstrated by Audibert and her colleagues (1981) who have reported on a synthetic peptide analogous to a region of diphtheria toxin (residues 188-201) which after attachment to bovine serum albumin, led to the production in guinea pigs of antibodies capable of neutralizing the dermonecrotic activity of diphtheria toxin.

This work was subsequently extended to additional synthetic peptides of the same region which were attached together with MDP to the synthetic carrier multichain poly-DL-alanine. The resulting totally synthetic immunogen, when administered in aqueous solution, provoked neutralizing antibodies in mice (Audibert et al, 1982).

We have used the synthetic approach in the case of two bacterial toxins - that of <u>vibrio cholera</u> and that of <u>Shigella dysenteria</u>. In both cases it was possible to provoke antibodies cross-reactive with native toxin that were capable of partially neutralizing its biological activity, as summarized in the following.

IV. CHOLERA TOXIN

The toxin of <u>Vibrio cholerae</u> is composed of two subunits, A and B. Subunit A activates adenylate cyclase, which triggers the biological activity, whereas subunit B is responsible for binding to cell receptors and expresses most of the immunopotent determinants. Antibodies to the B subunits are capable also of neutralizing the biological activity of the intact toxin. In vieew of its immunodominant role the B subunit was the obvious candidate for attempting the use of synthetic peptides for immunization. Moreover, in view of the high level of sequence homology between the B subunits of cholera toxin (CT) and the heat labile toxin of <u>E. coli</u> (LT) we have also investigated whether the same synthetic peptides will also cross-react with the <u>E. coli</u> toxin.

Six peptides, corresponding to sequences 8-20; 30-42; 50-64; 69-85; 75-85; and 83-97 (denoted CTP 1 to CTP 6, respectively) were synthesized and coupled to tetanus toxoid. All six conjugates elicited antibodies against the respective homologous peptides and four out of the six also reacted, to different extents, with the intact B subunit and with native cholera toxin (Jacob et al, 1983). The cross-reactivity was manifested in both ELISA and immunoblotting assays.

Of most interest among these peptides were CTP1 (residues 8 to 20) and CTP 3 (residues 50 to 64). Antisera against these two peptides exerted significant inhibition of the biological activity of cholera toxin. The toxic effect of CT can be demonstrated by skin vascular permeation and by fluid accumulation in ligated small intestinal loops, as well as on the biochemical level by the induction of adenylate cyclase. The inhibitory effect of the antipeptide sera was manifested in all the assays of the biological activity of the toxin with very good correlation between the biochemical level and the end biological effect of the toxin. In both cases inhibition values of 60 to 70% were obtained (Jacob et al, 1984a).

Simultaneous immunization with the two peptides, CTP1 and CTP3, was carried out in order to evaluate whether the combined immune response might lead to augmentation of the neutralizing effect achieved by each peptide separately. The peptides were administered to rabbits when bound together to the same tetanus toxoid carrier molecule, presented on separate carrier molecules, or in a polymerized form given without any carrier. However, when compared to the results obtained with the individual peptides, either as TT conjugate or in the polymerized form, there was no elevation in the anticholera toxin binding or augmentation of the neutralizing efficiency of the antibodies obtained as a result of the combined immunization. Similarly, simultaneous addition of the two antisera (anti-CTP1 and anti-CTP3) did not result in additivity of the inhibitory effect.

It is noteworthy that all synthetic peptides investigated, in their free form, as such do not induce any activation of adenylate cyclase as measured by cAMP production or fluid secretion into ligated ileal loops. This implies a low probability of undesirable biological effects as a result of their administration, which is a crucial consideration for the possible use of synthetic material for vaccination.

V. HEAT-LABILE TOXIN OF E. COLI

The heat labile toxin (LT) of pathogenic strains of E. coli is the causative agent of diarrhea in many tropical countries and due to its wide spread, presents probably a more serious health problem than cholera. As mentioned above, there is a high level of sequence homology between the B subunits of the LT and CT. Moreover, immunological relationship was demonstrated between the two toxins, with the existence of both shared and specific antigenic determinants (Lindhom et al, 1983). Using synthetic peptides we have corroborated this assumption. We demonstrated that indeed the antiserum elicited by CTP 3 (residues 50-64) is highly cross-reactive with the LT in both radioimmunoassay and immunoblotting. This is not surprising, since in this region the sequence homology between the two toxins is complete. The antiserum against CTP 1 (residues 8-20) was also cross-reactive with the two toxins, although to a much lower extent. This cross-reaction was observed with toxins of both human and porcine strains of E. coli. Moreover, antisera to both CTP 1 and CTP 3, which are inhibitory towards CT, were found equally effective in neutralizing the biological activity of the E. coli LT. This was manifested by significant inhibition of both adenylate cyclase induction and the fluid secretion into ligated ileal loops of rats (Fig. 1).

Of special interest was the observation that antisera against the two synthetic peptides cross-reacted with multiple strains of E. coli (Jacob et al, 1984b). All these strains were identified as LT-positive and most of them were cross-reactive with antisera against CT. However, a few of these strains exhibited higher reactivity with anti-CTP 3 than with the intact cholera toxin. It is thus possible that a synthetic peptide might prove a more efficient immunizing agent than an intact native protein, in particular when cross-reactivity is considered.

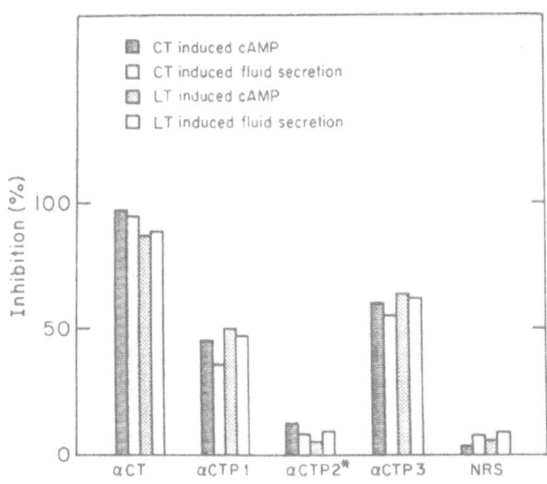

*Antisera to CTP 4, 5, 6-gave very similar results

Fig. 1. Inhibition of biological activities of cholera toxin and heat labile toxin of E. coli by various antisera. Correlation between inhibition of CT and LT by the antisera is manifested by inhibition of fluid secretion into ligated intestinal loops and inhibition of adenylate cyclase activation. Measurements of cAMP levels were made after 3 hr incubation of either 50 ng CT or 0.25 μg LT with 1:10 dilution of various antisera. Each bar represents the mean value obtained in three to five separate experiments.

VI. PRIMING IMMUNIZATION

The ability of synthetic peptides to prime the immune system to a secondary stimulus with the whole organism or native protein, is of great importance in vaccination. This approach for immunization might be of general practical value in endemic areas, where the population is constantly vulnerable to a low level of a particular infectious agent. This exposure has no influence on the unprimed immune system, but could serve as a booster in the case of individuals primed with an appropriate peptide, thus leading to a secondary immunological reaction. The basis for the priming effect lies within the heterogenic nature of the immune response. Antibodies which are formed against synthetic peptides most probably have various specificities, only a part of which may be relevant to the native protein. A secondary stimulus with the native protein serves as a specific booster only to those cells which produce antibodies that recognize the native protein or causative agent, thus enhancing the specific antibody population, selectively. Since the administration of the native protein serves as a booster to cells which are already sensitized, a very low dose of it should be sufficient for eliciting a secondary immune response against the intact protein. This phenomenon was indeed observed with several systems using synthetic peptides, although not in every case.

In our studies, rabbits were "primed" by one or two injections, at 4-week intervals, of 1 mg peptide-TT conjugate, followed by a single booster injection of a subeffective dose (1 μg) of CT. All cholera toxin peptides investigated were able to prime for a significant neutralizing anti-CT immune response as demonstrated by high levels of cross-reacting antibodies and by efficient inhibition of CT-induced cAMP (Jacob et al, 1986). The priming effect was achieved irrespective of whether the immunization with the peptide conjugate alone can lead to a neutralizing response or not. In fact, even conjugates of peptides that do not induce any CT-cross-reactive antibodies at all, also led to a priming phenomenon, though to a lower level. The priming effect was even more pronounced after a single immunization with the carrier-linked peptides, which as such did not result in any immune response, but after boosting led to a high level of anti CT antibodies. Moreover, a single dose of CTP3, by itself nonimmunogenic, primed rabbits to a vigorous immune response towards subsequent administration of single small subimmunizing doses of both cholera toxin and heat labile toxin of E. coli, as well as to neutralization of their biological activity.

VII. COMBINED USE OF SYNTHETIC PEPTIDES AND RECOMBINANT DNA

An alternative approach to the chemical synthesis of vaccines is the use of genetic engineering. We attempted to bridge the synthetic and recombinant DNA approaches with regard to the most effective cholera toxin peptide, namely CTP3. Our working hypothesis was that the expression of the CTP3 peptide by bacteria may provide an appropriate agent for induction of immunity towards cholera and E. coli toxins. Two plasmids containing a synthetic "gene" coding for the region 50 to 64 in the amino acid sequence of cholera toxin, denoted pAM1 and pAM2, were prepared (Jacob et al, 1985) and inserted in phase into the gene coding for E. coli β-galactosidase (Fig. 2).

The resulting fusion protein, which was expressed by the two vectors, was enzymatically active and reacted with antibodies against -galactosidase. It also reacted with antibodies against CTP3 and, though to a much lesser extent, with antibodies against intact CT. Immunization with the fusion protein by itself, did not lead to significant titer of antibodies recognizing CT. However, when followed by a booster injection of a minute amount (1 μg) of CT, too small to provide any immune response by itself, it

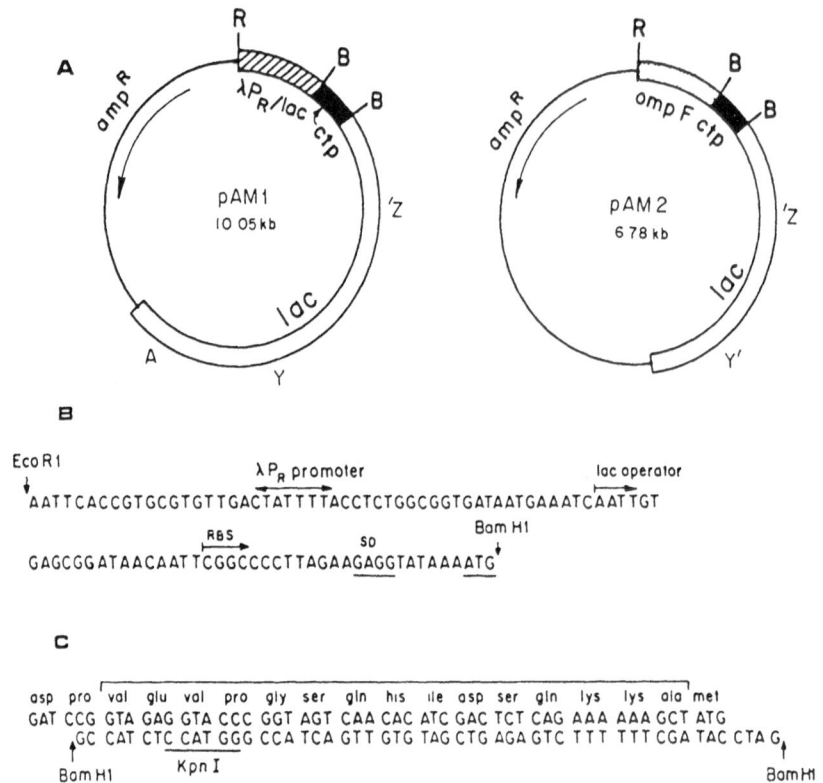

Fig. 2.   Structure of vectors for the expression of CTP3-β-galactosidase fusion products. (A) General structure of expression vectors. Hatched box, composite P_r/Lac promoter (in pAMI); dotted box, ompF promoter and amino-terminal coding sequence (in pAM2); filled boxes, CPT3 coding sequence (ctp); open boxes, parts of the lac operon; line, part of pBR322 sequence. Restriction sites: R, EcorR1, B, BamH1. (B) Sequence of the regulatory region in pAMI. Arrows designate cleavage sites by the indicated restriction enzymes. RBS, ribosome binding site; SD, Shine and Dalgarno sequence. (C) Sequence of the synthetic, 54-bp DNA duplex coding for CTP3 (ctp). The amino acid sequence corresponding to CTP3 is overlined. Arrows designate cleavage by BamH1.

led to a substantial level of neutralizing antibodies. Similar results were obtained after boosting with the heat-labile toxin of E. coli (Jacob et al, 1985). These findings demonstrate that when relevant antigenic determinants are defined, their corresponding synthetic oligonucleotides can be utilized, thus combining genetic engineering and peptide chemistry for the future synthetic vaccines.

VIII.   SHIGA TOXIN

Shiga toxin, the toxin isolated from Shigella dysenteriae 1 strains is one of the most potent of the lethal microbial toxins. It was initially classified as a neurotoxin due to the limb paralysis and death it causes when parenterally administered to sensitive animals. In addition, it is cytotoxic to certain tissue culture cells, and also exerts enterotoxic activity when applied to intestinal ileal loops in vivo. This toxin has been purified and shown to consist of A and B subunits of Mr 32,000 and Mr 6500.

Recently, the complete amino acid sequence of the B chain was determined, according to which the molecular weight of this subunit is 7,716 daltons (Seida et al, 1986). Incidentally, this sequence is identical to that of the B subunit of the shiga-like E. coli toxin, as deduced recently from its DNA sequence. Antibodies raised against the B subunit were shown to neutralize the cytotoxic effects of the toxin in the HeLa cell monolayers and also inhibit the binding of labeled toxin to these cells. In view of this, the B chain of shiga toxin was subjected to an evaluation of the synthetic approach to vaccination (Harari et al, 1988).

Computerized plots of hydrophilicity and surface residues identified the N-terminal part of the B chain as the most probable antigenic region. Hence, three overlapping peptides of this region, corresponding to sequences 5-18, 13-26 and 7-26 were synthesized and conjugated to several carriers or polymerized. Rabbit antisera raised against all the conjugates and polymers were highly reactive with the respective homologous peptides, and those manifesting the highest titer were further evaluated for their reactivity with the intact toxin. In the case of all three peptides high extent of cross reactivity with the Shiga toxin was demonstrated by both ELISA and immunoblotting tests. More significantly, the antisera showed considerable neutralizing capacity towards all three biological effects of the toxin. Thus, the anti-peptide antibodies at a 1:100 dilution led to about 70% inhibition of HeLa cell cytotoxicity, almost equal to that effected by antibodies against whole Shiga toxin in a similar dilution.

Antipeptide antisera also neutralized the enterotoxic activity as well as the neurotoxic activity of the toxin. In the case of the enterotoxic activity, antiserum against the peptide 5-18 showed the highest activity, exhibiting essentially the same level of neutralization (around 80%) as the anti-Shiga holotoxin. In the case of the neurotoxic activity the most efficient neutralization was achieved by antisera against the peptide 7-26. In these experiments, prior incubation of the toxin with the anti-peptide sera reduced its lethal effect by approximately 60%.

Not only were antisera induced by the three peptides capable of neutralizing the various activities of the toxin, but active immunization of mice with the conjugates of the three peptides led to the development of partial protection against the lethal effect of the Shiga toxin (Table 2). As shown, in non-immunized mice 96% mortality was reached 7 days after they were exposed to the lethal dose (2 LD_{50}'s) of the toxin. In contrast, in mice immunized with the conjugates of the three peptides and exposed to equal dose of toxin, the mortality rate was much lower. The best protection was obtained by immunization with a polymer of the peptide 13-26. In this case, the mortality was only 20%, with 80% long-term survival. We consider these results to be important for the rational design of peptide vaccines against shigellosis in the future.

CONCLUDING REMARKS

It is obvious that if the idea of synthetic peptide vaccines materializes, it will have tremendous advantages. Synthesis of materials that incorporate peptides consisting of the relevant antigenic determinants of several different viruses and/or bacteria in their structure may lead to multivalent vaccines that will be effective simultaneously towards all these pathogens. It will be possible to build adjuvanticity into the vaccines and thus overcome a serious problem in the production of existing vaccines. It may be possible to take advantage of our knowledge about the genetic control of the immune response and its linkage to the major histocompatibility complex for the design of synthetic vaccines tailored to individuals according to their HLA type.

Table 2. Protective Immunization against Neurotoxicity of Shiga Toxin

Immunizing agent	Number of Experiments	Total number of mice	Mortality (%)			% Protection
			Day 3	Day 5	Day 7	
None	9	100	15 ± 4.4	75 ± 8.0	96 ± 3.3	0
5-18 conjugates	7	90	19 ± 7.6	58 ± 7.8	59 ± 7.3	38
13-26 derivatives (all)	9	120	9 ± 4.8	33 ± 8.8	40 ± 9.4	58
13-26 polymer	5	60	4 ± 4.0	16 ± 7.4	20 ± 8.9	79
13-26-TT	4	50	15 ± 9.5	55 ± 9.5	65 ± 5.0	32
7-26 conjugates	3	40	13 ±13.3	53 ±12.6	60 ±11.5	37

[a]Numbers represent Mean ± S.E.

The key question is, however, whether synthetic peptides can form the basis for practical vaccines. The results obtained so far, both in our laboratory, as summarized above, and in other laboratories, show great promise. There are still many problems to solve of which a most difficult one is the development of an appropriate adjuvant. To this effect, the recent progress in this field, as evidenced by the cumulative information presented in this conference is certainly most relevant and encouraging.

REFERENCES

Ada, G.L., Leung, K.N. and Ertl, H., 1981, An analysis of effector T-cell generation and function in mice exposed to influenza A or Sendai viruses, Immun.Rev., 58:5.

Arnon, R., Maron, E., Sela, M. and Anfinsen, C.B., 1971, Antibodies reactive with native lysozyme elicited by a completely synthetic antigen, Proc.Natl.Acad.Sci.USA, 68:1450.

Arnon, R., Sela, M., Parant, M. and Chedid, L., 1980, Antiviral response elicited by a completely synthetic antigen with built-in adjuvanticity, Proc.Natl.Acad.Sci.USA., 77:6769.

Audibert, F., Jolivet, M., Chedid, L., Alouf, J.E., Boquet, P., Rivaille, P. and Siffert, O., 1981, Active antitoxic immunization by a diphtheria toxin synthetic oligopeptide, Nature, 289:593.

Audibert, F., Jolivet, M., Chedid, L., Arnon. R., Sela, M., 1982, Successful immunization with a totally synthetic diphtheria vaccine, Proc.Natl. Acad.Sci.USA, 79:5042.

Carelli, C., Audibert, F., Faillard, J. and Chedid, L., 1982, Immunological castration of male mice by a totally synthetic vaccine (LH-RH conjugated to MDP-lysine) administered in saline, Int.J. Immunopharmacol., 4:290.

Ellouz, F., Adam, A., Ciorbaru, R. and Lederer, E., 1974, Minimal structural requirements for adjuvant activity of bacterial peptidoglycan derivatives, Biochim.Biophys.Res.Commun., 55:1317.

Harari, I., Donohue-Rolfe, A., Keusch, G. and Arnon, R., 1988, Synthetic peptides of Shiga Toxin B subunit induce antibodies which neutralize its biological activity, Infect.and Immun., 56:1618.

Jackson, D.C., Murray, J.M., White, D.O., Fagan, C.N. and Tregear, G.W., 1982, Antigenic activity of a synthetic peptide corresponding to the loop region of influenza virus hemagglutin, Virology, 120:273.

Jacob, C.O., Sela, M. and Arnon, R., 1983, Antibodies against synthetic peptides of the B subunit of cholera toxin: cross reaction and neutralization of the toxin, Proc.Natl.Acad.Sci.USA, 80:7611.

Jacob, C.O., Sela, M., Pines, M., Hurwitz, S. and Arnon, R., 1984a, Adenylate cyclase activation by cholera toxin as well as its activity are inhibited with antibodies against related synthetic peptides, Proc.Natl.Acad.Sci.USA, 8:7893.

Jacob, C.O., Pines, M. and Arnon, R., 1984b, Neutralization of heat labile toxin of E. coli by antibodies to synthetic peptides derived from B subunit of cholera toxin, EMBO J., 3:2889.

Jacob, C.O., Leitner, M., Zamir, A., Salomon, D. and Arnon, R., 1985, Priming immunization against cholera toxin and E. coli heat-labile toxin by a cholera toxin short peptide-β-galactosidase hybrid synthesized in E. coli, EMBO J., 4:3339.

Jacob, C.O., Grossfeld, S., Sela, M. and Arnon, R., 1986, Priming immune response to cholera toxin induced by synthetic peptides, Eur.J.Immun., 16:1057.

Langbeheim, H., Arnon, R. and Sela, M., 1976, Antiviral effect on MS-2 coliphage obtained with a synthetic antigen, Proc.Natl.Acad.Sci.USA, 73:4636.

Laver, W.G., Air, G.M., 1979, "Structure and variation in influenza virus", Elsevier, North Holland, Amsterdam.

Lindhom, L., Holmgren, J., Wikstrom, M., Karlsonn, V., Andersson, K. and Lycke, N., 1983, Monoclonal antibodies to cholera toxin with special reference to cross-reactions with Escherichia coli heat-labile enterotoxin, Infect.Immun., 40:570.

Muller, G.M., Shapira, M. and Arnon, R., 1982, Anti-influenza response achieved by immunization with a synthetic conjugate, Proc.Natl.Acad.Sci.USA, 79:569.

Seidah, N.G., Donohue-Rolfe, A., Lazure, C., Auclair, A., Keusch, G.T. and Chretien, M., 1986, Complete amino acid sequence of Shigella toxin B-chain, J.Biol.Chem., 261:13928.

Shapira, M., Jibson, M., Muller, G.M. and Arnon, R., 1984, Immunity and protection against influenza virus by a synthetic peptide corresponding to antigenic sites of hemagglutinin, Proc.Natl.Acad.Sci.USA, 81:2461.

Shapira, M., Jolivet, M. and Arnon, R., 1985, Synthetic vaccine against influenza with built-in adjuvanticity, Int.J.Immunopharmacol., 7:719.

Wiley, D.C., Wilson, I.A. and Skehel, J.J., 1981, Structural identifications of the antibody binding sites of Hong Kong influenza hemagglutinin and their involvement in antigenic variation, Nature, 289:373.

THE DEVELOPMENT AND PRELIMINARY CLINICAL EVALUATION OF AN ANTIFERTILITY

VACCINE

P.D. Griffin

Special Programme of Research, Development and Research
Training in Human Reproduction, World Health Organization
1211 Geneva 27, Switzerland

INTRODUCTION

The World Health Organization's Special Programme of Research, Development and Research Training in Human Reproduction is a global programme of international technical cooperation initiated by WHO to promote, coordinate, support, conduct and evaluate research in human reproduction with special reference to the needs of developing countries, by:

(a) promoting and supporting research aimed at finding and developing new safe, effective and acceptable methods of fertility regulation as well as identifying and eliminating obstacles to such research and development;

(b) identifying and evaluating health and safety problems associated with fertility regulation technology, analysing the behavioural and social determinants of fertility regulation, and testing cost-effective interventions to develop improved approaches to fertility regulation within the context of reproductive health services;

(c) strengthening the training and research capability of developing countries to conduct research in the field of human reproduction;

(d) establishing a basis for collaboration with other programmes engaged in research and development in human reproduction, including the identification of priorities across the field and the coordination of activities in the light of such priorities (Barzelatto, 1988).

The WHO Special Programme's research activities are conducted through the Task Force mechanism. These Task Forces are multi-disciplinary, multi-national groups, with members from both developing and developed countries, which have been established in order to achieve certain objectives within a defined time-frame. In the case of one of these Task Forces, the Task Force on Vaccines for Fertility Regulation, these objectives are to develop safe and effective vaccines that will inhibit fertility without producing unacceptable side-effects. The advantages that such vaccines would have, include:

(a) lack of pharmacological activity;

(b) long-lasting action following a single injection or course of immunization;

(c) administration by a procedure associated with positive health benefits;

(d) low manufacturing costs and ease of delivery within existing health services.

Whilst virtually every step in the reproductive process is accessible to immunological attack, not all of these processes represent attractive targets for vaccine development. Immunization against some of the molecules essential for successful reproduction can result in a low level of anti-fertility efficacy and/or can produce unacceptable side-effects, ranging from minor disturbances in endocrine function to the more serious immuno-pathological sequelae of auto-immunity and immune-complex disease. In order to avoid these potential side-effects and hazards, it is necessary to select, for vaccine development, molecules that satisfy the following criteria:

(a) essential for successful completion of the reproductive process;

(b) accessible to immune attack;

(c) specific to the intended target and not represented in other body tissues or fluids;

(d) located in a site where a specific and controlled immune response will have no immunopathological or other undesirable consequences;

(e) present only transiently and in small amounts.

These criteria are met by some molecules in the sperm cell membrane, the zona pellucida of the ovum, and the membrane of the trophoblast cells of the peri-implantation embryo. In addition, some secreted products of these tissues also appear to be promising vaccine candidates. Prototype vaccines, incorporating naturally derived and synthetic preparations based on several of these tissue-specific immunogens, have been produced and evaluated by the Task Force since its formation in 1974. The work that has reached the most advanced stage, however, concerns the development of vaccines directed against the pregnancy hormone, human chorionic gonadotrophin (hCG) (Griffin, 1988). In addition to WHO, the National Institute of Immunology, New Delhi, India, and the Population Council in New York, USA, are also supporting research to develop anti-hCG vaccines. The preclinical and clinical studies supported by the Task Force, as well as a summary of its ongoing research programme in the anti-hCG vaccine area, are described below.

DEVELOPMENT OF THE ANTI-HCG VACCINE

Chorionic gonadotrophins are produced and secreted by the cells of the trophectoderm of the peri-implantation embryo of primates, including man, perhaps as early as 7 days post-ovulation. In a fertile cycle, the principal function of hCG appears to be to provide trophic support to the ovarian corpus luteum, which in turn produces progesterone needed for maintenance of the lining of the uterus in preparation for the implantation of the early embryo and the establishment of pregnancy. In a non-fertile cycle, no hCG is present and the corpus luteum regresses approximately 14 days post-ovulation, its production of progesterone falls, the uterine lining breaks down and is sloughed off in the natural process of menstruation. A vaccine directed against hCG, therefore, would inhibit the feedback function of the hormone and lead to a physiologically normal menstruation at or about the expected time.

Active immunization of infra-human primates with prototype vaccines incorporating whole hCG, its beta subunit or fragments thereof, have been shown to have a powerful antifertility effect without altering the menstrual cycles of the immunized animals (Stevens, 1975; Hearn, 1976). In theory, this antifertility effect could be caused by neutralization of the biological activity of hCG in the maternal circulation and by a direct antibody-mediated or cell-mediated immune attack on the trophectoderm cells of the peri-implantation embryo. Preliminary information obtained in studies with the anti-hCG vaccine developed by the Task Force, suggests that both of these mechanisms may be involved (Griffin, 1986). However, passive immunization studies using isologous antisera raised to hCG have shown that antibody alone is highly effective in terminating established pregnancies in the baboon. In developing its vaccine, the Task Force has taken the ability to elicit high levels of anti-hCG antibodies as the basis for selection of the various vaccine components. The studies to be described here have been carried out largely under the direction of Professor Vernon Stevens at Ohio State University, in Columbus, Ohio, USA.

Selection of the Immunogen Component

The hCG molecule consists of two subunits. One of these, the alpha subunit, is identical to that of a number of pituitary hormones with a similar quarternary structure, such as human follicle stimulating hormone (hFSH), human luteinizing hormone (hLH) and human thyroid stimulating hormone (hTSH). Immunization with the whole hCG molecule, therefore, produces a non-specific antibody response characterized by extensive cross-reactions with these other hormones. Furthermore, the beta subunit of hCG exhibits a substantial degree of sequence homology with the beta subunit of hLH, so that antibodies raised to beta-hCG cross-react with hLH. Fortunately, beta-hCG has a carboxyterminal extension of approximately 30 amino acids which is not present in beta-hLH nor in the other pituitary hormones (Morgan et al, 1973). The Task Force strategy, therefore, has been to develop a vaccine based on the apparently hCG-specific, carboxyterminal region of beta-hCG.

A large number of peptides of various lengths and positions within this C-terminal region of beta-hCG (CTP-beta-hCG) have been prepared and evaluated for their ability to generate antibodies which react specifically with hCG and which are also capable of neutralizing the biological activity of the native hormone. In addition, peptides with poly-proline 'spacers' at either the N- or C-terminus have been prepared in an effort to improve the peptide's presentation to the afferent arm of the immune response. Although suitably conjugated short peptides of ten or more amino acids can elicit substantial titres of anti-hCG antibodies, the peptides need to be 22 residues or more in length in order to elicit antibodies capable of neutralizing the biological activity of hCG in vivo and that have substantial hCG binding properties in vitro. The Task Force has found that the 37 amino acid peptide, corresponding to the natural sequence of the 109-145 region of beta-hCG, represents the optimal immunogen in terms of these two requirements whilst, at the same time, retaining the desired hCG specificity (Stevens et al, 1981a).

Selection of the Carrier Component

The 109-145 CTP-beta-hCG peptide is only weakly immunogenic in rabbits and baboons. In the human, which is immunologically tolerant to the hCG molecule, either as a result of exposure to the hormone in utero or because of its similarity to hLH, the peptide would be expected to be even less potent as an immunogen. To break tolerance to the peptide, the Task Force has evaluated a number of molecules for their abilities to act as carriers. Both natural and synthetic molecules have been evaluated for this purpose, with bovine gamma globulin (bGG) giving the best response in terms of antibody titre (Stevens et al, 1981a). The use of bGG for this purpose, however, was considered unattractive because of the risk of inducing cross reactive immunity to the endogenous gamma globulins of vaccine recipients and attention was focussed on the next most potent carriers, diphtheria toxoid and tetanus toxoid. Although there was little to choose between these two candidates in terms of their properties as carriers, diphtheria toxoid was considered the more attractive because of its reduced potency for eliciting hypersensitivity reactions as well as its high content of free-amino groups. This latter property offered a major advantage for conjugation to the 109-145 CTP-beta-hCG peptide, as described below.

Selection of the Peptide-Carrier Conjugation Procedure

The classical reagents used for the preparation of peptide-protein conjugates, such as carbodiimide, glutaraldehyde and the thiocyanates, are very efficient for the production of immunogenic preparations. The products obtained using these procedures, however, are very heterogeneous, consisting of peptide to peptide, protein to protein and a variety of peptide to protein conjugates. Whilst these preparations are adequate for experimental purposes, they would not satisfy the stringent requirements of the regulatory authorities for components of a vaccine intended for clinical use. The Task Force has therefore developed a conjugation procedure using a bifunctional reagent, 6-maleimido caproic acyl N-hydroxy succinimide ester (MCS), which permits the 109-145 CTP-beta-hCG peptide to be linked through the sulfhydral group of the cysteine residue at its N-terminus, to a free-amino group on the diphtheria toxoid carrier (Griffin, 1986). This reaction is predictable in that the conjugation only occurs in this manner and controllable in that the number of peptides linked to the carrier can be adjusted within a tolerance of +/- 2 peptides. The immunogenicities of the 109-145 CTP-beta-hCG:DT conjugates prepared in this way increase with increasing peptide loading up to a maximum of approximately 25 peptide molecules per 100 000 M_r of carrier. At this level of peptide loading, the conjugate consists of approximately 50% peptide and 50% carrier by weight. A conjugate of this composition has been selected by the Task Force for use in its prototype anti-hCG vaccine.

Selection of the Adjuvant Component

In order to enhance the immunogenicity of the 109-145 CTP-beta-hCG:DT conjugate, and to ensure an adequate level of stimulation of the immune system following the primary injection, the Task Force has evaluated several adjuvant compounds potentially suitable for clinical use. Using Complete Freund's Adjuvant (CFA) as the reference preparation, comparative evaluations were carried out in rabbits with a number of muramyl dipeptide (MDP) analogues. Two of these compounds, N-acetyl-nor-muramyl-L-alanyl-D-isoglutamine (Ciba-Geigy CGP 11,637) and N-glycol-nor-muramyl-L-alpha-abu-D-isoglutamine (Syntex DT-1), were found to be more potent than CFA in eliciting antibodies to the 109-145 CTP-beta-hCG:DT conjugate. There were essentially no differences in the immunopotencies of these two compounds, however, a considerable amount of preclinical toxicology had already been carried out on the former, CGP 11,637, and in view of the substantial savings in time and costs that this represented, this compound was selected as the adjuvant of choice by the Task Force (Stevens et al, 1981b).

Selection of the Vehicle Component

Whilst the magnitude and duration of the immune response elicited by an anti-fertility vaccine, and hence its duration of effect, will be influenced

190

by the genetic constitution of the recipient, the Task Force has estimated that such vaccines should provide protection from the risk of pregnancy for a period of 12-24 months, following a single injection or course of immunization, in order to be an attractive addition to the currently available methods of family planning. However, should effective immunity last longer than the planned 12-24 months and in some cases be predictably irreversible, the vaccine may still be attractive as a non-invasive alternative to surgical sterilization.

Injection of the 109-145 CTP-beta-hCG:DT conjugate and MDP adjuvant in saline elicits an anti-hCG response of short duration probably because of the rapid dissemination and clearance of the injected material. The Task Force has evaluated a number of simple and complex emulsions, formed from a variety of oil materials, in order to identify a clinically acceptable vehicle in which the vaccine components can be administered and that has the appropriate depot and delayed release characteristics to ensure the generation of an effective level of immunity of the desired duration. One such preparation, consisting of 4 parts squalene to 1 part mannide mono-oleate in 4 parts phosphate buffered saline, was found to be the most effective of the various formulations tested in terms of eliciting high levels of anti-hCG antibody levels and this vehicle formulation was selected for the prototype anti-hCG vaccine (Stevens et al, 1981b).

Preclinical Efficacy and Safety Studies

By 1980, after six years of intensive research, the Task Force had produced a complex but potent prototype anti-hCG vaccine formulation consisting of the 109-145 CTP-beta-hCG peptide conjugated to diphtheria toxoid, mixed with a MDP adjuvant, and suspended in a water-in-oil emulsion.

When administered to a group of fertile female baboons, this prototype vaccine was capable of reducing the fertility rate to less than 5% compared to 70% in a group of untreated control animals (Stevens et al, 1981c). These results were particularly encouraging in view of the fact that antibodies raised to the hCG peptide vaccine crossreacted by only 5-15% with the endogenous baboon CG.

In preparation for seeking approval to carry out a Phase I clinical trial of the vaccine in human volunteers, a series of acute and sub-acute toxicity and immunosafety studies were carried out in mice, rats, rabbits and baboons. The data obtained in these studies were submitted, together with a clinical trial protocol, as an Investigational New Drug Application to the United States Food and Drug Administration and to the Australian Department of Health. Approval to conduct the phase I trial was given by these regulatory bodies in the second half of 1985 and the clinical trial of this largely synthetic vaccine was initiated early in 1986.

PHASE I CLINICAL TRIAL OF THE ANTI-HCG VACCINE

The principal objectives of this study, carried out by Professor Warren Jones and his colleagues at Flinders Medical Centre in Adelaide, Australia, were to assess the nature and extent of any adverse side-effects associated with the use of the vaccine as well as to monitor the magnitude and duration of the anti-hCG antibody titres elicited by the vaccine.

Following a lengthy screening exercise, in which women volunteers were interviewed and examined to ensure that they satisfied the inclusion criteria as well as being fully informed of the nature and objectives of the clinical trial, a total of 30 subjects were recruited for initial study. All of the subjects had been previously, electively sterilized and were assigned to one of five dosage groups. In each group, subects 1, 3, 5 and 6 received the full vaccine preparation and subjects 2 and 4 received a 'placebo' preparation consisting of all of the vaccine constituents apart from the 109-145 CTP-beta-hCG:DT immunogen. All subjects received two injections of their respective preparations and doses at an interval of 6 weeks and follow-up was effected on both an in-patient and out-patient basis for a period of 18 months. In addition to routine physical examinations and laboratory analyses, a large number of biochemical, haematological and immunological tests relevant to this particular trial, were carried out.

Ten of the 30 subjects, principally in the higher dose groups, experienced transient arthralgia and myalgia following injection of the vaccine and were subsequently found to have developed poor anti-hCG antibody responses. These findings suggested that these subjects had probably received unstable emulsion formulations which broke down soon after injection and rapidly released the vaccine components. When replacement subjects were immunized with the vaccine in an emulsion which was tested for stability prior to injection, the incidence and severity of the side-effects were greatly reduced and the antibody responses greatly improved. A dose reponse effect was observed among the five groups in terms of the ability of the vaccine to elicit anti-hCG antibodies. The amount of hCG present in the maternal circulation at the time of implantation in the human, 135 mIU/ml, approximates to 0.26 nmol/l of hCG binding capacity in vitro. Preliminary data indicate that the biological neutralizing capacity of the anti-hCG antisera is approximately 50% of its in vitro binding capacity for the hormone, suggesting that an antibody titre of at least 0.56 nmol/l will be needed to achieve an antifertility effect in the human if the vaccine works by this mechanism alone. This threshold level of antibody production was exceeded in all of the dose groups, particularly in groups 4 and 5 where antibody levels 5-7 times greater than those expected to confer an antifertility effect were measured. For the majority of subjects in all dose groups, the anti-hCG antibodies remained above the threshold level for approximately 6 months and in the highest dose group for greater than 10 months (Jones et al, 1988).

Apart from the arthralgia and myalgia referred to earlier, no serious side-effects were reported by the subjects nor detected in any of the physical examinations and laboratory tests carried out during the clinical trial. Crossreactive immunity assays were carried out against a large panel of rat, baboon and some human tissues. Although no reactivity was observed with any of the pituitary hormones, especially hLH, some sera reacted with cells in the islets of Langerhans in the baboon pancreas, and further studies to identify the material responsible for this reactivity are currently underway.

DEVELOPMENT OF AN OPTIMIZED ANTI-HCG VACCINE

In parallel with the Phase I clinical trial, the Task Force has also been conducting studies to develop an optimized anti-hCG vaccine suitable for mass production and wide-scale clinical use. This research has included comparative evaluations of a number of alternative hCG peptides, carrier molecules and carrier peptides, adjuvants and vaccine delivery systems.

The development of a delivery system which would enable effective immunity to be elicited as the result of a single administration of vaccine, would have major advantages in those countries where health care services are hard put to meet the needs of their populations. Much of the Task Force work in this area has focussed on formulations that will release the anti-hCG vaccine continuously, or on a pulsed basis, over a long period of time, thereby mimicking the more conventional multi-injection schedule.

Preliminary results obtained in experiments using liposomes, suggest that these systems are capable of releasing the 109-145 CTP-beta-hCG:DT immunogen over a period of several weeks or months but that several injections are needed to obtain long-lasting immunity of the desired magnitude. Experiments using iscoms have only recently been initiated and insufficient data are available at the present time to draw any conclusions about the potential of this particular system.

The most encouraging data have been obtained with the 109-145 CTP-beta-hCG:DT immunogen and MDP incorporated into a number of different biodegradable/biocompatible microspheres prepared using a range of poly-glycolic/poly-lactic copolymers. By adjusting the ratio of the two polymers, microcapsules with different release rate characteristics can be prepared. Preliminary data show that theoretically effective levels of anti-hCG immunity lasting well in excess of 12 months, can be elicited following a single injection of two microcapsule formulations, one of which has a peak of vaccine release at 12 weeks and the other at 22 weeks (V.C. Stevens, unpublished information). Further Task Force funded work on these systems is currently underway.

CONCLUSIONS

The Phase I trial carried out with the prototype anti-hCG vaccine developed by the Task Force, has demonstrated that a synthetic peptide vaccine can be used clinically to elicit anti-hCG immunity devoid of unacceptable side-effects and of a magnitude and duration expected to confer an antifertility effect lasting in excess of 10 months. The trial has also demonstrated the need for an improved formulation of the vaccine in order to avoid the emulsion instability problems and to extend the duration of effective immunity. Ongoing research in this area promises to yield a safe, effective, long-lasting but reversible anti-hCG vaccine formulation in the near future that will be suitable for large-scale clinical use. In addition, the 'vaccine engineering' experience gained by the Task Force over the past 14 years will be of value to the projects that it is currently supporting to develop other fertility regulating vaccines as well as to the field of vaccine development in general.

REFERENCES

Barzelatto, J., 1988, World Health Organization Special Programme of Research, Development and Research Training in Human Reproduction, Biennial Report 1986-1987, W.H.O., 11.

Griffin, P.D., 1986, A fertility regulating vaccine based on the carboxyl-terminal peptide of the beta subunit of hCG, in: "Immunological Approaches to Contraception and Promotion of Fertility", G.P. Talwar, ed., Plenum Publishing Corporation.

Griffin, P.D., 1988, World Health Organization Special Programme of Research, Development and Research Training in Human Reproduction, Biennial Report 1986-1987, W.H.O., 177.

Hearn, J.P., 1976, Immunization against pregnancy, Proc.Roy.Soc.London (Biology), 195:149.

Jones, W.R., Bradley, J., Judd, S.J., Denholm, E.H., Ing, R.M.Y., Mueller, U.W., Powell, J.H., Griffin, P.D., Stevens, V.C., 1988, Phase I clinical trial of a World Health Organization birth control vaccine, Lancet, 8598:1295.

Morgan, F.J., Birken, S., Canfield, R.E., 1973, Human chorionic gonadotrophin: a proposal for the amino acid sequence, Molec. and Cell. Biochem., 2:97.

Stevens, V.C., 1975, Antifertility effects from immunization with intact, subunits and fragments of hCG, in:" Physiological Effects of Immunity Against Reproductive Hormones", R.G. Edwards, M.H. Johnson, eds., Cambridge University Press.

Stevens, V.C., Cinader, B., Powell, J.H., Lee, A.C., Koh, S.W., 1981a, Preparation and formulation of a human chorionic gonadotrophin antifertility vaccine: selection of a peptide immunogen, Amer.J.Reprod. Immun., 1:307.

Stevens, V.C., Cinader, B., Powell, J.H., Lee, A.C., Koh, S.W., 1981b, Preparation and formulation of a human chorionic gonadotrophin antifertility vaccine: selection of adjuvant and vehicle. Amer.J.Reprod. Immun., 1:315.

Stevens, V.C., Powell, J.E., Lee, A.C., Griffin, P.D., 1981c, Antifertility effects from immunization of female baboons with C-terminal peptides of human chorionic gonadotrophin, Fertility and Sterility, 36(No.1): 98.

RECENT PROGRESS WITH VACCINES AGAINST EPSTEIN-BARR VIRUS INFECTION

M.A. Epstein

Nuffield Department of Clinical Medicine
University of Oxford, John Radcliffe Hospital
Headington, Oxford OX3 9DU, UK

INTRODUCTION

For well over a decade there has been strong evidence indicating that Epstein-Barr (EB) virus, one of the six human herpesviruses, forms an essential link in the complicated chain of events leading to the development of two human cancers (reviewed in Epstein and Achong, 1979), endemic Burkitt's lymphoma (BL) (Burkitt, 1963) and undifferentiated nasopharyngeal carcinoma (NPC) (Shanmugaratnam, 1971). More recently, it has emerged that the virus also seems to be causally linked to the malignant lymphomas which arise with undue frequency in immunodepressed individuals (Purtillo, 1984).

Although endemic BL is not of much significance in world cancer terms, undifferentiated NPC certainly is since it is the most common tumor of men and the second most common of women amongst all populations of Southern Chinese origin (Shanmugaratnam, 1971) and also has a medium-high incidence across North Africa, in the Sudan, and in Kenya (Camoun et al, 1974; Clifford, 1970). In view of the importance of NPC it was proposed already in 1976 (Epstein, 1976) that prevention or reduction of EB virus infection in high-risk populations by an anti-viral vaccine might be expected to decrease the incidence of the cancer by removing the viral link in the causal chain, in just the same way as reducing cigarette smoking in a population decreases the incidence of bronchogenic carcinoma (Doll and Peto, 1976).

An important precedent for successful prevention of a naturally occurring cancer by anti-viral vaccination has been provided in the case of Marek's disease of chickens (Marek, 1907). With this herpesvirus-induced condition huge numbers of commercial chickens were regularly lost from malignant lymphomas (Payne et al, 1976) until the introduction of efficient anti-viral vaccines (Churchill et al, 1969; Okazaki et al, 1970). Live apathogenic virus vaccines are currently used, but it should be noted that experimental vaccines based on purified antigens from cells infected with Marek's disease herpesvirus have also given excellent protection (Kaaden and Kietzschold, 1974). For an analogous vaccine against EB virus, the EB virus-determined membrane antigen (MA) was chosen from the outset as immunogen (Epstein, 1976) because antibodies to it were known to be virus-neutralizing. Work in several laboratories demonstrated that MA consists of two, antigenically related, large glycoprotein molecules of 340Kd and 270Kd (MA gp340 and gp270) (reviewed in Epstein, 1984) and the larger of these was selected for study. Over the years, a prototype MA gp340-based vaccine was developed.

VALIDATION OF A PROTOTYPE VACCINE BASED ON MA gp340

In the early experiments MA gp340 was purified by a molecular weight-based procedure (Morgan et al, 1983), ensuring optimum yields at each step by monitoring with a quantitative radio-immunoassay (North et al, 1982). The product was rendered immunogenic by incorporation in artificial liposomes and the antibodies induced in banal laboratory animals were assessed by means of a highly sensitive ELIZA (Randle et al, 1984) and standard virus-neutralization tests (Moss and Pope, 1972; de Schryver et al, 1974). The only animal known to respond with lesions to experimental EB virus infection is the rare cotton-top tamarin (Saguinus oedipus oedipus) (Miller et al, 1977) and to provide animals for testing protection by the vaccine a successful breeding colony had to be set up (Kirkwood et al, 1983 and 1985). At the same time, it was necessary to determine the dose of virus for use as challenge after vaccination which would cause lesions in all unprotected animals, and to establish the nature of the lesions. The tumors which rapidly follow injection of such a 100% pathogenic dose have been shown by Southern blotting and immunoglobulin gene probing to be multiple, clonally-derived, malignant lymphomas, with each separate mass in any one animal arising from a separate malignant transformation event (Cleary et al, 1985).

Material prepared as outlined above has been used as a prototype vaccine in cotton-top tamarins and it has been shown that the immunized animals were protected against tumor induction when challenged with the 100% lymphomagenic dose of virus, thus demonstrating the efficacy of MA gp340 as a protective immunogen (Epstein et al, 1985). Indeed, very recent experiments indicate that it induces specific cellular immunological responses in addition to virus-neutralizing antibodies.

VACCINES AGAINST EB VIRUS FOR HUMAN USE

Following the successful experiments with the prototype vaccine (Epstein et al, 1985) efforts have been directed towards the preparation of a vaccine suitable for use in man. Various approaches to this goal are set out in Table 1 and each of these has already been addressed.

Table 1. Approaches to an MA gp340-Based Vaccine for Human Use

---

(1) Subunit production in genetically engineered cells
         bacterial
         yeast
         mammalian

(2) Use of recombinant viral vectors
         vaccinia
         varicella

(3) Improved purification and use of novel adjuvants
         immunostimulating complexes (Iscoms)
         muramyl dipeptide analogues

MA gp340 Production in Genetically Engineered Cells

The region of the EB virus genome carrying the gene coding for MA has been identified (Hummel et al, 1984) and the nucleotide sequence of this gene is known (Biggin et al, 1984). For the production of MA gp340 in genetically engineered cells the gene has been cloned and expressed in cultures of E. coli (Beisel et al, 1985), yeast (Schultz et al, 1987) and several types of mammalian cells (Whang et al, 1987; Conway et al, 1988). However, purification of the product has so far proved exceptionally difficult and little progress has been made with any of these systems.

Recombinant Vaccine Viruses Engineered to Express MA gp340

As regards the use of genetically engineered viral vectors, the MA gp340 gene has been inserted into recombinant strains of vaccinia virus. The first recombinant was made with the WR strain and the gene was expressed under the control of a vaccinia virus promoter during replication (Mackett and Arrand, 1985). The WR strain of vaccinia virus is relatively virulent and when it was used to immunize tamarins intradermally these small animals (of about 500 grams) developed skin lesions 4 or 5 cms in diameter with a secondary crop of further pustules one week later. Antibody titres to vaccinia virus ranging from 1:5000 to 1:10000 were induced, but administration of a second intradermal immunization with the WR recombinant virus did not increase the response. When challenged with the 100% lymphomagenic dose of EB virus, three out of four animals were protected and this result was considered of some interest. However, the WR strain of vaccinia virus is far too virulent for administration to man, a point not usually stressed in reported work on recombinant vaccinia virus vaccines. In view of this, a similar recombinant was made with the Wyeth (New York Board of Health) vaccine strain of vaccinia virus which is acceptable for human use. But this recombinant based on an attenuated vaccinia strain only induced minimal skin lesions in vaccinated tamarins and gave much lower titres of antibody to vaccinia, presumably because of the rather restricted level of virus replication; the vaccinated animals proved quite unresistant to challenge with EB virus (Morgan et al, 1988).

Another approach has been to use analogous genetic engineering techniques to insert the MA gp340 gene into varicella virus (Lowe et al, 1987) using from the outset the Oka vaccine strain (Takahashi et al, 1974) to prepare the recombinant. The Oka strain has been administered extensively over the years to vaccinate immunocompromised children and adverse reactions have not been reported (Weibel et al, 1984). Although there is at present no appropriate experimental animal for work with varicella virus, thought is being given to other possible ways of assessing the suitability of the new recombinant as a live viral vaccine.

Improved Purification of MA gp340 for use with Novel Adjuvants

While the above mentioned studies have been going forward, a new purification procedure has been elaborated for the easy, rapid and efficient preparation of MA gp340 from the membranes of infected cells (David and Morgan, 1988). This procedure is based on fast protein liquid chromatography (FPLC) and is suitable for scaling up for batch production. Material obtained by this FPLC method is routinely homogeneous and when it was injected into mice together with a synthetic muramyl dipeptide adjuvant formulation (Allison and Byers, 1986) it rapidly elicited very high titres of virus-neutralizing antibodies (David and Morgan, 1988). Preliminary experiments in cotton-top tamarins likewise using the muramyl dipeptide analogue have shown that vaccinated animals are protected against the standard 100% lymphomagenic dose of virus.

The FPLC-purified MA gp340 has also been used with immunostimulating complexes (Iscoms) (Morein et al, 1987) which combine the multimeric presentation of antigen with an inbuilt adjuvant, and which elicit both humoral and cell-mediated immune responses. When tamarins were vaccinated with gp340 incorporated into Iscoms and were subsequently challenged with the 100% lymphomagenic dose of virus, the animals were completely protected (Morgan et al, 1988).

PRIORITIES FOR HUMAN VACCINE TRIALS

Negotiations are currently being undertaken for the production of gp340 under conditions which will meet the requirements of the licensing authorities, for use as a first generation vaccine in man. It is envisaged that a small batch of FPLC-purified gp340 will be prepared for adminstration, together with the muramyl dipeptide analogue formulation, in a Phase I trial involving a small number of informed human volunteers who will receive the preparation to demonstrate its safety and its capacity to induce neutralizing antibodies in those who have not been infected by EB virus or to augment such antibodies in those who have been. Specific cytotoxic T cell responses will also be monitored using standard techniques (Rickinson, 1986).

Testing of the FPLC-purified First Generation Subunit Vaccine

For the second stage, studies will have to be undertaken on a larger scale and for this, EB virus-induced infectious mononucleosis (IM) provides an excellent experimental testing ground. Young adults who have escaped the usual silent primary infection of childhood can be readily detected (Pereira et al, 1969) and it is well known that such individuals are at risk for delayed primary infection which is accompanied in 50% of cases by the clinical manifestations of IM (Niederman et al, 1970); University Health Physicians and PHLS Laboratories, 1971). There are thus good grounds for applying such a screening programme to a group of university or college students and then undertaking a double-blind vaccine trial among volunteers who are seronegative and hence have never been infected by EB virus. Not only would the efficacy of the vaccine in preventing primary infection become evident in a relatively short time, but there would also be the added advantage that those who were successfully vaccinated would not have to face a possible attack of IM with its attendant disruption of courses. At the same time, there are strong ethical reasons for providing a vaccination programme for the rare individuals at risk for developing the genetically determined X-linked lymphoproliferative (XLP) syndrome after primary EB virus infection, since this condition is invariably a serious, and often a life-threatening, disease in affected males (Purtilo, 1984).

Once vaccine prevention of primary EB virus infection has been demonstrated with coincidental protection against IM and the grave manifestations of XLP syndrome, the way would be open for the final phase of trials. For this, the vaccine would have to be deployed in the field in an appropriate area where endemic BL has a high incidence. This would, by definition, be in some developing country (Burkitt, 1963) and special considerations would thus arise. Primary EB virus infection occurs at a very early age in the social conditions and standards of hygiene of the Third World (Henle and Henle, 1969; de The, 1979) so that vaccinations would have to be carried out during the first few months of life. However, such a schedule is exactly comparable to that required for hepatitis B virus (HBV) vaccination (Zuckerman, 1985; Deinhardt and Jilg, 1986) and the logistics for an EB virus vaccine trial against BL in the tropics are less complicated that those for the WHO thirty year prospective study of vaccination against HBV infection for the prevention of primary liver cancer (International Agency for Research on Cancer, 1985) which is a disease of adult life. In contrast, in BL the peak incidence is at about age 7 (Burkitt, 1963) and the influence of EB virus vaccination would therefore be apparent within ten years.

Possibilities for Second Generation Subunit Vaccines

As soon as it becomes clear that the first generation, FPLC-purified, gp340 subunit vaccine efficiently induces humoral and cellular protective responses capable of preventing or significantly modifying primary infection with EB virus, a strong incentive will have been provided for the development of cheaper and more efficient methods of gp340 production. As indicated above, genetically engineered bacterial, yeast, and mammalian cells have been induced to express gp340 and efforts to improve this approach so that the product could be readily harvested in an acceptable form would clearly be worthwhile.

Similarly, further development of recombinant viral vectors which could actually be used in man for vaccination should also be considered once the efficacy of the first generation vaccine has been established.

PREVENTION OF NPC

Vaccination against EB virus infection to prevent NPC is likely to require considerable long-term effort. This tumour occurs mostly in later life (Shanmugaratnam, 1971) and immunity would have to be maintained for many decades, longer even than for HBV vaccine protection against primary liver cell cancer. Nevertheless, progress with a vaccine against EB virus is continuing fast and what may appear somewhat daunting now is likely to prove possible sooner rather than later since success at each step in the demonstration of the efficacy of the gp340-based vaccine will generate acceleration in the rate of progress.

REFERENCES

Allison, A.C. and Byars, N.E., 1986, An adjuvant formulation that selectively elicits the formation of antibodies of protective isotype and cell mediated immunity, J.Immunol.Methods, 95:157.

Beisel, C., Tanner, J., Matsuo, T., Thorley-Lawson, D., Kezdy, F. and Kieff, E., 1985, Two major outer envelope glycoproteins of Epstein-Barr virus encoded by the same gene, J.Virol., 54:665.

Biggin, M., Farrell, P.J. and Barrell, B.G., 1984, Transcription and DNA sequence of the Bam HIL fragment of B95-8 Epstein-Barr Virus, EMBO.J., 3:1083.

Burkitt, D., 1963, A lymphoma syndrome in tropical Africa, in: "International Review of Experimental Pathology", G.W. Richter and M.A. Epstein, eds., Academic Press, New York.

Cammoun, M., Hoerner, G.V. and Mourali, N., 1974, Tumors of the nasopharynx in Tunisia: an anatomic and clinical study based on 143 cases, Cancer, 3:184.

Churchill, A.E., Payne, L.N. and Chubb, R.C., 1969, Immunization against Marek's disease using a live attenuated virus, Nature, 221:744.

Cleary, M.L., Epstein, M.A., Finerty, S., Dorfman, R.F., Bornkamm, G.W., Kirkwood, J.K., Morgan, A.J. and Sklar, J., 1985, Individual tumors of multifocal EB virus-induced malignant lymphomas in tamarins arise from different B cell clones, Science, 228:722.

Clifford, P., 1970, A review: On the epidemiology of nasopharyngeal carcinoma, Int.J.Cancer, 5:287.

Conway, M., Morgan, A. and Mackett, M., 1988, Expression of Epstein-Barr

virus antigen gp340/220 in mouse fibroblasts using a bovine papilloma
   virus vector, submitted to press.
David, E.M. and Morgan, A.J., 1988, Efficient purification of Epstein-Barr
   virus membrane antigen gp340 by fast protein liquid chromatography,
   J.Immunol.Methods, 108:231.
Deinhardt, F. and Jilg, W., 1986, Vaccines against hepatitis, Ann.Inst.
   Pasteur,Virol., 137E:79.
Doll, R. and Peto, R., 1976, Mortality in relation to smoking: 20 years'
   observation on male British doctors, Brit.Med.J., 2:1525.
Epstein, M.A., 1976, Epstein-Barr virus - is it time to develop a vaccine
   program? J.Nat.Cancer Inst., 56:697.
Epstein, M.A., 1984, A prototype vaccine to prevent Epstein-Barr (EB) virus-
   associated tumours, Proc.Roy.Soc.B.Lond., 221:1.
Epstein, M.A. and Achong, B.G., 1979, "The Epstein-Barr Virus", M.A. Epstein
   and B.G. Achong, eds., Springer, Berlin.
Epstein, M.A., Morgan, A.J., Finerty, S., Randle, B.J. and Kirkwood, J.K.,
   1985, Protection of cotton-top tamarins against Epstein-Barr virus-
   induced malignant lymphoma by a prototype subunit vaccine, Nature,
   318:287.
Henle, W. and Henle, G., 1969, The relation between the Epstein-Barr virus
   and infectious mononucleosis, Burkitt's lymphoma and cancer of the
   postnasal space, E.African Med.J., 46:402.
Hummel, M., Thorley-Lawson, D.A. and Kieff, E., 1984, An Epstein-Barr virus
   DNA fragment encodes messages for the two major envelope glycoproteins (gp350/300 and gp220/200), J.Virol., 49:413.
International Agency for Research on Cancer, 1985, An intervention study to
   evaluate the effectiveness of Hepatitis B vaccine for the prevention
   of hepatocellular carcinoma in a high risk population, IARC Working
   Paper 3/6:1.
Kaaden, O.R. and Dietzschold, B., 1974, Alterations of the immunological
   specificity of plasma membranes of cells infected with Marek's disease and turkey herpes viruses, J.Gen.Virol., 25:1.
Kirkwood, J.K., Epstein, M.A. and Terlecki, A.J., 1983, Factors influencing
   population growth of a colony of cotton-top tamarins, Lab.Animals,
   17:35.
Kirkwood, J.K., Epstein, M.A., Terlecki, A.J. and Underwood, S.J., 1985,
   Rearing a second generation of cotton-top tamarins (Saguinus oedipus
   oedipus) in captivity, Lab.Animals, 19:269.
Lowe, R.S., Keller, P.M., Keech, B.J., Davison, A.J., Whang, Y., Morgan,
   A.J., Kieff, E. and Ellis, R.W., 1987, Varicella-zoster virus as a
   live vector for the expression of foreign genes, Proc.Nat.Acad.Sci.
   USA, 84:3896.
Mackett, M. and Arrand, J.R., 1985, Recombinant vaccinia virus induces
   neutralising antibodies in rabbits against Epstein-Barr virus
   membrane antigen gp340, EMBO,J., 4:3229.
Marek, J., 1907, Multiple Nervenentzündung (polyneuritis) bei Hühnern,
   Deutsch.Tierärztl Wschr., 15:417.
Miller, G., Shope, T., Coope, D., Waters, C., Pagano, J., Bornkamm, G.W.
   and Henle, W., 1977, Lymphoma in cotton-top marmosets after inoculation with Epstein-Barr virus: tumor incidence, histologic spectrum, antibody responses, demonstration of viral DNA, and characterization of viruses, J.Exp.Med., 145:948.
Morein, B., Lovgren, K., Hogland, S. and Sundquist, B., 1987, The iscom:
   an immunostimulating complex, Immunol.Today, 8:333.
Morgan, A.J., Finerty, S., Lovgren, K., Scullion, F.T. and Morein, B.,
   1988, Prevention of Epstein-Barr (EB) virus-induced lyhmphoma in
   cottontop tamarins by vaccination with the EB virus envelope glycoprotein gp340 incorporated into Iscoms, J.Gen.Virol., in press.

Morgan, A.J., Mackett, M., Finerty, S., Arrand, J., Scullion, F. and Epstein, M.A., 1988, Recombinant vaccinia viruses expressing Epstein-Barr virus glycoprotein gp340 protect cottontop tamarins against EB virus-induced malignant lymphomas, J.Med.Virol., in press.

Morgan, A.J., North, J.R. and Epstein, M.A., 1983, Purification and properties of the gp340 component of Epstein-Barr (EB) virus membrane antigen (MA) in an immunogenic form, J.Gen.Virol., 64:455.

Moss, D.J. and Pope, J.H., 1972, Assay of the infectivity of the Epstein-Barr virus by transformation of human leucocytes in vitro, J. Gen. Virol., 17:233.

Niederman, J.C., Evans, A.S., Subrahmanyan, L. and McCollum, R.W., 1970, Prevalence, incidence and persistence of EB virus antibody in young adults, New Eng.J.Med., 282:361.

North, J.R., Morgan, A.J., Thompson, J.L. and Epstein, M.A., 1982, Quantification of an EB virus-associated membrane antigen (MA) component, J.Virol.Methods, 5:55.

Okazaki, W., Purchase, H.G. and Burmester, B.R., 1970, Protection against Marek's disease by vaccination with a herpesvirus of turkeys, Avian Dis., 14:413.

Payne, L.N., Fraxier, J.A. and Powell, P.C., 1976, Pathogenesis of Marek's disease, in: "International Review of Experimental Pathology", G.W. Richter and M.A. Epstein, eds., Academic Press, New York.

Pereira, M.A., Blake, J.M. and Macrae, A.D., 1969, EB virus antibody at different ages, Brit.Med.J., 4:526.

Purtilo, D.T., 1984, Immune deficiency and cancer. Epstein-Barr virus and lymphoproliferative malignancies, Plenum Medical Book Co., New York.

Randle, B.J. and Epstein, M.A., 1984, A highly sensitive enzyme-linked immunosorbent assay to quantitate antibodies to Epstein-Barr virus membrane antigen gp340, J.Virol.Methods, 9:201.

Rickinson, A.B., 1986, Cellular immunological responses to the virus infection in: "The Epstein-Barr virus: Recent Advances", M.A. Epstein and B.G. Achong, eds., William Heinemann Medical Books, London.

de Schryver, A., Klein, G., Hewetson, J., Rocchi, G., Henle, W., Henle, G., Moss, D.J. and Pope, J.H., 1974, Comparison of EBV neutralization tests based on abortive infection or transformation of lymphoid cells and their relation to membrane reactive antibodies (anti MA), Int.J. Cancer, 13:353.

Schultz, L.D., Tanner, J., Hofmann, K., Emini, E., Kieff, E. and Ellis, R.W., 1987, Expression and analysis of EBV gp350 in yeast Saccharomyces cerevisiae, in: "Epstein-Barr Virus and Human Disease", P.H. Levine, D.V. Ablashi, M. Nonoyama, G.R. Pearson and R. Glaser, eds., Humana Press, Clifton.

Shanmugaratnam, K., 1971, Studies on the etiology of nasopharyngeal carcinoma, in: "International Review of Experimental Pathology", G.W. Richter and M.A. Epstein, eds., Academic Press, New York.

Takahashi, M., Otsuka, T., Okuno, Y., Asano, Y., Yazaki, T. and Isomura, S., 1974, Live varicella vaccine used to prevent the spread of varicella in children in hospital, Lancet ii:1288.

de Thé, G., 1979, Demographic studies implicating the virus in the causation of Burkitt's lymphoma; prospects for nasopharyngeal carcinoma, in: "The Epstein-Barr Virus", M.A. Epstein and B.G. Achong, eds., Springer, Berlin.

University Health Physicians and PHLS Laboratories, 1971, Infectious mononucleosis and its relationship to EB virus antibody, Brit.Med.J., iv:643.

Weibel, R.E., Neff, B.J., Kuter, B.J., Guess, H.A., Rothenburger, C.A. Fitzgerald, A.J., Connor, K.A., McLean, A.A., Hilleman, M.R. and Buynak, E.B., 1984, Live attenuated varicella virus vaccine: efficiency trial in healthy children, New Eng.J.Med., 310:1409.

Whang, Y., Silberklang, M., Morgan, A., Munshi, S., Lenny, A.B., Ellis, R.W. and Kieff, E., 1987, Expression of Epstein-Barr virus gp350/220 gene in rodent and primate cells, J.Virol., 61:1796.

Zuckerman, A.J., 1985, Prevention of hepatocellular carcinoma by immunization against hepatitis B, in: "International Review of Experimental Pathology", G.W. Richter and M.A. Epstein, eds., Academic Press, Orlando.

MEASUREMENT OF ANTIBODY AFFINITY, CONCENTRATION AND ISOTYPE TO EVALUATE

ANTIGENS, ADJUVANTS, AND IMMUNIZATION REGIMENS IN VACCINE RESEARCH

Brian W. Hughes, John S. Kenney and Anthony C. Allison

Department of Immunology, Syntex Research, Palo Alto
CA 94304, USA

INTRODUCTION

The recent explosion in vaccine research has brought into focus the need for better methods of characterizing the serum antibodies produced in animals immunized with various adjuvants and antigens. The typical end-point ELISAs used to characterize sera fail to differentiate between antibody affinity and concentration and cannot quantitate antibody isotypes. Antibody affinity and concentration when measured directly can serve as invariant standards to judge the quality of immunization and are not subject to arbitrary interpretation that is applied to end-point titer ELISAs. Furthermore, antibody affinity determines the effectiveness of antibody in protection (reviewed in Steward and Steensgard, 1983). Antibody isotypes, which are often neglected in evaluating vaccines, have also been shown to have a marked effect on the generation of immunoprophylaxis (Wechsler and Kongshavn, 1986). We therefore suggest that measurements of antibody affinity, concentration, and isotype should be included in the evaluation of the efficacy of a vaccine. To measure these characteristics, we have developed a convenient solution-phase RIA for antibody affinity and concentration and an ELISA for antibody isotypes.

CRITICISMS OF THE END-POINT TITER ELISA

The solid-phase end-point titer ELISA has dominated serum characterization techniques. Typically, antigen is coated on a solid-phase (i.e. microplate well) and then reacted with various dilutions of serum, anti-immunoglobulin antibody-enzyme conjugate, and enzyme substrate. The resultant optical density is proportional to the amount of serum antibody that bound to the plate. Results are often expressed as a graph of the antibody dilution vs. O.D. or as the serum dilution that gives an arbitrary end-point value (such as twice background or 0.2 O.D. etc.). In practice, such graphical representation and arbitrary end-point values make it difficult to compare findings from different laboratories.

More serious objections can be raised against solid-phase ELISAs. First, even though an ELISA curve is a function of both affinity and concentration, it is impossible to distinguish the relative contribution of affinity and concentration by the shape of the curve. In vivo, antibody affinity and concentration vary independently of one another (Steward and Petty,

1972). Furthermore, mouse breeding studies indicate that antibody affinity and concentration are under independent genetic control (Steward and Petty, 1976). Second, the binding of protein antigens to the surface of ELISA wells can cause the protein to denature (Apple et al, 1984). Often, a substantial portion of the antibody-binding activity generated by immunization with native antigen is directed exclusively to denatured antigen, in part due to the adjuvant used (Wolberg et al, 1970; Kenney et al, 1989a). Hence, binding activity detected using a solid-phase antigen ELISA may not accurately reflect the native antigen-binding activity of the antibody (Gani et al, 1988). Third, a simple ELISA cannot quantitate antibody isotypes, which may be an important predictor of protection by complement-mediated and antibody-dependent cellular cytotoxic mechanisms.

ANTIBODY AFFINITY AND CONCENTRATION

Affinity is the measure of how tightly an antibody binds to its ligand. As antibody affinity increases, epitope recognition becomes more specific and non-specific cross-reactivity decreases. Because antibody affinity and concentration are intrinsic properties, they can be used directly to compare immunization regimens used by researchers in different laboratories. Antibody affinity measured in a serum constitutes an "average" affinity of all the antibodies present. In the assay we use, the average serum affinity appears to be a constant that can be reproducibly measured over a 20-fold range of serum dilutions (manuscipt in preparation). Therefore, our method of measuring antibody affinity and concentration directly offers a precise means for comparing two or more sera.

RIA FOR ANTIBODY AFFINITY AND CONCENTRATION

We have developed a simple radio-immunoprecipitation assay (RIA) for both antibody affinity and concentration (Kenney et al, 1989b) using the method described by Müller (1980) as the basis of our assay. Our assay is performed in Sarstedt 600 μl RIA vials in a 96-well microplate format. Antibody-antigen complexes are isolated using a second immunoprecipitating antibody. This makes the assay applicable to virtually any combination of antigen and antibody.

First, radio-labeled antigen is incubated with dilutions of serum to determine that dilution which binds approximately 40% of labeled antigen (Fig. 1, Part I). Second, this serum dilution is used in a competitive binding assay in the presence of a range of concentrations of unlabeled antigen (Fig. 1, Part II). The concentration of unlabeled antigen that inhibits the binding of 50% of the labeled antigen is then used to calculate the serum antibody affinity. From the antibody affinity and the dilution of antibody used in Part II, the concentration of the antibody can also be calculated (see Appendix).

APPLICATION OF RIA

Analysis of Adjuvants

In comparing adjuvant systems, we ran an end-point titer ELISA on sera pooled from mice immunized with human serum albumin (HSA) in either a mixture of Alhydrogel [Al(OH)$_3$] and N-acetylmuramyl-L-threonyl-D-isoglutamine (abbr. as AL/MDP) or Freund's adjuvant. In the ELISA, the Freund's sera produced the higher titer (Fig. 2A). However, when both adjuvants were tested in Part I of the RIA, the dilution curve of the AL/MDP sera was nearly identical to that of the Freund's sera (Fig. 2B). This difference

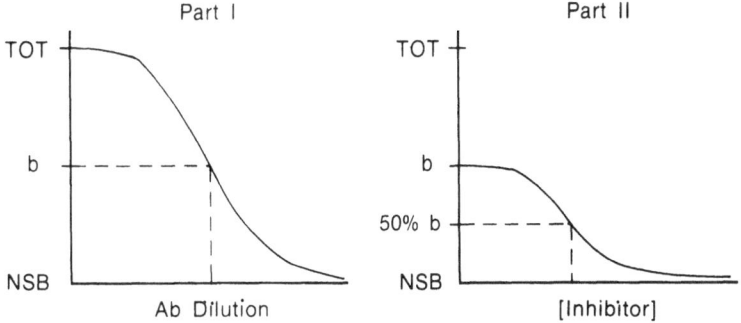

Fig. 1. Part I: Titration of antibody dilutions against radio-labeled antigen.

Part II: Titration of unlabeled antigen against antibody dilution that binds 40% of radio-labeled antigen. TOT, total radio-labeled antigen, cpm. NSB, non-specific binding of radio-labeled antigen, cpm. b, antibody dilution that binds 40% of radio-labeled antigen.

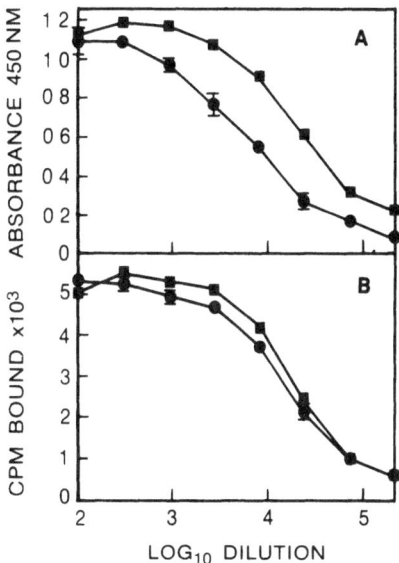

Fig. 2. (A) ELISA titration of pooled sera from mice immunized with HSA using Freund's adjuvant (■) or AL/MDP (●). (B) RIA titration of the same sera.

in the performance of the sera in the two assays is likely the result of higher antibody binding activity in the Freund's sera to denatured solid-phase antigen in the ELISA. Compared to AL/MDP, immunization with Freund's adjuvant results in a higher percentage of antibodies which bind exclusively to denatured antigen (Kenney et al, 1989a).

Antibody affinity and concentration were determined using the RIA. The AL/MDP sera had an antibody affinity of 11×10^9 1/mole and a concentration of 188 μg/ml while the Freund's sera had an antibody affinity of 4.9×10^9 1/mole and a concentration of 635 μg/ml. Although Freund's sera had more immunoglobulin, the AL/MDP sera had over twice the affinity for the antigen.

The RIA allowed us to establish that the AL/MDP adjuvant produced titers that were equivalent to the Freund's adjuvant in binding native antigen and produced higher affinity antibodies than Freund's. These results could not have been obtained if we had relied solely on ELISA for our evaluation.

Analysis of Antigens

The RIA was used to evaluate the ability of somatostatin and somatostatin analogues conjugated to thyroglobulin to induce antibodies against somatostatin in guinea pigs. Previous attempts to detect antibodies in guinea pig sera to unconjugated somatostatin and its analogues using an ELISA were unsuccessful. This may have been due either to surface denaturation of the peptides or to steric hindrance of antibody binding to coated peptides. On the other hand, we detected anti-somatostatin antibodies in the sera using the RIA. Three analaogues were found to induce antibody that cross-reacted with native somatostatin. Surprisingly, conjugated native somatostatin itself was a poor immunogen.

The RIA enabled us to distinguish antibody affinity and concentration in the sera. Affinity calculations showed that one analogue, analogue A, generated a serum antibody affinity of 2.9×10^9 1/mole (Table 1). Since the affinity of somatostatin for its receptor is in this range (e.g. 1.9×10^9 1/mole for rat pituitary receptor) this analaogue may be suitable for generating antibodies which could block somatostatin binding to its receptor.

The RIA was also used to investigate the relative affinity of the sera for the analogues as compared to their affinity for native somatostatin. In this analysis the analogues were substituted for native somatostatin as the unlabeled antigen inhibitor. When analogue A was used as the inhibitor with sera from animals immunized with the analogue A conjugate, we found analogue A to be 2-fold less potent that somatostatin as an inhibitor of radiolabeled somatostatin binding (data not shown). This suggests that analogue A, when conjugated to thyroglobulin, might present epitopes which were more like native somatostatin than the unconjugated analogue itself.

Table 1. Affinity and Concentration of Antibodies in Sera of Guinea Pigs[*] Immunized with Somatostatin Analogues

Immunogen	Serum Antibody	
	K_{aff} (x 10^9 1/mole)	[Ig] (ng/ml)
Analogue A	2.9	1,256
Analogue B	0.27	723
Analogue C	1.1	75
Native Somatostatin	**	**

*Guinea pigs were immunized three times with somatostatin or somatostatin analogue conjugated to bovine thyroglobulin in Syntex Adjuvant Formulation-1. Pooled sera (5 animal/group) were assayed for affinity and concentration using the RIA as described in the Appendix.

**Antibody not detected.

The use of the RIA in the analysis of peptide antigens has distinct advantages. First, it enables detection of antibodies to peptides where an ELISA may not. Second, the immunogenicity of peptides can be ranked according to the affinity and concentration of elicited antibodies. Third, affinity and concentration values can be related to the potential therapeutic utility of the antibodies.

ANTIBODY ISOTYPE

Structural variants of the immunoglobulin heavy chain constant region that are found in all individuals of a species are called isotypes. Each isotype makes its own unique contribution to host protection against infectious agents and tumors. For example, in the mouse the IgG2a isotype has been shown to be the most effective isotype in conferring protection. Different adjuvants and immunization regimens can influence the isotype distribution of antigen-specific immunoglobulins that are elicited by immunization (Kenney et al, 1989a). Determination of the isotype distribution of the antibody response should therefore be part of the evaluation of the efficacy of a vaccine.

ISOTYPE ASSAY

To determine the isotype profile of antigen-specific antibodies in serum, we developed an isotype ELISA (Fig. 3). Serum is incubated at a subsaturating dilution in antigen-coated wells. Bound antibody is detected using specific anti-isotype antibody-enzyme conjugates. To quantitate the amount of bound antibody, the absorbances of known concentrations of immunoglobulin isotype standards are measured by ELISA using wells coated with polyclonal anti-Ig instead of antigen. The relative concentrations of antigen-specific antibody isotypes are then determined from standard curves of absorbance versus the concentration of the isotype standard.

For the assay to be valid several conditions must be satisfied. The polyclonal anti-Ig used to capture the isotype standards must have the same affinity for all isotypes. The isotype standards must be pure and the anti-isotype antibody-enzyme conjugates highly specific for their particular isotype. Moreover, the assumption must be made that antibodies of different isotypes have the same average affinity for the antigen and that the distribution of isotypes for native and surface-denatured antigen is the same. In practice, the affinity of the capture antibody for the isotype standard may not be the same as the affinity of the antigen-specific antibody for the antigen. For these reasons, the isotype concentrations obtained by this method are relative values, rather than absolute values. However, the

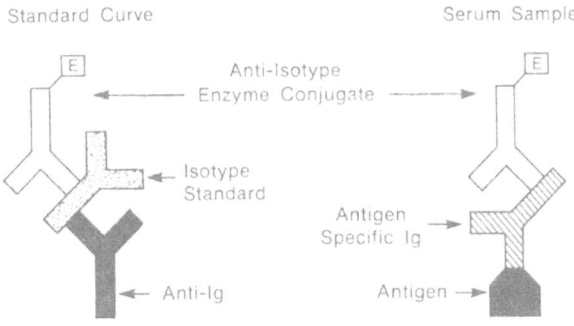

Fig. 3. Configuration of isotype ELISA.

relative concentration obtained in this assay can be compared with that obtained by the RIA for total immunoglobulin to calculate the concentration of a particular antibody isotype.

APPLICATION OF THE ISOTYPE ASSAY

We have used the isotype assay to evaluate the effect of adjuvants and the route of administration on the isotype distribution of antibodies (Kenney et al, 1989a). As shown in Table 2, sera from mice immunized by the s.c. or i.p. route gave different isotype profiles (Table 2). The s.c. serum had 91% IgG1, 7% IgG2a and 2% IgG2b while the i.p. serum had 78% IgG1, 20% IgG2a and 2% IgG2b. Mice immunized by the s.c. route also had higher concentrations of antibody as determined by the RIA (Table 2). Comparison of the calculated values for IgG2a were 13 and 23 µg/ml for the s.c.-immunized and i.p.-immunized mouse sera, respectively. Thus, the isotype assay allows evaluation of the effects of different immunization conditions such as the route of administration on the elicitation of antibodies of protective isotypes.

SUMMARY

Antibody affinity, concentration, and isotype are the key elements in the generation of effective humoral immunity. These elements are independently regulated and may be independently affected by the choice of immunogen, adjuvant and immunization regimen. We have described assay systems for determining antibody affinity, concentration and isotype. We have also described several applications of these assays to determine optimum conditions for immunization. The assays are useful tools to evaluate the potential of immunogens, adjuvants and immunization regimens in vaccine research.

Table 2. Isotype Distribution and Concentration of Antibodies in Sera of Mice Immunized with Human Serum Albumin in AL/MDP[a]

Route	Percent[b] Concentration (µg/ml)[c]				
	G1	G2a	G2b	G3	Total
s.c.	91 / 171	7 / 13	2 / 4	0 / 0	100 / 188
i.p.	78 / 90	20 / 23	2 / 2	0 / 0	100 / 115

[a]Mice were immunized with 5 µg human serum albumin in AL/MDP. Pooled sera (5 animals/group) were analyzed using the isotype assay and affinity and concentration assay as described in the Appendix. Antibodies of IgM and IgE isotype were not detected in the sera.

[b]Percent of total antibody as determined in the isotype assay.

[c]Concentration of antibody using the fraction of each isotype and the total antibody concentration determined by the RIA.

APPENDIX

Affinity and Concentration RIA

Assays are performed in RIA vials (#73.1055) in aluminum racks (#95.1012), both from Sarstedt (Princeton, NJ). Incubation buffer (0.1% non-fat dry milk/0.05% thimerosal/PBS) is used as the diluent for antibody and antigen. Incubation buffer is stored at 4°C and freshly prepared every few weeks. In addition to vials containing mixtures of labeled antigen (tracer) and test antibody the following controls are included: tracer alone (non-specific binding, NSB), tracer plus a concentration of antibody resulting in maximum tracer binding (total specific binding, TSB). The final concentrations of reactants (at the 75 μl incubation volume) are used in all calculations.

Part I - antibody titration. Starting at a dilution of 1:100, eight 3-fold serial dilutions are made of the test serum. To each vial (in triplicate) is added 25 μl of a test serum dilution and 50 μl of 5,000 cpm of tracer. The reactants are mixed using an orbital shaker, incubated overnight at 4°C and immunoprecipitated as described below. Results are plotted as the antibody dilution vs. cpm bound at each dilution (Fig. 1, Part I). The scale for cpm bound is set using NSB as the minimum value and TSB as the maximum value. The antibody dilution that binds approximately 40% of the tracer is estimated from the curve and used in part II (below).

Part II - antigen titration. The appropriate range of concentrations of unlabeled antigen (inhibitor) is determined as follows:

If $K = \dfrac{8}{3([I] - [T])}$ at 50% inhibition of tracer binding

(Muller, 1980) and [T] is considered to be negligible,

then $K = \dfrac{8}{3\,[I]}$ or $[I] = \dfrac{2.67}{K}$

where: K is the desired antibody affinity in 1/mole, and [I] and [T] are the inhibitor and tracer concentration, respectively, in mole/l.

Using the concentration of inhibitor calculated above as the midpoint, seven 3-fold dilutions of the inhibitor are prepared. An eighth sample containing only incubation buffer, no inhibitor (NIB), is included.

To each vial (in triplicate) is added 25 μl of a dilution of unlabeled antigen, 25 μl of 5,000 cpm of tracer (at exactly twice the concentration used in Part I), and 25 μl of a dilution of test serum giving approximately 40% binding as determined in Part I. The reactants are mixed using an orbital shaker, incubated overnight at 4°C and immunoprecipitated as described below. The results are plotted as the inhibitor concentration vs. cpm bound at each inhibitor concentration (Fig. 1, Part II). The scale for cpm bound is set using NSB as the minimum value and binding of tracer in the absence of inhibitor (NIB) as the maximum value (b). The inhibitor concentration resulting in 50% tracer bound is estimated from the curves and used to calculate affinity (see below).

Immunoprecipitation. Reactions are immunoprecipitated by adding 25 μl of an optimum dilution of normal carrier serum and 25 μl of an optimum dilution of precipitating anti-Ig second antibody (Pel-Freez, Rogers, AR), typically 10% and 20%, respectively. Vials are mixed using an orbital shaker and incubated 4 hours at 4°C. Then, 200 μl of cold (4°C) 2% polyethylene glycol (PEG)-6000 (Sigma #P-2139) (2% PEG/0.05% thimerosal/PBS) is added to the vials. The vials are centrifuged 600 x g, 20 minutes at 4°C using a microplate carrier. The supernatant is aspirated using an eight-channel manifold and a vacuum of 5 in. Hg, being careful not to draw up the pellet. Radioactivity in the pellet is determined using a gamma counter.

Calculation of antibody affinity (from Muller, 1980):

$$K = \frac{1}{(([I] - [T])\ (1 - 1.5b + 0.5\ b^2))}$$

where: K = the antibody affinity in 1/mole

$[I]$ = the inhibitor concentration which gives 50% inhibition of tracer binding in mole/1

$[T]$ = the tracer concentration in mole/1

$b = \dfrac{(NIB - NSB)}{(TSB - NSB)}$ all in cpm

Calculation of antibody concentration (from Muller, 1980):

$$[Ab_t] = b[T] + \frac{b}{(1-b)K}$$

where: $[Ab_t]$ = total antibody binding sites in mole/1.
By assuming that each IgG has two binding sites and a M_r = 150,000

$$[Ig]g/ml = (\text{Serum dil.})\frac{(150,000g)}{mole}\frac{(0.001)}{ml}\frac{([Ab_t])}{2}$$

where: Serum dil. = the serum dilution in the 75 µl reaction volume in Part II.

ISOTYPE ASSAY

Optimization of assay conditions. Before performing the isotype assay, the dilution of serum that is sub-saturating for coated antigen must be determined in a separate assay. Eight 3-fold serial dilutions, starting at 1:100, of the test serum are incubated on antigen-coated wells and detected with polyclonal anti-Ig (pan specific) peroxidase-conjugate using the conditions described below. From a plot of absorbance vs. serum dilution, the sub-saturating dilution is determined. In addition, the range and concentration of the standard curve dilutions for each isotype standard and the optimum dilution of each isotype-specific peroxidase-conjugate must be determined to give the proper working range.

Coating of microplate wells. For the isotype standard, 3 x 8 wells on each plate are coated with 50 µl of polyclonal anti-Ig (Southern Biotech. Assoc., Birmingham, AL) diluted to 10 ug/ml in PBS. These wells will be used for the standard curve. The remaining wells, which are to be used for the serum samples, are coated with 50 ul of antigen in PBS (10 µg/ml). The plate is then incubated at 4°C in an humidified environment. The wells are emptied, washed 3 times with wash buffer (PBS/0.05% thimerosal) and blocked with 100 µl of 5% non-fat dry milk/PBS 0.05% thimerosal. Following an incubation for one hour at room temperature, the wells are emptied and washed 3 times with wash buffer.

Preparation and addition of isotype standard. Seven serial 3-fold dilutions of isotype standard (Southern Biotech. Assoc.) are prepared in incubation buffer (1% non-fat dry milk/PBS/0.05% thimerosal). To all anti-Ig coated wells except the eighth set is added (in triplicate) 50 μl of each dilution of isotype standard. To each well of the eighth set of the standard curve wells is added 50 μl of incubation buffer. These wells are the background control for non-specific binding by the anti-isotype antibody-enzyme conjugate.

Addition of test sera. To each antigen coated well is added (in triplicate) 50 μl of a sub-saturating test serum dilution. The plates are incubated for 2 hours at room temperature with shaking. The wells are then emptied and washed 4 times with wash buffer.

Detection of bound antibody. To each well is added 50 μl of anti-isotype peroxidase-conjugate (Southern Biotech. Assoc.) diluted in incubation buffer. The plates are incubated for two hours at room temperature with shaking. The wells are emptied and washed five times with wash buffer. To each well is added 50 μl of orthophenylenediamine (OPD) substrate solution (1 mg/ml OPD (Zymed Labs., So. San Francisco)/0.03% H_2O_2/0.1 M citrate buffer pH 4.9), and the plates are incubated for 30 minutes. The absorbance at 450 nm is read with a microplate spectrophotometer using the background control well as blank.

Data analysis. The standard curve is plotted and used to evaluate the isotype concentration for the test sera. The total immunoglobulin concentration for all of the isotypes is computed to determine their relative percent concentration. If available, the total immunoglobulin concentration calculated from the affinity RIA can be used to determine the absolute immunoglobulin concentration for each isotype.

Acknowledgements

We thank Noelene Byars and Mary Welch for providing the guinea pig anti-somatostatin sera.

REFERENCES

Apple, R., Knauper, B., Pesce, A. and Michael, G., 1984, Shared antigenic determinants of native and denatured bovine serum albumin are recognized by both B- and T-cells, Mol.Immunol., 21:901.

Gani, M., Coley, J. and Porter, P., 1987, Epitope masking and immuno-dominance - complications in the selection of monoclonal antibodies against HCG, Hybridoma, 6:637.

Kenney, J.S., Hughes, B.H., Masada, M.P. and Allison, A.C., 1989a, Influence of adjuvants on the quantity, affinity, isotype, and epitope specificity of murine antibodies, J.Immunol.Methods, in press.

Kenney, J.S., Hughes, B.H. and Allison, A.C., 1989b, Determination of antibody affinity and concentration by solution-phase micro-radioimmunoassay, in: "Focus on Laboratory Methods in Immunology", H. Zola, ed., CRC Press, Boca Raton.

Muller, R., 1980, Calculation of average antibody affinity in anti-hapten sera from data obtained by competitive radioimmunoassay, J.Immunol. Methods, 34:345.

Steward, M.W. and Petty, R.E., 1972, The use of ammonium sulphate globulin precipitation for determination of affinity of anti-protein antibodies in mouse serum, Immunology, 22:747.

Steward, M.W. and Petty, R.E., 1976, Evidence for the genetic control of antibody affinity from breeding studies with inbred mouse strains producing high and low affinity antibody, Immunology, 30:789.

Steward, M.W. and Steensgaard, J., 1983, Antibody affinity: thermodynamic aspects and biological significance, CRC Press, Boca Raton.

Wechsler, D.S. and Kongshavn, P.A.L., 1986, Heat-labile IgG2a antibodies affect cure of Trypanosoma musculi infection in C57BL/6 mice, J.Immunol., 137:2968.

Wolberg, G., Liu, C.T. and Adler, F.L., 1970, Passive hemagglutination. II. Titration of antibody against determinants unique for aggregated denatured bovine serum albumin and further studies on gelatin, J.Immunol., 105:797.

RECOMMENDATIONS FOR THE ASSESSMENT OF ADJUVANTS (IMMUNOPOTENTIATORS)

D.E.S. Stewart-Tull

Microbiology Department, Alexander Stone Building
Garscube Estate, Bearsden, Glasgow G61 1QH, UK

INTRODUCTION

During the last decade there has been a significant expansion in the
field of adjuvant research in many countries due, in part, to the success of
the techniques of genetic manipulation, nucleotide sequencing and peptide
synthesis. The potential exists to formulate totally synthetic vaccines
(Jolivet et al, 1987) but it is also realized that the successful use of the
new technologies will require a continued research effort to find adjuvant
substances which do not have adverse biological properties (Vane and
Cuatrecasas, 1984; Paquet et al, 1986; Stewart-Tull, 1988).

There are conflicting views about the beneficial or detrimental characteristics of the adjuvant substances themselves or the suspending media used
in the formulation of experimental vaccines. Numerous contraindications
have been cited in published papers (WHO 1976; Edelman, 1980; Stewart-Tull,
1985, 1986; Allison et al, 1986; Gisler et al, 1979) and researchers involved in vaccine development must have cognisance of these criticisms.
However, it is equally important to refute continued criticism where the
argument is ill-founded. Unfortunately, few scientists are prepared to
withdraw an adverse comment even if there is experimental evidence to null-
ify the complaints. In addition, newcomers to the field lack the experience
of the established adjuvant researcher and often through sub-optimal testing
they obtain erroneous results or fail to do crucial experiments. The presence of such a large majority of the world's adjuvant researchers, apart
from some French and Japanese workers, at this conference encouraged me to
seek a consensus agreement for some sensible guidelines for the standard-
ization of the assessment of adjuvanted vaccine formulations.

THE OPPOSITION TO ADJUVANTS

There is an understandable tendency for individual researchers to obtain support for their own adjuvant and to decry those of other groups. We
have seen these jousting knights attempting to seek the favours of participants at this conference so that they will be tempted to use one defined adjuvant. However, it is likely that the choice of adjuvant may depend on the
chemical characteristics of the antigen molecules. For example, there would
be less advantage in the use of Freund complete adjuvant (FCA) or incomplete
adjuvant (FIA) if the antigen molecule lacks a hydrophobic sequence as this

is essential for holding it in the oil at the oil-water interface, some antigens may lack the charge required to effect the adsorption to aluminium hydroxide and different responses may be obtained with different MDP derivatives. It is important to test a variety of adjuvants in the experimental systems before selecting a particular one for inclusion in a vaccine.

OIL-ADJUVANTED FORMULATIONS

Many investigations were reported in the early 1940s on the search for an oil-adjuvanted influenza vaccine (Friedewald, 1944a, b; Henle and Henle, 1945) and these directed attention to the search for suitable oil components and emulsifiers (reviewed by Stewart-Tull, 1983). Some 22,000 Army personnel were given mineral oil-adjuvanted influenza vaccine in clinicial trials (Davenport, 1968) and attempts are being made to follow up their immunological history (R. Edelman, personal communication).

In 1976, we had investigated the reason why some oils produced very severe reactions leading to hemorrhagic lesions and loss of integral structure within the guinea-pig footpad in some Japanese experiments. The answer to these contraindications was very clearcut, mineral oils consisting of short-chain hydrocarbons, $\underline{n}C_6H_{14} - \underline{n}C_{10}H_{22}$ were responsible. Mineral oils with long-chain, fully saturated hydrocarbons, $\underline{n}C_{16}H_{34} - \underline{n}C_{19}H_{40}$ did not cause severe problems. After publication of these data one company manufacturing an oil-adjuvanted veterinary vaccine informed me that three of five 50 gallon drums were found to be contaminated with short-chain hydrocarbons. It has been my aim to persuade other adjuvant workers to include GLC analyses in papers which are critical of oil-adjuvanted formulations. However, the argument fails to be resolved because in two studies on cholera and tetanus, with a limited number of vaccinees, adverse reactions were recorded (MacLennan et al, 1965; Ogonuki et al, 1967) but we have no knowledge of the quality of oil used.

Similarly, Arlacel A (mannide mono-oleate: ICI America) was shown to be a carcinogen and a co-carcinogen in male Swiss-Webster mice (Murray et al, 1972) but to my knowledge there is no report of this effect in any other strain of mouse. These findings need to be substantiated in parallel with accurate chemical analyses of the materials before they become the yardstick of non-acceptability of oil-adjuvanted formulations.

ADJUVANT DOSAGE

This is another perennial problem encountered in experimental adjuvant studies (Stewart-Tull, 1980, 1984, 1985a, b, 1988; Stewart-Tull and Parant, 1986). If the dose of adjuvant is too low, the substance may appear to be adjuvant-inactive; if it is too high an immunodepressive effect may be found. Numerous researchers may have floundered because of the lack of a dose response curve, but where this has been done the optimal dose of adjuvant is often quite apparent (Räsänen et al, 1982).

CONTRAINDICATIONS

A suitable vaccine adjuvant must not induce hypersensitivity to host tissue nor to the adjuvant itself. It should not possess cross-reactive antigens. Carcinogenic, teratogenic and abortogenic activities must be absent. The adjuvant should induce neither an autoimmune disease nor a polyarthritis nor an allergic reponse to extraneous antigens (e.g. food antigens with oral vaccines) in the particular experimental system used.

BENEFICIAL ADJUVANT PROPERTIES

A candidate adjuvant substance should produce a humoral and/or cell-mediated immune response while potentiating the immunogenicity of the vaccine antigen. The adjuvant should be non-toxic, non-pyrogenic and preferably biodegradable.

THE COMPARISON OF ADJUVANTS

It will be clear from these few examples that there is a need for some clear guidelines to test new adjuvants for contraindications and to enable comparisons to be made between one adjuvant and another (Tables 1 and 2).

CONSENSUS VIEW OF COMPARATIVE TESTS FOR ADJUVANTS

Glassware

All the bottles, pipettes and glass syringes should be cleaned with E-TOXA-CLEAN (210-3: Sigma Chemical Company, a cleaning agent for glassware prior to inactivation of endotoxins by steam sterilization and dry heating).

Toxicity Test

After discussion of the various available tests (a, tissue culture cytopathogenic effect in human MRC-5 or mouse fibroblasts; b, mouse LD_{50} tests; c, intracutaneous toxicity test in rabbits; d, weight gain test in mice) it was agreed to use the creatine phosphokinase assay (Kit No. 520 for colorimetric determination in serum or plasma; Sigma Chemical Company). The phosphorylation of ADP by phosphocreatine to give ATP and free creatine is catalyzed by this enzyme. Hughes (1962) used the assay to measure the

$$\text{ADP + phosphocreatine} \xrightarrow{\text{creatine phosphokinase}} \text{ATP + creatine}$$

$$\text{Creatine} + \alpha\text{-naphthol} + \text{diacetyl} \longrightarrow \text{colored complex.}$$

enzyme in serum; no appreciable changes occur if the serum sample is stored at 2-6°C or frozen for 5 days. The amount of color is directly proportional to enzyme activity and from a calibration curve International Units litre^{-1} can be determined. According to Gray et al (1974) levels of creatine phosphokinase <2000 IU 1^{-1} are acceptable but >3000 IU 1^{-1} would be an indicator of toxicity.

Pyrogenicity Tests

1. The diluents used in the preparation of experimental vaccines should be tested for their activity in the Limulus polyphemus assay.

2. The rabbit pyrogenicity test is the recognized procedure in the European and United States' Pharmacopoeias for the detection of endotoxin. Rabbits are injected subcutaneously with variable doses of the vaccine adjuvants in 1.0 ml pyrogen-free phosphate-buffered saline. The rectal temperature is recorded hourly and the increase should be <2.0°F for the preparation to be deemed non-pyrogenic.

Choice of Antigen

The majority of the early studies with FCA were done with chicken egg albumin (see Stewart-Tull, 1983) as this is a poor immunogen when injected

Table 1. Scheme of Adjuvant Testing

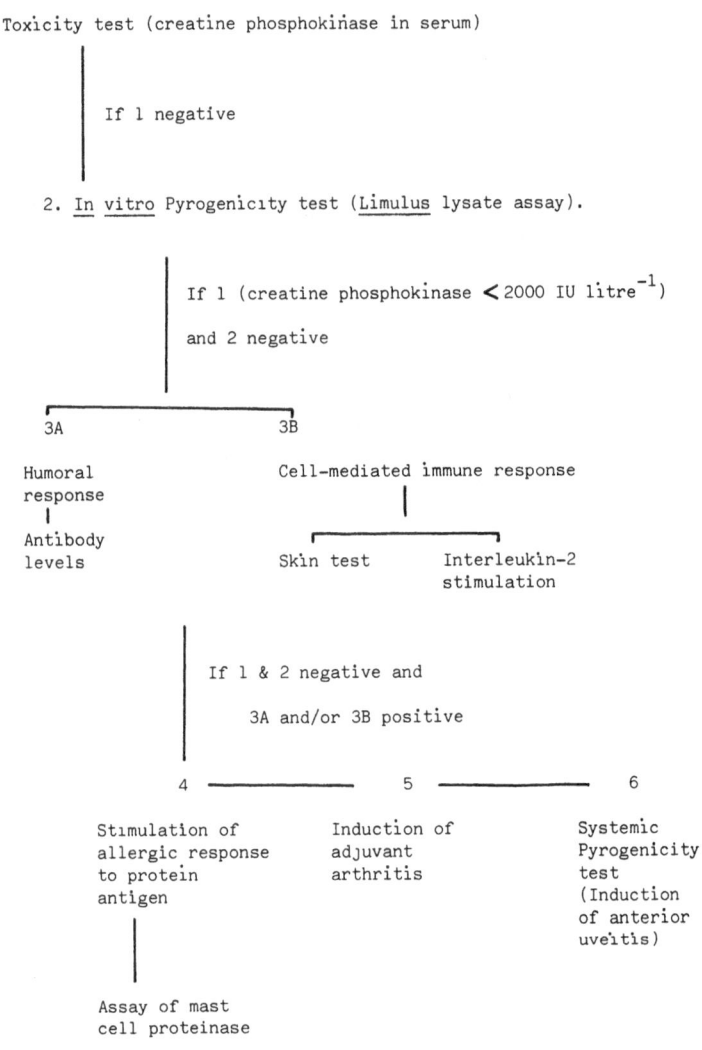

1. Toxicity test (creatine phosphokinase in serum)

   If 1 negative

2. In vitro Pyrogenicity test (Limulus lysate assay).

   If 1 (creatine phosphokinase $<$ 2000 IU litre^{-1})

   and 2 negative

3A | 3B

Humoral   Cell-mediated immune response
response

Antibody
levels   Skin test   Interleukin-2
                                 stimulation

If 1 & 2 negative and

3A and/or 3B positive

4 ——————— 5 ——————— 6

Stimulation of   Induction of   Systemic
allergic response adjuvant   Pyrogenicity
to protein   arthritis   test
antigen   (Induction
                                  of anterior
                                  uveitis)

Assay of mast
cell proteinase

---

alone. Participants agreed that one of the standard antigens should be OVALBUMIN (A5503, Grade V crystallized and lyophilized from Sigma Chemical Company). This protein antigen is soluble in water but it is imperative that the lyophilized powder should be dried in a desiccator over NaOH pellets and conc H_2SO_4 in separate containers as the preparation may absorb water to nearly 50% of its weight!

In addition, there was approval for the use of an amphipathic antigen molecule, INFLUENZA H_3N_2 TYPE A HEMAGGLUTININ (obtainable from Dr. F. Brown, Wellcome Biotechnology, Langley Court, Beckenham, Kent BR3 3BS).

Standard Adjuvants

There is no doubt that the most potent adjuvant known to adjuvant researchers is FREUND'S COMPLETE ADJUVANT. At present the Statens Serum Institute, Copenhagen is producing a classical formulation which is marketed by Superfos a/s, Frydenlundsvej 30, P.O. Box 39, DK-2950, Vedbaek, Denmark:-

```
     FCA 85% (v/v) Marcol 52              FIA 85% (v/v) Marcol 52
         15% (v/v) Arlacel A                  15% (v/v) Arlacel A
                   (mannide mono-oleate)
     2.0 mg/ml Mycobacterium
               tuberculosis
```

The mineral oil mixture containing M. tuberculosis should be sonicated briefly before use to ensure the even dispersion of the bacterial organisms. For the guinea-pig the optimal injection mixture is composed of 10.0 mg ovalbumin in 0.5 ml physiological saline plus 0.5 ml FCA containing 1.0 mg M. tuberculosis; the mixture is repeatedly drawn up into a 1.0 ml glass syringe with Luer fitting and expelled until emulsified. It is important to check this by placing a drop of the mixture onto the surface of water. If it disperses, the mixture is oil-in-water, if it stays as a blob it is a water-in-oil emulsion. The injection dose (0.2 ml) contains 2.0 mg antigen and 200 μg M. Tuberculosis (see Stewart-Tull, 1983). For the dose response curve, injection doses should be prepared containing 25, 50, 100, 200 and 400 μg of M. tuberculosis.

Aluminium Hydroxide (Alhydrogel)

Participants agreed that the standard preparation should be Alhydrogel produced by Superfos a/s; it is a stable, viscous homogenous material retaining its properties even after storage for several years. Alhydrogel is a sterile pyrogen-free, pure product which complies with the specifications of the British Pharmacopoeia. The gel stability and re-suspension characteristics of vaccines containing alhydrogel remain satisfactory even for 3-4 years. The chemical analysis is as follows:-

Dry matter as Al_2O_3	2.0%
Equivalent to $Al(OH)_3$	3.0%
Conductivity (max)	0.5 milli Siemens
Nitrate as N (")	0.005%
Free Sulphate (")	0.05%
Total Sulphate (")	0.1%
pH	6.0-7.0

Alhydrogel has a positive charge under biological conditions and therefore it will adsorb proteins which are negatively charged; note that basic proteins are poorly adsorbed. The degree of adsorption is influenced by 1) the concentration and nature of antigen, salts and buffering ions and 2) pH of the resulting mixture. The protein binding capacity is approximately 17.0 mg protein (as HSA) ml^{-1} but this should be checked by the method of Weeke et al (1975) although we have found it is satisfactory to use the quantitative single radial immunodiffusion method of Mancini, Carbonara and Heremans (1965) instead of rocket immunoelectrophoresis.

From the studies of Rethy (1965) with Shigella and Salmonella antigens and Schmidt (1967) with influenza vaccine it is apparent that increasing amounts of aluminium hydroxide with standard amounts of antigen leads to a gradual loss of toxicity. It is suggested that the maximum adsorption of protein (1.0 ml) to Alhydrogel (1.0 ml) should be determined. For example 1.0 ml physiological saline containing the amount of protein optimally adsorbed to Alhydrogel (e.g. 17.0 mg HSA), 1.0 ml Alhydrogel 2% i.e. final concentration of aluminium hydroxide 1.5%, 0.2 ml injection dose or 1.7 mg protein/0.2 ml mixture.

For the dose response curve the 1.0 ml physiological saline containing the protein (e.g. 17.0 mg HSA) is kept constant but the amount of Alhydrogel is varied e.g. 0.2, 0.4, 0.5, 0.8 ml and the volume adjusted to 2.0 ml with

Table 2.   Summary of Recommended Guidelines of Tests for Adjuvant Comparisons

1.   Standard antigens:

   (i) Ovalbumin (soluble antigen: Sigma Chemical Company)

   (ii) Influenza Hemagglutinin (particulate amphipathic molecule:
   obtainable from Wellcome Biotechnology, Beckenham, Kent BR3
   3BS).   Prepared from H_3N_2 type A strain.

2.   Standard adjuvants:

   A.   Freund complete adjuvant (prepared by the State Serum
   Institute, Copenhagen and marketed by Superfos Biosector a/s,
   P.O. Box 39, 2950 Vedbaek, Denmark).

   B.   Aluminium hydroxide gel ('Alhydrogel', Superfos Biosector a/s).

3.   Vaccine mixture components:

   a)   Aqueous diluents - a Limulus lysate assay should be done on
   aqueous diluents and it is advisable to use glassware cleaned
   in E-TOXA-CLEAN for the preparation of the experimental
   vaccine.

   b)   Oil formulations - i) GLC analysis to determine the hydrocarbon
   chain length in the sample to be used.   These data ought to be
   provided by the manufacturer supplying the oil.

   ii) conductivity test on the final emulsion.

   iii) viscosity tests on the final emulsion.

   c)   Aluminium hydroxide -the binding capacity of the Alhydrogel for
   the antigen should be determined.   This ought to be the
   responsibility of the manufacturers.

4.   Animal species: Guinea-pig (White Hartley strain 300-400g in
   weight), 10 animals/group.

5.   Injection route: Intramuscular only.

6.   Injection schedule:

   Primary immune response -   Day 0   inject adjuvant mixture
   i.m. into hindlimb.

   Day 1 & 2 bleed animal for
   creatine phosphokinase
   activity in serum.

   Day 10-12 palpate the draining
   inguinal lymph nodes.

| | Day 19 | Skin test for delayed-type hypersensitivity to the antigen. |
| | Day 21 | Bleed animal. |

Secondary immune response - Day 0-12 as above.

	Day 28	booster injection.
	Day 44	skin test for delayed-type hypersensitivity.
	Day 46	bleed animal.

7. Procedures

a) The vaccine antigen should be tested for innate adjuvant activity against the standard antigens suspended in Freund incomplete adjuvant.

b) The experimental adjuvant should be tested against the standard adjuvants and standard antigens to determine the relative potency. The adjuvant mixtures should be injected by the i.m. route.

c) Pyrogenicity test, the experimental vaccine should be tested for pyrogenic activity in at least two rabbits.

d) Toxicity test, the experimental vaccine should be tested for the stimulation of guinea-pig serum creatine phosphokinase activity.

e) Measurement of humoral antibody response
   i) Enzyme-linked immunosorbent assay (ELISA)
   ii) Quantitative precipitin test (μg Ab N ml^{-1}) or suitable alternative.

f) Measurement of delayed-type hypersensitivity to the protein antigen. Guinea-pigs should be skin-tested with the vaccine antigen preparation on days 19 or 46.

g) Measurement of lymphokine production: Guinea-pig peripheral blood monocytes should be tested for the presence of increased interleukin-2 (IL-2) synthesis in response to specific antigen.

h) The vaccine formulation should be tested for:-
   i) Induction of allergy, ie. stimulation of antigen specific IgE against the vaccine antigen and standard antigens) in guinea-pigs.

(Cont'd.)

Table 2. (Cont'd.)

---

ii) <u>Induction of adjuvant arthritis</u>, the vaccine

formulation should be tested for the induction of

adjuvant polyarthritis in Lewis strain rats after

injection at the base of the tail.

(iii) <u>Systemic pyrogenicity test</u>, stimulation of an anterior

uveitis in rabbits.

---

physiological saline. To exceed the optimal binding capacity, dissolve the antigen in 0.5 ml physiological saline and add 1.5 ml Alhydrogel.

<u>Injection of guinea-pigs</u>: It was agreed to use 10 animals/group. However, this is an arbitrary number of animals which does not take into consideration the inherent precision (λ) and the width of confidence limits which would be regarded as acceptable (Fig. 1).

Analysis of the results obtained with the dose response curves for the standard and test adjuvants may be done by the six-point parallel-line assay, described in detail by Wardlaw (1985). In principle, the log antibody titre could be plotted against the log dose of adjuvant and values for a low, medium and high dose selected from the linear portion of the curves for the standard and test adjuvant. These values could be used to plot the dose-response lines of a six-point parallel-line assay. However, it would be more accurate to test these three doses of adjuvant in further groups of guinea-pigs.

In the comparison of different adjuvants there might be a lack of parallelism and other procedures may be used, e.g. slope-ratio assay where the straight-line, dose-response curves of the standard and test are not parallel but meet at the focal point of zero dose. The result is calculated as described by Wardlaw (1985).

A. <u>Economy of antigen</u>. As some of the vaccine antigens may be expensive it is worthwhile to produce a dose response curve of <u>variable antigen</u> against <u>constant adjuvant</u>. It is suggested that this should be done using FCA (as the most potent adjuvant) containing 200 µg <u>M. tuberculosis</u>/0.2 ml injection dose and Alhydrogel (as an adjuvant licenced for human use) at the optimal binding capacity but with variable doses of antigen.

B. <u>Comparison of adjuvants in experimental vaccines</u>. Each animal is injected with 0.2 ml of the experimental vaccine or with the two standard adjuvant mixtures prepared as described above under <u>Standard adjuvants</u>, intramuscularly into the right hind-limb.

The draining inguinal palpable lymph nodes are graded from 3.0 (pea-sized) to 0.5 (millet seed) in each animal, injected with the mixtures of group A or B mentioned above, after 10 days.

TEST FOR INDUCTION OF DELAYED-TYPE HYPERSENSITIVITY TO THE VACCINE ANTIGEN

Skin Test

This test should be done on day 19 in the primary response and day 44 in the secondary response. The hair on the flanks of the animals is removed and the bare skin is injected with 0.1 ml saline containing 100 µg antigen.

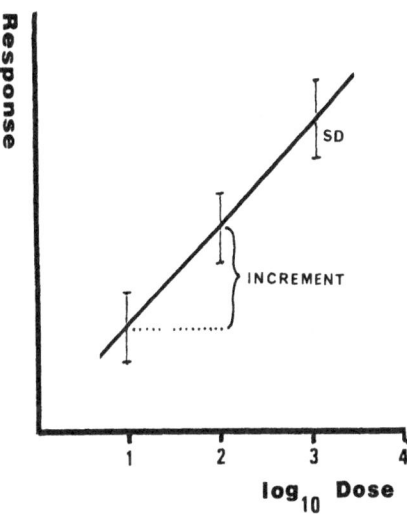

Fig. 1.   Theoretical relationship between response, dose and number of
          animals in the group.

Note that λ, $\dfrac{SD}{slope}$ should be as low as possible; slope =

$\dfrac{\text{increment of responses}}{10\log_{10}}$; λ^2 is proportional to N (number of animals,

provided that each animal yields one observation).

Skin reactions are measured (diameter of erythema in mm) at 24 and 48 hr;
the mean value $\pm$ standard error is calculated for each group.

Measurement of Interleukin 2 (IL-2) Production

This is not an essential test for the comparison of adjuvants but it is
useful to monitor the cell-mediated response after re-exposure to the antigen in the vaccine. The method involves the isolation of peripheral blood
monocytes from immunized and control guinea-pigs, exposure of the cells to
different concentrations of antigen in tissue culture medium and assay of
the supernatant medium by uptake of ^{3}H-thymidine using an IL-2-dependent
cytotoxic lymphocyte cell line (Byars and Allison, 1987). In addition, it
was suggested that a proliferative in vitro response should be examined in a
standard T-cell transformation test.

Measurement of Antibody Levels in Guinea-pig Sera

Animals are easily bled by cardiac puncture immediately after terminal
anaesthesia with carbon dioxide gas on day 21 for primary response or day 46
for secondary response. The sera should be tested against the test antigen
in an ELISA assay, chequerboard titration of antigen and antiserum to determine affinity between antigen and antibody.

It is my opinion that considerable variation exists in the determination of an end-point in an ELISA test for antibody titre. With many sera
some residual activity, with $OD_{492\ nm}$ values just above baseline, is observed at low antibody concentrations. This may be due to many factors, e.g.
raw plastic being exposed to antibody or conjugate and/or the ratio of enzyme to secondary antibody and the time of exposure to the substrate. It is
suggested that the following protocol should be adopted during the trial
period and that a baseline value of 0.3 $OD_{492\ nm}$ should be used.

PROTOCOL:

Volume in well

First reagent	.35 ml	Ovalbumin (10 μg). Wash three times
Second reagent	.30 ml	Guinea-pig anti-ovalbumin dilutions. Wash three times
Third reagent	.25 ml	Antibody: Enzyme (anti Guinea-pig: horseradish peroxidase) Wash three times
Fourth reagent	.20 ml	Enzyme substrate
Stopping reagent	50 μl	12.5% H_2SO4

Read Plate in spectro-
photometer

COATING BUFFER: 0.05 M carbonate buffer pH 9.6.
    1.59 g Na_2CO_3 (sodium carbonate anhydrous, 105.99 MW)
    2.93 g $NaHCO_3$ (sodium hydrogen carbonate, 84.01 MW)
    Dissolve in distilled water, total volume one litre.

WASHING BUFFER:
    Phosphate buffered saline   8.0 g NaCl
                                 0.2 g KH_2PO_4
                                 2.8 g $Na_2HPO_4.12H_2O$
                                 0.2 g KCl
    Dissolve in distilled water, total volume one litre.
    Make to 0.05% with Tween 20.

SUBSTRATE: O-phenylenediamine in citrate/phosphate buffer, pH 5.0.
    Solution A, 0.1 M citric acid (21.01 g/litre)
    Solution B, 0.2 M phosphate (35.6 g $Na_2HPO_4.2H_2O$/litre)
    49.0 ml of A + 51.0 ml B, pH 5.0

    (WEAR MASKS AND GLOVES WHEN WEIGHING OUT OR USING O-PHENYLENEDIAMIME.)

    Add 34.0 mg O-phenylenediamine to 100 ml of citrate/phosphate buffer pH
5.0 and add 20 μl of hydrogen peroxide (20 vol.) just before use. Place in
a brown glass reagent bottle. Keep in the dark and use on the day only.

INDUCTION OF ALLERGIES

Stimulation of IgE Antibodies

    Wistar rats, 9-11 weeks old weighing 200-300 g are fed and watered nor-
mally until 24 hr prior to the administration of the test adjuvant. At this
time the animals are given ovalbumin (Grade II; Sigma Chemical Co.) mixed to
a paste with water. The adjuvant dose is administered orally and the ovalbumin diet is continued for a further 12 hr. Twenty-four days later the
rats are challenged intravenously with ovalbumin in 0.75 ml saline. At 0.5,
1.0 and 4.0 hr after challenge the rats are anaesthetised and bled from the
carotid artery. The sera are collected from the clotted blood, freeze-dried

and stored in sealed tubes (Miller et al, 1983). The amount of rat mast-cell proteinase II (ng), a chymotrypsin-like proteinase, in the sera is determined by an ELISA antibody capture technique, see below.

Measurement of Mast Cell Proteinase 2

This assay was suggested since it measures the end-point of IgE production with mast cell degranulation. An ELISA kit for this assay is available from the Moredun Institute, 408, Gilmerton Road, Edinburgh, EH17 7JH, Scotland. After sensitization of rats by oral administration of ovalbumin (A5253, Grade II, Sigma) and experimental adjuvant the animal is left for 24 days and re-exposed to the antigen. The serum levels of mast cell proteinases increases within 5 hr of exposure to antigen and is assayed by ELISA.

Induction of Adjuvant Arthritis in Rats

The experimental adjuvant should be dissolved or suspended in phosphate-buffered saline (pH 7.2) to give an appropriate range of concentrations. The adjuvant preparation is mixed with an equal volume of FIA and emulsified to produce a water-in-oil emulsion.

Female Lewis inbred rats are injected with 0.1 ml of the water-in-oil emulsion either into the inguinal lymph nodes or at the base of the tail. The animals are examined daily for 4 weeks after the injection for signs of adjuvant arthritis development (Pearson and Wood, 1959; Waksman, Pearson and Sharp, 1960). There is usually evidence of swollen tail joints, arthritis in the feet and sometimes nodular lesions on the ears in positive reactions.

Systemic Pyrogenicity Test: Induction of Vascular Leakage as a Measure of Intraocular Inflammation (Uveitis)

As shown in the scheme of testing adjuvants (Tables 1 and 2), if the in vitro tests for toxicity and pyrogenicity produce negative results, it is possible to proceed to this test which measures the inflammatory effect of pyrogenic substances after systemic (i.v.) administration of 0.05-10 mg kg^{-1} body weight of the rabbit (Waters et al, 1986). The protein ml^{-1} aqueous humor after 2.5 hr gives an indication of vascular leakage. However, the test is expensive and wasteful of animals (three rabbits/dose of adjuvant) so it will be important to monitor how many adjuvants will be tested by this method during the two years trial period.

COMMENT

There was considerable discussion about the choice of standard antigens and adjuvants. As a particulate antigen, sheep erythrocytes (srbc), cannot be incorporated into liposomes (Gregory Gregoriadis) nor into ISCOMS (Bror Morein). The use of tetanus toxoid was favoured by Lajos Rethy but as Anthony Allison pointed out this is a potent antigen on its own and doesn't really require an adjuvant. Robert Hunter asked 'what would be the effect of a TNP hapten group on the ovalbumin?' He also cautioned that amphiphilic membrane proteins would react quite differently from ovalbumin and srbc's. There were also queries about the need for T-dependent or T-independent antigens.

With the standard adjuvants, other choices for experimental vaccines should also be considered. Kristian Dalsgaard stated that Quil A works well with amphipathic molecules and MDP with soluble antigens.

From the vaccine producers' point of view Simon Bartoling stated that researchers interested in FMDV would want to observe the adjuvant effect on this antigen and not on the others mentioned. This is relevant when we have a series of efficient safe adjuvants and it must be stressed that these guidelines are for the comparison of adjuvants. Further work would be required to select the best adjuvant for a particular vaccine antigen. Similar criticisms might be levelled at the choice of one route of adminis- tration because a good adjuvant might be missed e.g. the oral route may require adjuvants with different characteristics. Bror Morein added that intranasal immunization might be required and one benefit of subcutaneous injection was that any local reaction could be immediately observed. However, it was concluded that only the intramuscular route was required to make a basic comparison of adjuvants.

The adjuvant effect was originally measured on the primary immune res- ponse and FCA is known to depress the immune response after repeated admin- istration (see Stewart-Tull, 1983). However, Bror Morein said that some adjuvants were more suited to boosting the secondary immune response. The consensus view was that both responses should be examined.

A lively debate took place about the choice of animal. Some urged the use of at least two inbred strains of mouse because of MHC restriction and the fact that cell-mediated responses of outbred strains are often too low to measure, whereas some urged the use of outbred strains and others stated that results would be irrelevant with mice. In view of the limitations it was agreed to use the white guinea-pig.

Regulatory agenices may demand different toxicity tests to the levels of creatine phosphokinase in serum adopted in the guidelines. For instance, the EEC Pharmacopoeia uses the intradermal tolerance test (Berlin test). Six guinea-pigs are injected with 0.1 ml of a product at three different sites. The animals are weighed daily and the thickness of the skin fold and the diameter of a possible zone of erythema are measured. There is still the requirement for an LD_{50} test in mice; negative compounds cause no mor- talities, no peritonitis and normal weight gain of the mice.

At the meeting there was no consensus enthusiasm for including tests for cancer immunotherapy or induction of non-specific protection against bacterial pathogens as these are not essential comparative tests. Nevertheless, it would be important to know whether a potential candidate adjuvant possessed these activities. Similarly, it was decided to ignore the induction of experimental auto-immune disease, e.g. allergic encephalo- myelitis in guinea-pigs, possibly because of the artificiality of this test.

It must be remembered that these guidelines for testing adjuvants are in no way to be read as regulations, nor do we expect the regulatory bodies to adopt them before trial. Experience will show whether the experienced adjuvant researchers have chosen the most suitable series of tests. It may be in two years time that we shall need to rethink our strategy but at present whenever an adjuvant is to be used in an experimental vaccine the aim is to utilize these guidelines.

Some researchers might have difficulty in obtaining some of the spec- ified reagents required to complete some of the tests. Therefore, I will endeavour to publish in the European Adjuvant Newsletter results for the two standards and other researchers' results for different adjuvants. This database will serve as an essential aid for companies interested in the use of adjuvants and as a means of reducing the need for repetitive tests in laboratory animals.

Acknowledgements

I am indebted to the participants at this conference for their patience in helping me to formulate these guidelines. I should like to thank NATO and the Wellcome Foundation for their support which enabled me to attend the conference and Robert Bomford and Bror Morein for their active participation in the workshop. The statistical advice of Alastair Wardlaw is gratefully acknowledged.

REFERENCES

Allison, A.C., Byars, N.E. and Waters, R.V., 1986, "Advances in Carriers and Adjuvants for Veterinary Biologics", R.M. Nervig. P.M. Gough, M.L. Kaeberle and C.A. Whetstone, eds., Iowa State University Press, Ames, Iowa.

Byars, N.E. and Allison, A.C., 1987, Adjuvant formulation for use in vaccines to elicit both cell-mediated and humoral immunity, Vaccine, 5:223.

Davenport, F.M., 1968, Seventeen years' experience with mineral oil adjuvant influenza virus vaccines, Ann. Allergy, 26:288.

Edelman, R., 1980, Vaccine adjuvants, Rev.Infect.Dis., 2:370.

Friedewald, W.F., 1944a, Adjuvants in immunization with influenza virus vaccines, J.Exp.Med., 80:477.

Friedewald, W.F., 1944b, Enhancement of the immunizing capacity of influenza virus vaccines with adjuvants, Science, 99:453.

Gisler, R.H., Dietrich, F.M., Baschang, G., Brownbill, A., Schumann, G., Staber, F.G., Tarcsay, L., Wachsmuth, E.D. and Dukor, P., 1979, "Drugs and Immune Responsiveness", J.L. Turk and D. Parker, eds., Macmillan Press, London.

Gray, J.E., Weaver, R.N., Moran, J. and Feenstra, E.S., 1974, The parenteral toxicity of clindamycin 2-phosphate in laboratory animals, Toxicol. Appl.Pharmacol., 27:308.

Henle, W. and Henle, G., 1945, Effect of adjuvants on vaccination of human beings against influenza, Proc.Soc.Exptl.Biol.Med., 59:179.

Hughes, B.P., 1962, A method for the estimation of serum creatine kinase and its use in comparing creatine kinase and aldolase activity in normal and pathological sera, Clin.Chim.Acta, 7:579.

Jolivet, M.E., Audibert, F.M., Gras-Masse, H., Tartar, A.L., Schlesinger, D.H., Wirtz, R. and Chedid, L.A., 1987, Induction of biologically active antibodies by a polyvalent synthetic vaccine constructed without carrier, Infect.Immun., 55:1498.

MacLennan, R., Schofield, F.D., Pittman, M., Hardegree, M.C. and Barile, M.F., 1965, Immunization against neonatal tetanus in New Guinea. Antitoxin response of pregnant women to adjuvant and plain toxoids, Bull.WHO, 32:683.

Mancini, G., Carbonara, A.O. and Heremans, J.F., 1965, Immunochemical quantitation of antigens by single radial immunodiffusion, Immunochemistry, 2:235.

Miller, H.R.P., Woodbury, R.G., Huntley, J.F. and Newlands, G., 1983, Systemic release of mucosal mast-cell protease in primed rats challenged with Nippostrongylus brasiliensis, Immunology, 49:471.

Murray, R., Cohen, P. and Hardegree, M.C., 1972, Mineral oil adjuvants: biological and chemical studies, Ann.Allergy, 30:146

Ogonuki, H., Hashizume, S. and Abe, H., 1967, Histophathological tests of tissues in the sites of local reactions caused by the injection of oil-adjuvant cholera vaccine, Symp.Ser.Immunobiol.Stand., 6:125.

Paquet, A., Raines, K.M. and Brownback, P.C., 1986, Immunopotentiating activities of cell walls, peptidoglycans and teichoic acids from two strains of <u>Listeria monocytogenes</u>, <u>Infect.Immun.</u>, 54:170.

Pearson, C.M. and Wood, F.D., 1959, Studies of polyarthritis and other lesions induced in rats by injection of mycobacterial adjuvant. I. General clinical and pathologic characteristics and some modifying factors, <u>Arthritis and Rheumatism</u>, 2:440.

Rasanen, L., Mustikkamaki, U.P. and Arvilommi, H., 1982, Polyclonal response of human lymphocytes to bacterial cell walls, peptidoglycans and teichoic acids, <u>Immunology</u>, 46:481.

Rethy, L., 1965, The production and use of a stable aluminium hydroxide adsorbent, <u>Prog.Immunobiol.Stand.</u>, 2:80.

Schmidt, G., 1967, The adjuvant effect of aluminium hydroxide in influenza vaccine. International Symposium on Adjuvants of Immunity, Utrecht, 1966, <u>Sym.Ser.Immunobiol.Stand.</u>, 6:275.

Stewart-Tull, D.E.S., 1980, The immunological activities of bacterial peptidocylcans, <u>Ann.Rev.Microbiol.</u>, 34:311.

Stewart-Tull, D.E.S., 1983, Immunologically important constituents of mycobacteria: adjuvants, <u>in</u>: "Biology of Mycobacteria", Vol. 2, C. Ratledge and J. Stanford, eds., Academic Press, London.

Stewart-Tull, D.E.S., 1985a, Immunopotentiating conjugates, <u>Vaccine</u>, 3:40.

Stewart-Tull, D.E.S., 1985b, Immunopotentiating activity of peptidoglycan and surface polymers, <u>in</u>: "Immunology of the Bacterial Cell Envelope", D.E.S. Stewart-Tull and M. Davies, eds., John Wiley & Sons, Chichester.

Stewart-Tull, D.E.S., 1988, Immunostimulation with peptidoglycan or its synthetic derivatives, <u>Prog.Drug Res.</u>, 32, in press.

Stewart-Tull, D.E.S. and Parant, M., 1986, Immunopotentiators in vaccines: a dream to a reality, <u>Ann.Immunol.Hung.</u>, 26:197.

Vane, J. and Cuatrecasas, P., 1984, Genetic engineering and pharmaceuticals, <u>Nature</u>, 312:303.

Waksman, B.H., Pearson, C.M. and Sharp. J.T., 1960, Studies of arthritis and other lesions induced in rats by injection of mycobacterial adjuvants. II. Evidence that the disease is a disseminated immunologic response to exogenous antigen, <u>J.Immunol.</u> 85:403.

Wardlaw, A.C., 1985, "Practical Statistics for Experimental Biologists", John Wiley & Sons, Chichester.

Waters, R.V., Terrell, T.G. and Jones, G.H., 1986, Uveitis induction in the rabbit by muramyl dipeptides, <u>Infect.Immun.</u>, 51:816.

Weeke, B., Weeke, E. and Lowenstein, H., 1975, The adsorption of serum proteins to aluminium hydroxide gel examined by means of quantitative immunoelectrophoresis, <u>Scand.J.Immnol.</u>, 4:149.

World Health Organization, 1976, Immunological adjuvants, <u>Techn.Rep.Ser.</u>, 595:3.

INTERACTION BETWEEN INDUSTRY AND THE BASIC RESEARCHER IN THE UNIVERSITY

DURING VACCINE DEVELOPMENT

D.E.S. Stewart-Tull

Microbiology Department, Alexander Stone Building
Garscube Estate, Bearsden, Glasgow G61 1QH, UK

Generalizations about the relationship between the academic basic re-searcher and the pharmaceutical industry are difficult to make and as the university-based researcher on the panel I can only present a personal view. The relationship will vary from country to country so attitudes in the U.K. may not be coincident with those elsewhere.

My own position is slightly unusual as I had been in gainful employ-ment for two years prior to taking up my offered place at university. Subsequently, my passage was typical of students graduating during the expansionist period within the British university system: primary degree, doctorate and "tenured" lecturership. However, I retained part of the work ethic from the pre-university period and have continuously devoted part of my time to industrially orientated projects.

We have been urged 'to perceive the practical while pursuing the possible' (Hilleman, 1985). It could be argued that the relationship between the university and industry depends on a perceived need for sound basic research which may subsequently be exploited by industry.

During the last decade successive British governments have stipulated a contraction in university research funding and the present government ex-pects a greater financial input from industry. To the politicians this may seem reasonable when one considers the cost of training graduates - the universities cannot <u>sell</u> their products to industry! Whatever the economic views of one party or another some enlightened thought is required to cement good relations in this mixed marriage. The partners themselves may have widely separated opinions with reference to their perceived roles in re-search and development.

We need only to perceive the view of the temple of Poseidon and the island across the bay, from the verandah of this conference hotel, to ob-tain an image that some may have of each party.

THE ACADEMIC

The Temple of Poseidon, the superior academic in his ivory tower looks down on industry on the island.

An accurate picture or not? Some years ago a selected group of British university staff were questioned about their willingness to collaborate with industry. A proportion of them indicated quite vehemently that they were not employed to work for industry. A shortsighted approach you might believe but one needs to remember that the modern senior academic wears many different hats during the course of a year - teacher, administrator and basic research worker. Subsequently, the present government added to the list, industrial entrepreneur; the supposition was that academic scientists would exploit their own discoveries and enhance their salaries by forming companies.

Administration involves the running of both the university and the individual department with consequent membership of many varied committees and subcommittees eg. finance, academic development, biological and radiation hazards, library, management of residences, amenities, external relations, planning new buildings etc. etc. It never ceases to amaze me that the acquisition of a primary degree equips the academic to assume the roles of, for instance, financial and personnel manager or design consultant. Many years ago, after poring for endless evenings over the plans of a new building, I once asked a senior colleague how much of my time the university bought and he replied "not more than twenty-four hours of each day." As a researcher more and more time is spent chasing less and less money and I doubt the kudos for anyone in an "unfunded α-rated grant application."

The main criteria of academic excellence are still research council grants and refereed publications. There is minimal emphasis placed on industrial research and development with concomitant requirements for secrecy and non-publication of results until patent rights have been explored by legel departments.

From my own part I enjoy the urgency of industrial projects whether they be troubleshooting, contract research or R & D. There is a different type of challenge and pressure - I well remember the effect when twenty years ago I was asked whether a fungal contaminant would destroy the whisky in a bonded warehouse, the value of which was £147,000,000; certainly concentrates the mind!

Recently, we have taken out a patent and the lawyers assumed an immediate knowledge of existing patents and an ability to phrase the claims in acceptable legal language. Detailed answers to some twelve cited patents had to be prepared before some relief was obtained when the patent was granted. This was followed by the preparation of non-disclosure and commercial licencing agreements with the associated problems of 'intellectual property rights' and finance. An American lawyer 'phoned me from his golf-club one day to ask "did I know that a clause of our agreement contravened clause 85 of the Treaty of Rome?" Of course, every sientist is conversant with this treaty!

THE INDUSTRIALIST

<u>The prosperous industrial organization contemplates the financially starved university on the island and waits for exploitable knowledge.</u>

During this conference, there seems to be an attitude developing that basic research which is not capable of immediate scale-up to an industrial process is invalid or irrelevant. I think this attitude is reasonable during the pre-commercialization period after the basic research phase but not before. As I have already indicated basic research is only applied research waiting to be exploited. There is a great need for more university-based fundamental research in the vaccine field:-

1. to determine mechanisms of pathogenesis
2. to identify protective antigens (epitopes)
3. to examine new and existing adjuvants with a view to understanding their biological activities
4. to develop experimental vaccines and test these in the laboratory.

One might ask whether such studies are all necessary before recombinant or peptide vaccines can be launched? This reminds me of a conversation with a leading immunologist in the early sixties during the initial difficulties with transplant rejection. He indicated that the problem was that the technical skill of the surgeon Dr. Christian Barnard was fifteen years ahead of the understanding of transplantation immunology. Might there be greater risks of adverse effects if a product reaches the market-place before the researchers have completed the basic studies?

It is significant that during one or two of the sessions, delegates from industry have indicated that company lawyers would not permit them to reveal what they were doing. Some time ago (1986), I surmised that the day would come when university patent lawyers would inspect our papers before publication for patentable discoveries, the lawyer replied 'sooner rather than later' (Szczuka, 1986). An industrialist said to me that the problem with academics is that they do not want to maintain secrecy. Financial implications may dictate the slowing-down of the release of innovative ideas from universities. This would be a loss because, as one senior UK politician calculated, the earnings directly attributable to university discoveries amounts to more than £500,000,000/annum.

There is a need for the industrialist to educate the basic researcher and much of the talk by George Poste is extremely useful in this respect. The academic does need to adopt a realistic approach in the final development of an experimental vaccine, especially the cost of a vaccine preparation. In many developing countries $1 per year/individual is spent on healthcare. We have heard others express opinions about "breaking the cold chain" and the method of administration to the patient. These points were brought home to me by a doctor working for Save The Children who criticized speakers at a conference because they failed to realise the problems in transporting and administering a vaccine in the remoter parts of Africa.

A STABLE MARRIAGE?

Bridging the gap between the Temple of Poseidon and the island.

The crucial stage in the relationship between university and industry is during the courtship - ie. the pre-commercialization stage.

It has been my experience that during this period it is not possible 'to go Dutch' (pay equal shares) and sometimes I feel I am neither on the hill nor on the island but hanging over the gulf in between. Research councils see 'vaccine development' and argue that industrial support should be obtained. Industry does not see immediate exploitation and sometimes appears unwilling to finance what they still consider to be basic research, albeit innovative and exploitable basic research. The financial returns from a vaccine often do not justify such expenditure and, in 1985, I suggested that the government would need to accept a greater share of R & D expenditure. However one approaches this problem there is a need to encourage links between the university and industry, otherwise many academics will decide to conserve their energies for basic research funded by research councils, as Aeschylus wrote 'the force of necessity is irresistible'!

ADDENDUM

At the end of the session, there was no discussion but several Europeans told me that they found it difficult to forge links with industry during the pre-commercialization phase of vaccine development.

Acknowledgements

I should like to thank NATO and the Wellcome Foundation London for enabling me to attend this conference. The figures were drawn by my son, Iain and the cartoon was contributed by Dr. David Katz.

REFERENCES

Hilleman, M.R., 1985, Newer directions in vaccine development and utilization, J.inf.Dis., 151:407

Stewart-Tull, D.E.S., 1985, Vaccines: Who will buy? The SGM Quarterly, 12:109

Stewart-Tull, D.E.S., 1986, Research work to patent, The SGM Quarterly 13:1

Szczuka, J.T., 1986, A practical introduction to patents, The SGM Quarterly, 13:2.

Participants of the NATO Advanced Studies Institute "Immunological Adjuvants and Vaccines" held at Cape Sounion, Greece during 24 June-5 July, 1988. The organizing committee included Anthony C. Allison (ASI Co-director), Ruth Arnon, Gregory Gregoriadis (ASI Director and Chairman), John H.L. Playfair and George Poste.

CONTRIBUTORS

Allison, A.C., Department of Immunology, Institute of Biological Sciences, Syntex Research, Palo Alto, CA 94304, USA

Alving, C.R., Department of Membrane Biochemistry, Walter Reed Army Institute of Research, Washington, DC 20307-5100, USA

Arenzana-Seisdedos, F., Laboratoire d'Immunologie Virale, Institut Pasteur, 75724 Paris Cedex 15, France

Arnon, R., Department of Chemical Immunology, The Weizmann Institute of Science, Rehovot, 76100, Israel

Bennett, B., Department of Pathology, Emory University, Atlanta, Georgia 30322, USA

Bomford, R., Wellcome Biotechnology, Langley Court, Beckenham, Kent BR3 3BS, UK

Brade, H., Forschungsinstitut Borstel, Institut für Experimentelle Biologie und Medizin, Parkallee 22, D-2061 Borstel, FRG

Brade, L., Forschungsinstitut Borstel, Institut für Experimentelle Biologie und Medizin, Parkallee 22, D-2061 Borstel, FRG

Brown, F., Wellcome Biotechnology, Langley Court, Beckenham, Kent BR3 3BS, UK

Buus, S., University of Copenhagen, Copenhagen, Denmark

Buynitzky, S., Cytrx Corporation, Norcross, Georgia 30368, USA

Byars, N.E., Department of Immunology, Institute of Biological Sciences, Syntex Research, Palo Alto, CA 94304, USA

Check, I.J., Department of Pathology, Emory University, Atlanta, Georgia 30322, USA

van Dam, J.E.G., Laboratory of Microbiology, Faculty of Medicine, Utrecht University, Utrecht, The Netherlands

Epstein, M.A., Nuffield Department of Clinical Medicine, University of Oxford, John Radcliffe Hospital, Headington, Oxford OX3 9DU, UK

Flad, H-D., Forschungsinstitut Borstel, Institut für Experimentelle Biologie und Medizin, Parkallee 22, D-2061, FRG

Gregoriadis, G., Medical Research Council Group, Academic Department of Medicine, Royal Free Hospital School of Medicine, Pond Street, London NW3 2QG, UK

Grey, H.M., Cytel, San Diego, CA, USA

Griffin, P.D., Special Programme of Research, Development and Research Training in Human Reproduction, World Health Organization, 1211 Geneva 27, Switzerland

Hayre, M.D. Department of Membrane Biochemistry, Walter Reed Army Institute of Research, Washington, DC 20307-5100, USA

Heath, A.W., Department of Immunology, University College and Middlesex School of Medicine, Arthur Stanley House, 40-50 Tottenham Street, London W1P 9PG, UK

Hockmeyer, W.T., Department of Immunology, Walter Reed Army Institute of Research, Washington, DC 20307-5100, USA

Höglund, S., Institute of Biochemistry, Biomedicum, Box 576, S-751, 23 Uppsala, Sweden

Howerton, D., Department of Pathology, Emory University, Atlanta, Georgia 30322, USA

Hughes, B.W., Department of Immunology, Syntex Research, Palo Alto, CA 94304, USA

Hunter, R.L., Department of Pathology, Emory University, Atlanta, Georgia 30322, USA

Kenney, J.S., Department of Immunology, Syntex Research, Palo Alto, CA 94304, USA

Kraaijeveld, C.A., Laboratory of Microbiology, Faculty of Medicine, Utrecht University, Utrecht, The Netherlands

Loppnow, H., Forschungsinstitut Borstel, Institut für Experimentelle Biologie und Medizin, Parkallee 22, D-2061, FRG

Lövgren, K., Department of Virology, National Veterinary Institute, Biomedicum, Box 585, S-751, 23 Uppsala, Sweden

Mogensen, S.C., Institute of Medical Microbiology, University of Aarhus, Aarhus, Denmark

Morein, B., Swedish University of Agricultural Sciences, College of Veterinary Medicine, Department of Microbiology, Section of Virology, Biomedicum, Box 585, S-751 23 Uppsala, Sweden

Nakano, G., Department of Immunology, Institute of Biological Sciences, Syntex Research, Palo Alto, CA 94304, USA

Playfair, J.H.L., Department of Immunology, University College and Middlesex School of Medicine, Arthur Stanley House, 40-50 Tottenham Street, London W1P 9PG, UK

Rhodes, J., Wellcome Research Laboratories, Langley Court, Beckenham, Kent BR3 3BS, UK

Richards, R.L., Department of Membrane Biochemistry, Walter Reed Army
Institute of Research, Washington, DC 20307-5100, USA

Rietschel, E.T., Forschungsinstitut Borstel, Institut für Experimentelle
Biologie und Medizin, Parkallee 22, D-2061, Borstel, FRG

van Rooijen, N., Department of Histology, Medical Faculty, Free University,
P.O. Box 7161, 1007 MC Amsterdam, The Netherlands

Schade, U., Forschungsinstitut Borstel, Institut für Experimentelle Biologie
und Medizin, Parkallee 22, D-2061 Borstel, FRG

Sette, A., Cytel, San Diego, CA, USA

Seydel, U., Forschungsinstitut Borstel, Institut für Experimentelle Biologie
und Medizin, Parkallee 22, D-2061 Borstel, FRG

Snippe, H., Laboratory of Microbiology, Faculty of Medicine, Utrecht
University, Utrecht, The Netherlands

de Souza, J.B., Department of Immunology, University College and Middlesex
School of Medicine, Arthur Stanley House, 40-50 Tottenham Street, London
W1P 9PG, UK

Stewart-Tull, D.E.S., Microbiology Department, Alexander Stone Building,
Garscube Estate, Bearsden, Glasgow G61 1QH, UK

Su, D., Department of Microbiology, Fujian Medical College, Fuzhou, P.R.
China

Tan, L., Medical Research Council Group, Academic Department of Medicine,
Royal Free Hospital School of Medicine, Pond Street, London NW3 2QG, UK

Verheul, A.F.M., Laboratory of Microbiology, Faculty of Medicine, Utrecht
University, Utrecht, The Netherlands

Virelizier, J.L., Laboratoire d'Immunologie Virale, Institut Pasteur, 75724
Paris Cedex 15, France

Welch, M., Department of Immunology, Institute of Biological Sciences,
Syntex Research, Palo Alto, CA 94304, USA

Wirtz, R.A., Department of Entomology, Walter Reed Army Institute of
Research, Washington, DC 20307-5100, USA

Xiao, Q., Medical Research Council Group, Academic Department of Medicine,
Royal Free Hospital School of Medicine, Pond Street, London NW3 2QG, UK

Zähringer, U., Forschungsinstitut Borstel, Institut für Experimentelle
Biologie und Medizin, Parkallee 22, D-2061, FRG

INDEX